T0203996

Gene Expression Data Analysis

Gene Expression Data Analysis

A Statistical and Machine Learning Perspective

Pankaj Barah

Dhruba Kumar Bhattacharyya

Jugal Kumar Kalita

CRC Press
Taylor & Francis Group
Boca Raton London New York

CRC Press is an imprint of the
Taylor & Francis Group, an **informa** business

A CHAPMAN & HALL BOOK

First edition published 2022
by CRC Press
6000 Broken Sound Parkway NW, Suite 300, Boca Raton, FL 33487-2742

and by CRC Press
2 Park Square, Milton Park, Abingdon, Oxon, OX14 4RN

© 2022 Taylor & Francis Group, LLC

CRC Press is an imprint of Taylor & Francis Group, LLC

Library of Congress Cataloging-in-Publication Data

Names: Barah, Pankaj, author. | Bhattacharyya, Dhruba K., author. | Kalita, Jugal Kumar, author.
Title: Gene expression data analysis : a statistical and machine learning
    perspective / Pankaj Barah, Dhruba Kumar Bhattacharyya, Jugal Kumar Kalita.
Description: First edition. | Boca Raton : Chapman & Hall/CRC Press, 2022.
    | Includes bibliographical references and index. | Summary: "The book
    introduces phenomenal growth of data generated by increasing numbers of
    genome sequencing projects and other throughput technology-led
    experimental efforts. It provides information about various sources of
    gene expression data, and pre-processing, analysis, and validation of
    such data"-- Provided by publisher.
Identifiers: LCCN 2021012278 (print) | LCCN 2021012279 (ebook) | ISBN
    9780367338893 (hardback) | ISBN 9781032055756 (paperback) | ISBN 9780429322655 (ebook)
Subjects: LCSH: Gene expression--Statistical methods. | Gene
    expression--Data processing. | Machine learning.
Classification: LCC QH450 .B38 2022 (print) | LCC QH450 (ebook) | DDC 572.8/65--dc23
LC record available at https://lccn.loc.gov/2021012278
LC ebook record available at https://lccn.loc.gov/2021012279

ISBN: 978-0-367-33889-3 (hbk)
ISBN: 978-1-032-05575-6 (pbk)
ISBN: 978-0-429-32265-5 (ebk)
DOI: 10.1201/9780429322655

Typeset in CMR10 font
by KnowledgeWorks Global Ltd.

Dedication

This book is dedicated to my beloved father, Late Sidheswar Barah, who had the biggest contribution in shaping my life and career. He was an inspiring school teacher, a prolific public orator, author and social activist.

Pankaj Barah

This book is dedicated to my mother (Maa), Ms. Runu Bhattacharyya, and mother-in-law (Xahu), Ms. Bina Sarma, who constantly encouraged, inspired and motivated me in the successful completion of this humble work.

Dhruba Kumar Bhattacharyya

This book is lovingly dedicated to my father (Deuta), Benudhar Kalita, an accomplished author who passed away in 2012, and of whom I have fond memories, and my mother (Maa), Nirala Kalita, a school teacher, who has been gone for eighteen years. They are surely looking down from heaven. It is also dedicated to my 8-year-old daughter, Ananya Lonie, who I hope will grow to be a smart, thoughtful, compassionate and accomplished woman.

Jugal Kumar Kalita

Contents

Acknowledgments

This humble work would not have been possible without the constant support, encouragement and constructive criticism of a large number of people. Special thanks and sincere appreciation are due to the following dedicated faculty members: Dr. Hasin A. Ahmed, Dr. Rosy Sarma, Dr. Nazrul Haque, Dr. Priyakshi Mahanta, Dr. Swarup Roy, Dr. R. C. Baishya, and Dr. Bhabesh Nath. We are extremely grateful for the constant support extended by our sincere students and research scholars: Mr. Hussain A. Chowdhury, Ms. Chinmoyee Baruah, Mr. Akash Das, Mr. Ankur Sahoo, Ms. Pallabi Patwary, Ms. Koyel Mandal, Ms. Rachayita Bharadwaj, Mr. Pranjal Kakaty, and Mithil Gaikwad.

We are grateful to the panel of reviewers for their constructive suggestions and critical evaluation. The constant support and cooperation received from our colleagues and students during the period of writing this book is sincerely acknowledged.

Pankaj Barah
Dhruba Kumar Bhattacharyya
Jugal Kumar Kalita

Authors

 **Pankaj Barah** is an assistant professor in Molecular Biology and Biotechnology at Tezpur University. He received his M.Sc. degree in Bioinformatics (2006) from University of Madras in India and his PhD in Computational Systems Biology (2013) from the Norwegian University of Science and Technology (NTNU), Trondheim, Norway. He worked as a bioinformatics scientist in the division of Theoretical Bioinformatics at the German Cancer Research Center (DKFZ) in Heidelberg, Germany between 2015 and 2017. His research areas include computational systems biology, bioinformatics, evolutionary systems biology, next-generation sequencing (NGS), Big data analytics and biological networks. He has authored 20 research articles, edited two books and written five book chapters. He received the Ramalingaswami Re-entry Fellowship from the Department of Biotechnology, Government of India. Dr. Barah is currently a member of the Indian National Young Academy of Sciences.

 **Dhruba Kumar Bhattacharyya** is a professor in Computer Science and Engineering at Tezpur University. He teaches machine learning, network security, cryptography, and computational biology in UG, PG and PhD classes at Tezpur University. Professor Bhattacharyya's research areas include machine learning, network security, and bioinformatics. He has published more than 280 research articles in leading international journals and peer-reviewed conference proceedings. Dr. Bhattacharyya has authored five technical reference books and edited nine technical volumes. Under his guidance, 20 students have successfully completed their PhD. in the areas of machine learning, bioinformatics, and network security. He is PI of several major research grants, including the Centre of Excellence of Ministry of HRD of the Government of India under FAST instituted at Tezpur University. He has been serving as a guest editor of a special issue of *Springer*

Nature Journal of Computer Science. Professor Bhattacharyya is a Fellow of IETE and IE, India. He is also a Senior Member of IEEE. More details about Dr. Bhattacharyya can be found at   http://agnigarh.tezu.ernet.in/~dkb/index.html.

Jugal Kumar Kalita teaches computer science at the University of Colorado, Colorado Springs. He received MS and PhD degrees in computer and information science from the University of Pennsylvania in Philadelphia in 1988 and 1990, respectively. Prior to that, he received an MSc from the University of Saskatchewan in Saskatoon, Canada, in 1984 and a BTech from the Indian Institute of Technology, Kharagpur, in 1982. His expertise is in the areas of artificial intelligence and machine learning, and the application of techniques in machine learning to network security, natural language processing, and bioinformatics. He has published 130 papers in journals and refereed conferences. He is the author of a book on Perl titled *On Perl: Perl for Students and Professionals.* He is also a co-author of a book titled *Network Anomaly Detection: A Machine Learning Perspective* with Dr. Dhruba K Bhattacharyya. He received the Chancellor's Award at the University of Colorado, Colorado Springs, in 2011, in recognition of lifelong excellence in teaching, research, and service. More details about Dr. Kalita can be found at http://www.cs.uccs.edu/~kalita.

Preface

Bioinformatics is an exciting field of scientific and technological innovation. This evolving field integrates several disciplines, including computer science and informatics, modern biology, statistics, applied mathematics and artificial intelligence to provide solutions to crucial biological problems at the molecular level. With the rapid growth in data capture and machine learning techniques, it has become possible to obtain, organize, analyze and interpret biological data with an eye to uncovering hidden, non-trivial, and interesting patterns of valuable consequence. One major area within bioinformatics is analysis of gene expression data of disparate kinds such as microarrays, RNAseq, gene ontologies, protein-protein interactions, and various flavors of genome sequence data or combinations. A major advantage of gene expression data produced from the microarray and sequencing technologies is that they provide dynamic information about cell function, whereas the genome provides only static information. The measurement of the activity (expression) of thousands of genes at once so as to create a global picture of cellular function is known as gene expression profiling. Analysis and interpretation of such gene expression data using machine learning or statistical methods can help extract intrinsic patterns or knowledge, which may be of use towards uncovering causes of critical diseases.

Unlike most other computational biology books, this book focuses on gene expression data generation, characteristics of such data, and preprocessing to handle noise and redundancy, to help improve performance of analysis algorithms. This book will help readers learn about the various types of gene expression data such as microarray, MiRNA, RNAseq and ScRNA data in detail enabling them to develop data-centric algorithms towards interesting biological problem solving. The book also includes discussion of a long list of statistical and machine learning methods and techniques applied in finding interesting coexpressed, differentially coexpressed or differentially expressed patterns from microarray, MiRNA, RNAseq and ScRNA data to provide a clear understanding of the state of the art. Further, to provide hands-on experience, the book discusses a large number of tools and systems for creation, extraction and analysis of gene classes, groups or network modules, and their effectiveness from statistical as well as biological perspectives. A detailed pros and cons analysis of these tools will help provide a clear understanding of the ability of each tool. Inclusion of case studies to demonstrate the practical use of the approaches in critical disease handling through interesting biomarker identification makes our book different from most other books. Finally, the reader is exposed to current issues and challenges that need to be addressed to provide appropriate solutions for several known open biological problems.

Pankaj Barah
Dhruba Kumar Bhattacharyya
Jugal Kumar Kalita

MATLAB® and Simulink are registered trademarks of The MathWorks, Inc. For product information, please contact:
The MathWorks, Inc.
3 Apple Hill Drive
Natick, MA 01760–2098 USA
Tel: 508–647–7000
Fax: 508–647–7001
E-mail: info@mathworks.com
Web: www.mathworks.com

Chapter 1

Introduction

1.1 Introduction

An exciting area of significant scientific and technological innovation of recent times is bioinformatics. The field integrates diverse disciplines, including computer science and informatics, biology, statistics, applied mathematics and artificial intelligence to provide solutions to crucial biological problems at the molecular level. With the help of machine learning techniques and statistical methods, it has become possible to organize, analyze and interpret voluminous biological data with an eye to uncovering interesting patterns of great consequence. One major area within bioinformatics is analysis of gene expression data of disparate kinds such as microarrays, gene ontologies, protein-protein interactions and various flavors of genome sequence data or combinations. The genome provides only static information whereas gene expression data analysis produced from the microarray and sequencing technologies provide dynamic information about cell function. The measurement of the activity (expression) of thousands of genes at once so as to create a global picture of cellular function is known as gene expression profiling. Analysis and interpretation of such gene expression data using appropriate machine learning or statistical methods can help extract intrinsic patterns or knowledge, which may be of use towards uncovering causes of critical diseases.

Bio-medical science has been battling against many deadly diseases including cancer for many years, and grand successes have been promised but have been limited, in general. The number of humans affected by such deadly diseases is increasing day by day. Early detection and treatment using modern medical technology has been beneficial in combating the scourge of such

DOI: 10.1201/9780429322655-1

diseases and increasing survival rates. Machine learning is an exciting area of research and practice, which has been applied successfully in bioinformatics to uncover many interesting, yet previously unknown patterns towards identification of biomarker genes for such critical diseases.

1.2   Central Dogma

Genes are the primary factors that control traits of various characteristics in an organism. These characteristics may be associated with certain diseases or normal development processes. There are two major phases associated with the pathways through which genes control characteristics of an organism. In the first phase, genetic code is transferred from genes to proteins through a phenomenon called the Central Dogma.

The Central Dogma of Molecular Biology describes the formation of a protein molecule inside a living organism, as shown in Figure 1.1. The double-stranded DNA molecule is partially unzipped and an enzyme called RNA polymerase copies the gene's nucleotides one by one into an RNA molecule, called the messenger RNA or mRNA. This process is called *transcription*. The mRNA is a small, single-stranded sequence of nucleotides which moves out of the nucleus. Outside the nucleus, another set of proteins reads the sequence of the mRNA and gathers free floating amino acids to fuse them into a chain. The nucleotide sequence of the mRNA determines the order in which an amino acid is incorporated into the growing protein. The process of translating the mRNA sequence into a protein sequence is called *translation*.

The Central Dogma explains the biological process that results in the flow of genetic information into proteins from the information encoded in nucleotide sequences of DNA segments or genes. A protein is a biological macromolecule associated with almost all biological processes, typically governing the traits of various phenotypic and non-phenotypic characteristics in an organism. Hence, as shown in Figure 1.2, genes are the keys that drive protein structure and all biological processes, and thus traits of various characteristics in an organism.

1.3   Measuring Gene Expression

The magnitude of expression of a gene depends on a number of factors, including inter-gene regulatory relationships. The expression level of a gene

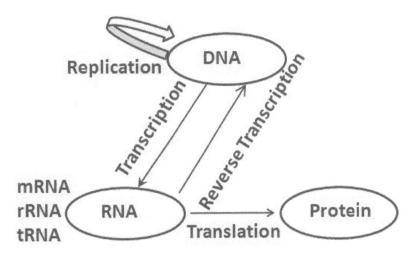

Figure 1.1: Central Dogma: An illustration.

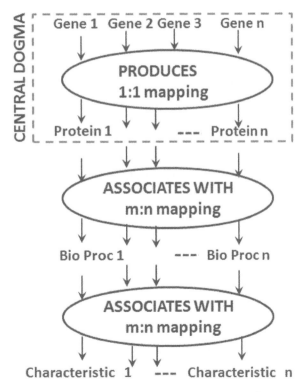

Figure 1.2: Flow of control from genes to traits in an organism.

is a major determinant of the presence of the corresponding governed characteristics in an organism. A number of technologies developed, including revolutionary microarray and sequencing technologies, help determine the expression levels of thousands of genes in a single experiment. Gene expression data generated by these technologies provide an ample resource from which useful biological knowledge can be extracted. Computational analysis of such data can be of great use to biologists. Figure 1.3 shows the exponential growth in the quantity of gene expression data collected over a period of ten years [1]. Figure 1.4 is another example, showing the growth of protein data over a period of ten years [2].

Using appropriate microarray or sequencing technology, it is possible to simultaneously examine the expression levels of thousands of genes across developmental stages, clinical conditions or time points. The real-valued gene expression data are obtained in the form of a matrix where the rows refer to the genes and the columns represent the conditions, stages or time points. Genes, which are the primary repository of biological information, help the growth and maintenance of an organism's cells. Required activities include construction and regulation of proteins as well as other molecules that determine the growth and functioning of the living organism, and ultimately to the transfer of genetic traits to the next generation.

RNA-Seq is a recent and a robust sequencing technology to measure the expression levels of nucleotide sequences corresponding to genes [147]. Determination of how nucleotides are strung together in a DNA molecule is called DNA sequencing. The term *next-generation sequencing* refers to a number of modern advanced high-throughput DNA sequencing techniques [293]. Pyrosequencing [14], DNA colony sequencing [216], massively parallel signature sequencing [68], illumina sequencing [418], DNA nanoball sequencing [444], and heliscope-single-molecule sequencing [334] are some examples of this family of techniques. In RNA-Seq technology, mRNA molecules are sequenced to short nucleotide base sequences. These sequences are then aligned with known nucleotide sequences corresponding to genes to determine expression levels of the genes.

1.4   Representation of Gene Expression Data

The widespread use of the technologies mentioned above has led to generation of an enormous amount of gene expression data that are witnesses to numerous biological phenomena in living organisms. Various types of gene

---

[1]http://www.ncbi.nlm.nih.gov/geo/
[2]http://www.rcsb.org/

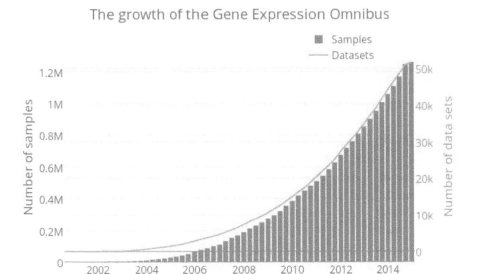

Figure 1.3: Growth statistics of gene expression data.

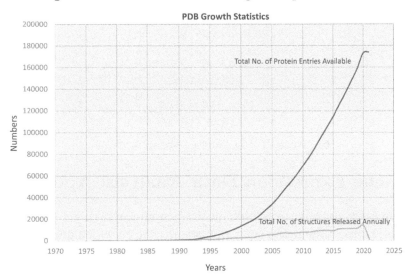

Figure 1.4: Growth statistics of protein data.

expression data correspond to how measurements are carried out and represented. If expression levels of genes are detected in multiple samples collected from different organisms, a two-dimensional gene expression dataset is produced where rows correspond to genes and columns correspond to samples or vice versa [329]. Certain expression datasets store expression levels of genes in one or more samples at various time points. Such specialized gene expression data are called *time-series gene expression data* [270]. A special form of time series gene expression data contains expression levels of multiple samples at multiple time points to form a three-dimensional structure. Such data are called *gene sample time* (GST) expression data [269]. Figure 1.5 presents the structure of a two-dimensional gene expression dataset, whereas three-dimensional GST expression is shown in Figure 1.6. There are numerous online repositories that store and maintain ever-growing gene expression datasets. ArrayTrack [407], ArrayExpress at EBI and Gene Expression Omnibus-NCBI [66] are three very widely used repositories of gene expression data.

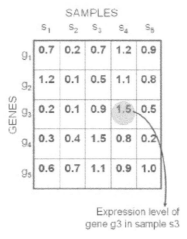

Figure 1.5: 2-D gene expression data.

1.5   Gene Expression Data Analysis: Applications

Gene expression data witness biological phenomena taking place in an organism and hence, they represent a raw resource from which ample biological knowledge can potentially be unearthed. Proper analysis of such data extracts information about underlying biological phenomena. Some problems that can be addressed by gene expression data analysis are briefly discussed below.

(a) **Prediction of Gene Function:** Predicting functions of genes based on their expression values requires designing computational techniques that operate on gene expression data [188].

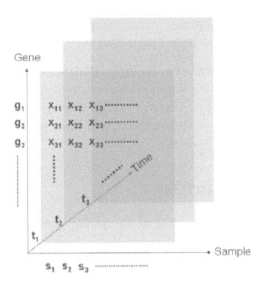

Figure 1.6: 3-D gene expression data.

(b) **Biologically Enriched Network Module Extraction:** Extracting a set of modules (i.e., groups of closely associated genes) with high biological significance from a biological (co-expression or regulatory) network supports identification of interesting behavior of a set of participating driver or causal genes across the states of a disease.

(c) **PPI Complex Identification:** Extracting clusters or complexes from protein-protein interaction (PPI) data supports analysis and interpretation of the association or interaction patterns across the states of a disease in progression. It is a well-accepted observation that in PPI analysis, a PPI cluster or complex consists of two regions, namely, a core, i.e., the central dense region and a periphery, i.e., a relatively sparse region around the core. Both regions impact our understanding of the unique and common properties among the proteins across the states of a disease. So, identification of both these regions with high precision using statistical or machine learning techniques is essential.

(d) **Disease Biomarker Identification:** Identifying causal or driver genes or interesting biomarkers for a given disease across conditions with high precision is the initial step in isolating genetic causes of diseases. To identify such biomarkers, the process exploits co-expression, differential co-expression, or differential gene expression analysis technique(s) that use statistical or machine learning approaches to analyze gene expression data.

(e) **Driver Gene Selection:** Identifying an optimal subset of driver genes for a given disease involves using appropriate statistical or machine learning techniques on gene expression data.

(f) **Inference of Regulatory Relationships:** Constructing regulatory networks by uncovering appropriate association types between a target gene and transcription factor (TF), or between two different TFs, or within the same TF, is necessary to generate interesting inferences for effective diagnosis of a disease.

1.6   Machine Learning

Machine learning is a major field of computer science that deals with the process of inferring interesting but non-trivial unknown patterns from data. During the past several decades, there has been enormous progress in this rapidly evolving field. Three basic machine learning techniques, data clustering, association mining and classification, have been widely used to analyze gene expression data. Extensive work in clustering, classification and association mining have been performed in the context of gene expression analysis. Publications in leading journals and conference proceedings justify this statement. Some of these developments are evolutionary enhancements of previous work; others are revolutionary, introducing new concepts and methods [59], [396]. Research in practical aspects of machine learning often addresses issues that benefits society.

Classification is the process of learning from a set of instances that have assigned class labels to extract intrinsic properties of classes based on attribute or feature values which can be used in the future to assign class labels to new objects or instances without class labels [59]. Thus, it is a process of empirically learning how to assign instances or data objects to a set of pre-defined classes. In other words, classification identifies knowledge in terms of patterns of feature values from the known objects or instances. Classifier development requires designing an inducer module which is able to learn from a set of objects with known class labels. The set of labeled instances or objects is called a training set. The trained classifier is assessed by using it to assign labels to a set of objects called testing samples. Decision trees [129], Naive Bayes [355], K-nearest neighbors [396] and Artificial Neural Networks are some well-known classification techniques.

A classifier can learn only when a set of known objects or instances with class labels are available. However, such training samples may not always be available at the time of learning. Clustering is an unsupervised learning technique that groups unlabeled instances or objects into clusters based on their similarity with respect to feature values without any prior knowledge of classes [162]. Clustering partitions data instances to form a set of disjoint groups, called clusters, so that objects within a cluster or group exhibit high similarity to each other, while objects in separate clusters show high dissimilarity [395]. Therefore, it is clear that an important aspect of cluster analysis is the notion of similarity among objects. A large number of similarity and dissimilarity measures are available. These include Manhattan distance, Euclidean distance, cosine distance, and Pearson correlation co-efficient to measure the proximity between objects. Some examples of clustering techniques are K-means [265], DBSCAN [118] and EM clustering [108].

Association rule mining detects associations among variables that occur in a set of tuples, each called a *transaction*, with a certainty value [162], [395]. It finds interesting associations among a large set of data items using a support-confidence framework. Association rule mining works in two major phases: (i) frequent itemset finding with reference to a user-defined threshold, called *minimum support*, and (ii) rule generation based on frequent itemsets (generated in the previous phase) with reference to another user-defined threshold, called *confidence*. The *support* value of a rule summarizes how frequently an antecedent occurs in the transactions. The *confidence* value of a rule measures how often the rule is found to be true. So, in this support-confidence framework, *minimum support* and *confidence* play a crucial role during association finding and rule generation. Usually, association rule mining is described in terms of market basket data, which is represented as a 0-1 matrix where each row in the matrix represents a transaction, and each column represents a set of items (1 represents the fact that the item is present whereas 0 represents that the item is absent). An association rule is of the form $X \to Y$ where X is a set of items called *antecedent* and Y is another set of items called *consequent*. $X \to Y$ implies that if items in X are bought in a transaction, the items in Y are also bought in the same transaction. Apriori [6] and FP-growth [163] are examples of association rule mining techniques.

The machine learning techniques mentioned here have already been successfully applied to solve a great variety of problems in bioinformatics to address issues that arise in applications.

1.7   Statistical and Biological Evaluation

It is a well-known fact that nothing is perfect, and thus there are often limitations to any technique or approach. An evaluation or assessment of quality or accuracy of a system, mechanism or method is usually performed as a snapshot in time. With the passage of time after a system is initially built, the environment or constraints change and new situations arise, and accordingly, the evaluation needs to be performed again after parameter tuning. However, the information obtained during one evaluation process plays a significant role in subsequent evaluations as well as in the validation and establishment of the final outcome of a process. A number of validity indices have been proposed [59] to assess the quality of outcomes generated by the computational techniques proposed for gene expression data analysis. To assess the significance of a gene (or gene groups) identified or extracted by machine learning and co-expression network analysis approaches, one can use both biological as well as statistical measures to validate the results. Based on the nature and operational aspects of these validity measures, these two categories of indices, i.e., statistical and biological, have their own advantages as well as limitations. None of the measures alone is sufficient for the purpose, and hence, often a combination of these measures is used.

Biological measures rely on external standard repositories or databases, which are used as references during validation of the results. Statistical measures attempt to assess the performance of the methods based on relevance and correctness with reference to some ground truths. *P*-value estimation helps evaluate the quality of the groups of genes or proteins predicted or extracted by a statistical or machine learning method with respect to the groups of genes defined by gene annotation in the Gene Ontology repository. *P*-value is computed using the hypergeometric test or Fisher's Exact Test [413], [270], [16]. This test is based on the null hypothesis that the list of genes in a group is a random sample from the population. Fisher's Exact Test computes *p*-values from the contingency table using the hypergeometric distribution. The *p*-value signifies how well a group of genes matches different GO categories. A low p-value of a group of genes in a cluster or module indicates that the genes belong to enriched functional categories defined in annotations of Gene Ontology and are biologically significant. There are also validity indices to assess the performance of a computational method in terms of metrics such as *precision, recall, F-measure*, and *accuracy*. These match the results against the pre-existing standards or ground truths. Precision is the fraction of the relevant instances or groups of genes or proteins among the retrieved instances or groups of genes/proteins.

On the other hand, recall represents the fraction of relevant instances or groups of genes/proteins that has been generated or extracted by a statistical or machine learning method over the total number of relevant instances or groups. Precision is also known as *positive predictive value* and recall is also called *sensitivity*. *F*-measure is the harmonic mean of precision and recall. It is represented as $2 \times (precision \times recall)/(precision + recall)$. It is also known as the traditional *F*-measure, which was found to be biased and hence updated to introduce other variants, such as F_2 (where *recall* is more emphasized) and $F_{0.5}$ (where *precision* is more emphasized). *Accuracy* is another commonly used measure to evaluate the performance of supervised and unsupervised learning techniques. Accuracy of a method is assessed in terms of positive predictive value (*ppv*) and sensitivity (S_n). It is defined as a geometric mean of *ppv* and S_n. *Co-localization score* is another measure to evaluate a method in terms of predicted groups with reference to a standard localization dataset. It is defined for a group c_i as $colocalz(c_i)=max(v_p/V)$, where v_p is the total number of objects in c_i which reside in location p, and V is the total number of objects in c_i. In addition to the above indices, one can also use pathway databases or repositories to establish the effectiveness of a method. A pathway represents a series of related biochemical reactions occurring in a living body. The number of common pathways found to be shared by the members of a predicted group or cluster with established driver or causal genes for a given disease is another effective validation for a computational method. There are several tools and standard repositories to support such validation activities. PANTHER [282] and Genecard [342] are some examples of tools and repositories.

1.8 Gene Expression Analysis Approaches

The three most prominent approaches for gene expression data analysis for extraction of interesting genes, patterns or biomarkers to support effective handling of a disease are: (a) co-expression analysis, (b) differential co-expression analysis, and (c) differential expression analysis.

Co-expression analysis attempts to find groups of genes that exhibit similar trends in expression values over a group of samples [275]. It helps find the association of a gene with unknown functions to a disease. It also discovers associations of a gene of unknown function with a biological process. Another interesting application is the prioritization of candidate disease genes. Four commonly used co-expression analysis approaches are: (a) clustering, biclustering or triclustering to find co-expressed groups of genes, (b) network module extraction, (c) intra- and inter-cluster hubgene-finding, and (d) guilt by association.

Differential co-expression analysis helps discover disease mechanisms and underlying regulatory dynamics, which cannot be discovered using co-expression or differential expression analysis alone [275], [16]. This approach identifies a group of genes which show varying co-expression patterns over different conditions such as developmental stages, tissue types and disease states. These genes are the regulatory genes responsible for different phenotypic variations. Disease-related genes have tissue-specific expression. So, this analysis can help identify genes that show co-expression in both similar and different types of tissues.

Differential expression (DE) analysis is another commonly used approach for transcriptome profiling to help with effective disease diagnosis. DE analysis aims to find differentially expressed genes, i.e., to identify a set of genes which exhibit significant differences in terms of expression levels across the states or conditions of a disease, such that the differences are greater than what would be expected due to natural random variations only [88], [351]. The two major tasks of DE analysis are: (a) to find the magnitude of differences in expression levels between two or more conditions based on read counts from replicated samples, i.e., to calculate the fold changes in read counts, taking into account the differences in sequence depths and variability, and (b) to estimate the significance of the differences using appropriate statistical tests.

Each of these approaches has its own purposes, advantages and limitations. However, all these approaches require some amount of preprocessing to transform the data into an adequate form for unbiased and effective analysis. Next, we discuss the preprocessing techniques usually applied on gene expression data followed by a brief discussion of each of the three gene expression data analysis approaches.

1.8.1   Preprocessing in Microarray and RNAseq Data

Widely used preprocessing techniques for both microarray and RNAseq data include missing value estimation, discretization, batch effect removal, normalization, data transformation, and removal of outlying entries or with low read counts. Microarray and RNA-seq technologies obtain average gene expression values of cells. This may not be useful in discovering information present in individual cells because there is a chance of losing much information. To address this important issue with both the aforesaid technologies, another advanced technology has been recently introduced, referred to as scRNA-seq technology, where gene expressions are quantified at a cell level in an experiment. The main difference between scRNAseq and RNA-seq is that the sequencing library contains a single cell.

These three technologies require different types of preprocessing to improve downstream analysis. In contrast to microarray data, in case of RNA-seq and scRNA-seq count data, often very few reads are mapped to a large number of genes. Such genes are not informative for downstream analysis. Zero inflation is an inherent bias in RNA-seq and scRNA-seq data. A bias called *dropout events*, is present only in scRNA-seq data. Imputation is required to address these two biases. Researchers assume that microarray data follow a normal distribution, whereas RNA-seq and scRNAseq data follow discrete data distributions. So, sometimes transformation of data is also required to use the methods, which were originally developed for microarray proper analysis.

Discretization

Discretization is a commonly used preprocessing task, especially for microarray data to facilitate effective gene expression data analysis. Use of appropriate discretization techniques can result in significant improvement in both classification and cluster analysis. In gene expression data analysis, discretization as a preprocessing task helps overcome the issues that arise due to the possible presence of noise, missing values and systematic variations during experiments with microarray technology [271]. Discretization attempts to convert real gene expression data into a typically small number of finite values without compromising the variation trend in the original data. Rules generated based on discrete values are usually compact, simple and convenient to use, and hence can lead to better predictive accuracy. Further, a significant portion of the noise is removed in the discretization process.

Discretization methods can be categorized as *supervised* and *unsupervised* [271]. A supervised discretization method transforms the raw data into discretized form using prior knowledge, while unsupervised methods do not use any such knowledge. Most commonly used discretization methods are unsupervised. Unsupervised discretization methods can be further subdivided into two distinct categories [58]: (i) using expression absolute values and (ii) using expression variations between time points. Discretization using expression absolute values discretizes the gene expression data directly using absolute gene expression values. These techniques can be further divided into row discretization, column discretization and matrix discretization. In row discretization, only the values in the rows are considered during discretization. In column discretization, values are discretized based on the values of the columns. In matrix discretization, both row and column values are considered during discretization.

Discretization using expression variations between time points can be, in general, applied to time series gene expression data. It computes variations between a pair of consecutive time points to discretize these variations. All these techniques can be further divided into three categories, namely, binary, ternary, and multilevel. A binary discretization technique discretizes gene expression data to two discretized values, whereas a ternary technique produces three discretized values. A multilevel discretization technique generates more than three discrete values in the discretized gene expression matrix.

Missing Value Estimation in Microarray Data

Microarray gene expression data are generally peppered with missing values due to inadequate processing, data recording, or environmental conditions. Such missing values can significantly deteriorate the performance of downstream analysis, and hence it is expected that the datasets are completed by inserting estimated values. Among the causes, inconsistent data recording techniques play a major role in the occurrence of missing values. Such inconsistencies may be the result of both human as well as machine errors. We categorize missing value imputation techniques into two categories: local and global.

A local imputation technique identifies a group of genes with the highest association or relevance (using appropriate proximity measure) to the target gene to estimate missing values. One can use the k-Nearest-Neighbors algorithm [438], [396] or its variants such as iterative kNN, sequential kNN (SKNN), least squares adaptive (LSA), local least squares (LLS), iterative LLS (ILLS), or sequential LLS (SLLS) for local imputation. For global imputation, singular value decomposition and Bayesian principal components analysis are two effective approaches. Between these two approaches, missing value estimation using local imputation has been found to be more effective.

In contrast to microarray data, bulk RNA-seq and scRNA-seq datasets generally contain large numbers of zeroes (zero inflation). scRNA-seq data also suffer from dropout events, which may mislead downstream analysis. Generally, the amount of mRNA present in a cell is very low and up to 1 million amplification is necessary. Low starting points lead to missing many mRNAs during cDNA production using reverse transcription, and thus they

may not be detectable in later steps of sequencing. This also happens due
to dropout events when a gene expressed highly or moderately in one cell
may not be expressed in another [245]. To address such problems, imputation is required. Local and global imputation techniques found effective in
microarray data, can also be used for bulk RNA-seq and scRNA-seq data.
However, some methods may not be applicable directly on RNA-seq and
scRNA-seq data because of the discrete distribution of count data. scRNA-
seq quantifies the gene expression profile of a single cell, and does not average
like other technologies working with gene expression data [177]. There are
some high-precision imputing methods for both microarray and RNAseq
data, but they require high computational resources. On the other hand,
there are cost-effective supervised imputing techniques for both categories
of data, but due to non-availability of learning samples, their use is limited.
So, special care is needed to address these issues.

Normalization

Normalization aims to transform raw gene expression data into an unbiased
form for enhanced inference generation in downstream analysis. It makes
gene expression levels comparable across samples and genes to generate valid
inferences. The major objective of normalization is to eliminate biases that
occur due to biological differences. A number of normalization techniques
have been proposed. These techniques are dependent on five major factors:
sequencing depth, size of transcript, sequencing error rate, GC-content, and
insert size [4]. From the results of empirical studies, it is evident that quantile normalization can significantly improve the RNA-seq data quality including those with low amounts of RNA [165]. Similarly, the package called
EDASeq by Bioconductor [346], can reduce GCcontent bias [52]. The NVT
package [113] can identify the best normalization approach for an RNA-
seq dataset by analyzing and evaluating multiple methods via visualization,
based on a user-defined set of uniformly expressed genes. Zyprych-Walczak
et al. [464] provide a procedure to find the optimal normalization for a specific dataset. They warn that an inappropriate normalization method affects
DE analysis.

Elimination of Low Read Counts

This preprocessing task attempts to handle the occurrence of a large abundance of 0s in the count data matrix, which often misleads both (differential)
co-expression and differential expression analysis. This may cause failure of

any correlation measure in identifying a pair of related genes, even with perfect correlation. Due to the presence of a very low number of mapped reads for a significant number of genes, downstream analysis may lead to a biased result. So, preprocessing is essential to overcome such issues. One possible solution is to discard such entries. Prior to performing (differential) co-expression or differential expression analysis, it is recommended that we select only differentially expressed genes. Several effective tools or algorithms have been developed with in-built pre-processing support to identify such differentially expressed genes, in the presence of a large number of 0 counts or low read counts.

Feature Selection

A feature selection algorithm aims to identify an optimal subset of relevant features, which ensure best possible classification accuracy. Feature selection, also sometimes referred to as variable selection, dimensionality reduction, attribute selection or variable subset selection, is integral to knowledge discovery and machine learning processes. It helps build a cost-effective high precision predictive model for a dataset. This multi-step pre-processing task can help extract interesting patterns. Feature selection can be supervised as well as unsupervised, depending on the use of knowledge during feature relevance estimation and evaluation.

An efficient feature selection method can improve the performance of a learning model in three different ways: (a) it can alleviate the effect of a large number of dimensions, (b) it can enhance the generalization capability as well as learning speed, and (c) it helps understand the data by discovering important features and associations among them. Feature selection has been the focus of interest for quite some time and substantial work is available. With the creation of huge databases and the consequent requirements for good machine learning techniques, new problems have arisen and novel approaches for feature selection are being proposed. Based on the nature of the algorithm, feature selection can be of four types: filter, wrapper, embedded and hybrid.

1.8.2   Co-Expressed Pattern-Finding Using Machine Learning

Gene expression data are generated from high-throughput microarray or sequencing technologies, and they are generally presented as matrices of expression levels of genes under different conditions. One objective of gene expression data analysis is to identify groups of genes with similar expression patterns over the full space or a subspace of conditions. This may reveal natural structures and identify interesting patterns in the underlying data. A cluster of genes can be defined as a set of biologically relevant genes which are similar based on a proximity measure. The goal of co-expressed pattern-finding, using supervised or unsupervised learning, is to assign appropriate class labels to the instances with or without prior knowledge. Another goal is to expand classes or groups, so that two criteria are satisfied: *homogeneity*—-elements in the same cluster are highly similar to each other, and *separation* —- elements from different classes or clusters have low similarity to each other. In supervised methods, the dataset is partitioned into disjoint classes using a class attribute. A classifier model is built based on the training data (data sample plus correct class labels) and later the model is used for predicting the class of an unknown sample. The goal of classification is to analyze the training set to develop an accurate description or model for each class using the attributes present in the data. Many classification models [162], [396], [395], [59] have been developed, including neural networks, genetic models, decision trees, KNN, and support vector machines, to cater to the learning needs of various application domains.

In unsupervised methods, the dataset is partitioned into disjoint groups called clusters with high intra-cluster similarity and low inter-cluster similarity. A similarity or distance measure is an important metric used to build clusters. To a large extent, the quality of clustering depends on the appropriateness of the similarity measure used for the data set or the domain of application. Numerous similarity and dissimilarity measures are available, such as Euclidean distance, cosine distance, and Pearson correlation co-efficient, to measure the amount of proximity between a pair of objects. A number of techniques are available for the purpose of clustering gene expression data. Partitioning methods, hierarchical methods, density-based methods, grid-based methods and model-based methods are some well-known clustering techniques.

Various traditional clustering techniques such as k-means [265] and CLINK [99] have been used in gene expression data analysis to identify groups of co-expressed genes, which share similar functionality. Some specialized clustering techniques such as CLICK [367] and DHC [196] have been proposed for gene expression data analysis. Genes may not always be similar across a whole set of features due to the fact that a biological phenomenon may take place on a subset of samples in a gene expression dataset [328]. This observation has led to the emergence of a variant of clustering that attempts to find groups of clusters that maximize similarity across subsets of features corresponding to subsets of samples. This variant of clustering is called biclustering [16]. The emergence of three-dimensional GST gene expression data presented a challenge to researchers to develop another variant of clustering to work in three-dimensional space. This variant of clustering is called triclustering [269]. There are very few triclustering techniques that extract genes which have similar expression patterns across a set of samples over a set of time points from GST expression data. A challenge is to take into account both inter-temporal and intra-temporal gene coherence in the process of tricluster mining. Other challenges in triclustering are avoidance of time-dominated and sample-dominated results and detection of time-latent triclusters.

In gene function prediction, one is concerned with the clusters that have high overlap with known biologically curated groups of genes. Genes which are included in such clusters, but do not belong to the biological groups with which the clusters have high overlap, are genes that have not yet been found to be associated with the biological phenomena corresponding to the biological groups. Such predictions can further be confirmed by biologists. In disease diagnosis, genes associated with a disease can also be predicted in a similar way. The disease can be diagnosed based on expression patterns of these genes along with the known genes. Weighted Gene Correlation Network Analysis (WGCNA) is a widely used package for finding such groups (or modules) of genes [235]. It applies hierarchical clustering on a co-expression network to find modules. This tool can be applied to both RNA-seq and microarray as well as to scRNA-seq data. This was the first tool applied to RNAseq data, and it can effectively find modules which are responsible for biological processes. cMonkey2 is a tool to detect co-regulated gene modules for any organism [345]. ConGEMs is a tool for extracting biomarker modules from transcriptomic data by detecting co-expression modules using association rule mining and weighted similarity scores [274]. FGMD is based on a hierarchical clustering algorithm [198]. It was developed to capture actual biological observations. It has shown excellent performance on different datasets for most functional pathway enrichment tests. This algorithm has been applied to both microarray and RNA-seq datasets, and modules constructed from these two types of datasets have many common genes.

The basic difference between classification and clustering is that classification assumes prior knowledge of class labels, while clustering does not assume any knowledge of classes. Supervised and unsupervised methods have their own biases and cannot always provide an accurate analysis of high-dimensional data. Therefore, ensemble methods are often used to improve the overall prediction accuracy by combining the outputs of several algorithms. We can also combine ensemble approaches to gain further improvement by building an ensemble of ensembles, known as a *meta-ensemble*.

Co-expression analysis enables researchers to identify groups of tightly correlated genes associated with biological processes or pathways, and it can be used to illustrate the correlation between genes in a graphical form. It shows which genes in a group of samples have the tendency to exhibit similar expression. A gene expression network generally has a scale-free topology and differential gene connectivity analysis of a network helps identify genes that are involved in different phenotype variations [88]. Network analysis extracts complex relationships between the genotype and the phenotype, and reveals the biological mechanism of complex diseases. Figure 1.7 shows an example biological network.

Figure 1.7: Biological network: An example.

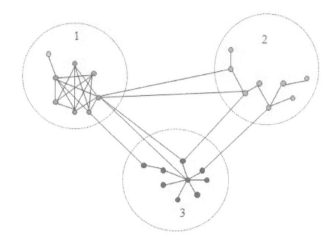

Figure 1.8: co-expression network modules: An example

1.8.3   Co-Expressed Pattern-Finding Using Network-Based Approaches

A co-expression network reveals genes responsible for a trait. To obtain the specific process responsible for a biological problem, co-expression network analysis is performed to extract a smaller network with modules of high biological significance from the co-expression networks of genes present in expression datasets. Figure 1.8 shows an example with network modules extracted from a co-expression network.

RNA-seq co-expression networks [39] enable the discovery of the association of a gene with unknown functions to a disease. It also enables researchers to associate a gene of unknown influences to a biological process. Another interesting application is prioritization of candidate disease genes. co-expression analysis can be used to focus on splice variants and non-coding RNAs [416].

Limitations of co-expression network analysis include: (a) causality information or any distinction between regulatory and regulated genes cannot be found from gene co-expression networks, (b) tissue or condition-specific modules (i.e., differentially expressed modules) may not be detectable in a tissue- or condition-independent co-expression network because the correlation measure used during computation of pairwise correlation of genes may show lower correlation between a pair of genes due to the lack of a proper correlation signal in other conditions or tissues for the same pair of genes, (c) modules which show different co-expression patterns across different tissues or conditions are not detectable in co-expression analysis. This limitation can be overcome by differential co-expression analysis, which can detect such modules.

1.9 Differential Co-Expression Analysis

This approach helps extract interesting intrinsic patterns, associations, or behaviors of molecular mechanisms which are non-trivial and cannot be discovered by co-expression or differential expression analysis alone [416]. Differential co-expression analysis can identify groups of genes which show varying co-expression patterns over different conditions such as developmental stages, tissue types and disease states. It can identify genes which are responsible for regulating different phenotypic variations. Disease related genes have tissue-specific expressions. So, this analysis helps identify genes that show co-expression in both similar and different types of tissues. The genes which have different co-expression patterns in different sample groups, are the probable regulatory genes responsible for phenotypic variations observed in different conditions. A large number of network-based methods have been developed using this approach to identify interesting disease biomarkers for critical diseases. RNAseq enables researchers to study the expression profiles within or across the sample groups [325]. This approach has already been established to perform well for disease sub-classification towards identification of prognostic markers and to investigate datasets with unknown sub-populations, e.g., large-scale single-cell RNA-seq data analysis. This approach is sensitive to noise.

1.10 Differential Expression Analysis

DE analysis is a common approach for transcriptome profiling to support effective disease diagnosis. The main aim of DE analysis is to find differentially expressed genes, i.e., to determine those genes for which gene expression-level differences are greater than what would be expected due to natural random variations only. It is a well-established approach for both microarray and RNAseq data. Many network-based gene expression data analysis methods have been introduced to identify interesting disease biomarkers, where DE analysis is used as an initial task to identify the input set of DE genes for subsequent co-expression analysis. Such analysis helps compare transcriptomes across developmental stages, and across disease states. Several powerful tools have been developed based on this approach, especially for RNAseq data analysis.

The basic tasks of DE analysis tools are: (a) to estimate the magnitude of variations in expressions between two or more conditions based on read counts from replicated samples, and (b) to compute the significance of differences using statistical tests. To determine the genes whose read count differences between two conditions are greater than expected by chance, DGE tools make assumptions about the distribution of read counts [88]. One of the most popular choices to model read counts is the Poisson distribution because it provides a good fit for counts arising from technical replicates, and it is suitable when a small number of biological replicates are available. However, this distribution predicts smaller variations than those seen between biological replicates. Robinson et al. [351] propose the use of the Negative Binomial (NB) distribution to model counts across samples and capture the extra-Poisson variability, known as over-dispersion. The NB distribution has parameters, which are uniquely determined by mean and variance. With the introduction of a scaling factor for the variance, NB outperforms Poisson, and is a widely used model for feature counting. However, the number of replicates in datasets of interest is often too small to estimate both parameters, mean and variance, reliably for each gene. Several empirical studies and evaluations of performance of DE analysis methods can be found in [88].

1.11 Tools and Systems for Gene Expression Data Analysis

Computational biologists have developed a number tools using (diff) co-expression and differential expression analysis to support microarray and RNAseq data. These tools, under each of the above three approaches, can be further classified based on the computational technique used. Due to much overlap in the algorithms, tools belonging to co-expression and differential co-expression analysis categories are grouped into one category. Machine learning techniques play a dominant role in this category of tools. For DE analysis tools, statistical techniques are more dominant.

1.11.1 (Diff) Co-Expression Analysis Tools and Systems

The most widely used programs for differential co-expression analysis are WGCNA   [235], DICER   [305], DiffCoEx   [403] and DINGO   [159]. These programs first identify co-expressed modules considering all samples, and then compare these modules with a defined set of samples, such as samples representing disease states or tissue types. For example, if a pair of genes show high positive correlation in a normal condition and a negative correlation in disease condition, there is a high chance that these genes are responsible for the disease. Siska et al. [375] developed a tool named Discordant and reported that Spearman's correlation is the best method for differential correlation analysis in RNA-seq data. Most differential co-expression methods perform analysis on the full space in samples of each sample group. These approaches may miss patterns that only cover a subset of samples, i.e., a sub-space in one or more sample groups. Fang et al. [122], [88] performed sub-space-based differential co-expression analysis and proposed a method to find the sub-space of differential co-expression patterns. They also found that discovered sub-space patterns are functionally enriched in terms of pathways and are related to microRNA and TFs. Many tools for (differential) co-expression analysis have been reported and most were initially developed for microarrays and later adapted for RNA-seq data analysis.

1.11.2 Differential Expression Analysis Tools and Systems

Many algorithms have been introduced for the identification of DEGs from RNAseq data. Most techniques accept the read-count data matrix as input [88]. The Read count is the number of reads mapped to each gene in the samples. Some methods directly accept count data as input, but others transform count data before taking them as input. The methods that directly work on count data can be classified into two groups, (a) parametric and (b) non-parametric. Parametric methods are generally influenced by outliers, but this limitation can be overcome by non-parametric methods. To control FDR (False Discovery Rate), the Benjamini-Hochberg procedure is used by most tools. RNA-Seq 2G [461] is a web portal containing many DE analysis tools. Some tools support RNA-seq analysis workflows, and these are treated as global tools. Most tools are implemented in R, making them platform independent.

1.12 Contribution of This Book

The following are the major contributions of this book.

(a) A background discussion of the Central Dogma, machine learning and statistical techniques, and statistical and biological enrichment as well as validation methods.

(b) An in-depth discussion of gene-gene co-expression, differential co-expression and differential expression analysis for microarray, MiRNA, RNAseq and scRNA data.

(c) A systematic presentation of statistical, machine learning, knowledge-based and soft computing methods for identification of interesting co-expressed, differentially co-expressed and differentially expressed genes in microarray, MiRNA, RNAseq and scRNA data.

(d) Discussion of a large number of practical tools, systems and repositories that are useful for computational biologists to create, analyze and validate biologically relevant co-expressed, differentially co-expressed and differentially expressed gene expression patterns.

(e) A practical wishlist and a list of important issues and research challenges.

1.13   Organization of This Book

This book discusses gene expression data analysis from both machine learning and statistical perspectives. Recently, there has been tremendous progress in technologies for gene expression data generation. The data are voluminous (both from dimensionality and number of instances) and are generated continuously. To analyze such data for identification of interesting patterns that are relevant and enriched for a given disease query, efficient machine learning algorithms are necessary. The development of such algorithms is possible only when the science of the biological data is clearly understood. This book presents several such useful algorithms under the categories of machine learning, statistical, knowledge-based and soft computing-based approaches. To understand the effectiveness of such an algorithm, method or system, we also discuss performance metrics (both statistical and biological) that can be used in a real-life as well as in a simulation environment. Further, this book also discusses a large number of tools, systems and repositories that are commonly used in analyzing and validating results for microarray, MiRNA, RNAseq and ScRNA data. The book is composed of four major parts.

Part I of our book contains one chapter. In Chapter 1, we provide an introduction to the Central Dogma and gene expression data types. This chapter also briefly introduces the purpose of various methods and techniques introduced in recent years to analyze such data. It highlights the distinguishing features of the book and the major contributions.

Part II of the book contains four chapters to provide background for learners, practitioners and researchers of this field. Chapter 2 delves deeper into issues related to the Central Dogma of modern biology. Chapter 3 discusses the generation of gene expression data and their characteristics. In Chapter 4, we provide background on supervised and unsupervised learning techniques. Similarly, to develop a background on evaluation and validation of the outcomes on gene expression data analysis, Chapter 5 presents necessary statistical and biological methods and techniques.

Part III is composed of three long chapters. Chapter 6 presents an overview of co-expressed pattern analysis, followed by an in-depth discussion of approaches, methods and techniques for co-expressed pattern-finding for microarray, MiRNA and RNAseq data, and scRNA data. In Chapter 7, we discuss popular and impactful methods and techniques for differentially co-expressed patterns for gene expression data. Similarly, in Chapter 8, we provide a generic framework for differentially expressed genes, followed by a large number of approaches, methods and techniques for differentially expressed pattern-finding in microarray, MiRNA and RNAseq data, and scRNA data.

Part IV of the book includes two chapters. Chapter 9 presents a number of tools, systems and repositories commonly used by computational biologists. The last chapter, Chapter 10, concludes the book with a discussion of research issues and challenges in this evolving field of research.

Chapter 2

Information Flow in Biological Systems

2.1 Concept of Systems Theory

Complex Systems consist of many interrelated and interdependent components or parts, connected to each other. To understand the global properties of complex systems, one cannot rely on a single rule. In other words, the characteristics of a complex system arise not due to properties of one individual component, but are the manifestations of the total interactions among the constituent parts. Thus, a complex system exhibits properties that emerge from the interactions among its individual components, which cannot be predicted from the properties of the individual constituent parts. Systems science is a rapidly evolving field studying how parts of a system give rise to the collective behavior of the system, and how the system interacts with its environment [126].

2.1.1 A Brief History of Systems Thinking

The idea of modern Systems Theory originated from Karl Ludwig von Bertalanffy's General Systems Theory (GST), described in his book titled *General System Theory: Foundations, Development, Applications* [55]. Many other contemporary experts such as Kenneth E. Boulding, William Ross Ashby and Anatol Rapoport, later adopted the concept proposed by Bertalanffy. They used the concept in diverse areas such as in mathematics, psychology, biology, game theory and social network analysis. However, the concept of systems thinking can be traced back to the days of Aristotle (384–322 BC).

DOI: 10.1201/9780429322655-2

The most famous quote about complex systems has come from Aristotle who said that "The whole is more than the sum of its parts" [376].

2.1.2   Areas of Application of Systems Theory in Biology

As Systems Theory says, high levels of organization exhibit emergent properties [126]. Biological systems are composed of highly interconnected elements, arranged in a hierarchical manner from the molecular to the whole organism and the ecosystem [258, 302]. The relationship between complexity and physiological stability has been observed among all biological systems. Interactions among modular components of a complex biological network can facilitate predictions of behavior under environmental perturbations. Temporal dynamics of parameters related to biological processes such as photosynthesis, metabolism, enzymatic reactions and a broad class of fluxes could be associated with a greater capacity of system homeostasis and successive adaptation and genome evolution.

2.2   Complexity in Biological Systems

2.2.1   Hierarchical Organization of Biological Systems from Macroscopic Levels to Microscopic Levels

Biological systems are organized in a hierarchical manner from the microscopic sub-atomic level to the macroscopic level, comprising ecosystems and communities. Lower levels of organization are gradually integrated to create higher levels. Atoms, as the primary building blocks of any matter, are also the foundations of life's organizational structure. A number of atoms interact among themselves through chemical bonds to get organized in a particular spatial arrangement to form a molecule. Examples of commonly known classes of biomolecules include DNA, RNA, protein, metabolites, and hormones. Molecules come together to form organelles, which are highly organized structures within a cell that carry out specific cellular functions. Examples of organelles are nucleus, endoplasmic reticulum, Golgi apparatus, mitochondria, ribosomes, chloroplast, and cell wall. The next level of hierarchy in biological organization is the cell, which is the functional unit of life. Living organisms can either be single-cellular, or multicellular. In multicellular organisms, similar cells that perform a common function come together to form tissues. Different types of tissues work together to form another level of functioning unit known as organs. Several organs and related tissues are integrated into organ systems. For example, the heart, blood, arteries, and veins are the interrelated tissues of the cardiovascular system. Members of a single type of organism (such as a species) living in a defined area form a population. When different species interact with the populations of other species, it becomes a community. All living organisms from diverse biological domains together with environmental factors form an ecosystem. The macroscopic organization may further be extended up to the biosphere that contains all abiotic environments on the planet Earth and beyond.

2.2.2 Information Flow in Biological Systems

Early chemists believed that the secrets of life resided in its chemical composition. Linus Pauling said, "Life is a relationship between molecules, not a property of any one molecule." Information flow is the core of all interactions within and among hierarchical levels of a biological system. Interactions within molecules in the form of chemical bonds can be seen as a flow recording attractions among atoms, ions or molecules. Information for any biological system representing a life form is encoded in genetic material or in its genome. Sophisticated molecular machineries can read the information to decode it to govern life processes through tightly regulated molecular interaction networks. Molecular networks such as gene-protein, protein—protein, and metabolic interactions work as the interconnected grid for the flow of genetic information, embedded in the genotype-to-phenotype continuum.

Technological advancements in the last several decades have enabled scientists to generate enormous amounts of high-throughput omics data such as genomic, proteomic, transcriptomic, metabolomic, phenomic, interactomic, and ionomic. The suffix -omics in these terms comes from the word -ome which means "in total." Omics are high-throughput technological means to characterize and quantify the functions of and interactions among a large number of biological molecules that maintain the structure, function, and dynamics of an organism. With such omics data, along with robust bioinformatics and data mining tools, scientists can now explore relevant correlations and construct mathematical or statistical models describing physiological states. Such models of the various cellular components such as enzyme activities and complexes, gene expression, metabolite pools or pathway flux modes can help gain insights into the complexity of biological systems at every level of structural and functional organization.

2.2.3   Top-Down and Bottom-Up Flow

The flow of biological information is bidirectional, i.e., from top (macroscopic) to bottom (microscopic) and vice versa. Let us consider an example to understand this statement. Interactions of living communities with the living and non-living environments are crucial for the sustainability of the living communities. A living organism adapts and evolves, adjusting with the local environment. Those that are not capable of achieving such adaptations are eliminated in the course of time. While integrating multidimensional heterogeneous data from omics experiments into consistent models, it is possible to describe and predict the behavior of biological systems, for example with respect to endogenous or environmental changes. The environment triggers pressure on the organism in a top-down manner, which gradually gets incorporated in the genetic codes at the bottom level of the hierarchy, subsequently propagating to next generations, establishing ways for genome evolution. On the other hand, mutations occur in the genome all the time. The mutations can alter the original genetic information encoded in the genome, which can be decoded and propagated through molecular interaction networks to the phenotypic level. Those phenotypes that can tolerate and thrive in the environmental conditions survive, and the mutations may propagate to the next generation (if the mutation occurs in the germ cells). In this case, the information flow is from the bottom to the top level of the hierarchy in the biological organization. Figure 2.1 illustrates this hierarchical organization of biological systems.

2.3   Central Dogma of Molecular Biology

In nature, living organisms from the unicellular level to the complex multicellular level have specific phenotypes. These phenotypes may either be directly visible, such as skin color, hair color, skin patterns, the presence or absence of feathers and height, or may be invisible such as various aspects of metabolism. Specific proteins control all phenotypes in living systems at the molecular level. The question arises: What exact mechanism causes the cellular systems to synthesize specific proteins responsible for or controlling a specific phenotype? What are the underlying mechanisms for the intra-species variability of the phenotypic characteristics? Answers to such questions can be found in the Central Dogma.

Francis Crick first proposed the Central Dogma in 1958. He is well known for his discovery of the DNA double helical structure. The Central Dogma can be stated as postulating the flow of genetic information from hereditary

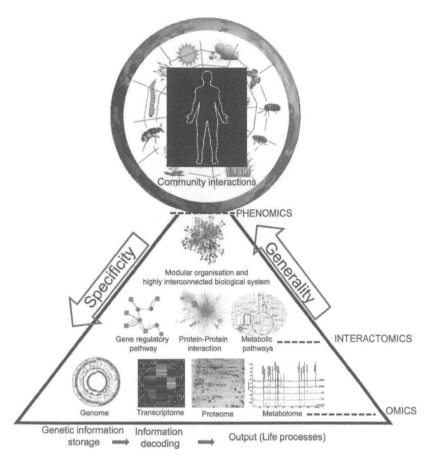

Figure 2.1: Hierarchical organization of complex biological systems from macroscopic to microscopic level showing top-down and bottom-up flow of information.

DNA macromolecules into functional products as protein molecules. The steps in the Central Dogma are the following.

- *Replication:* Making new DNA molecules from existing DNA molecules.

- *Transcription:* Making RNA molecules from existing DNA molecules.

- *Translation:* Making functional protein molecules from RNA molecules.

2.3.1 DNA Replication

The basic mechanism of replication is similar in all organisms. A semiconservative mode of replication produces two daughter macromolecules from one parent DNA macromolecule. The sequential steps in replication are shown in Figure 2.2.

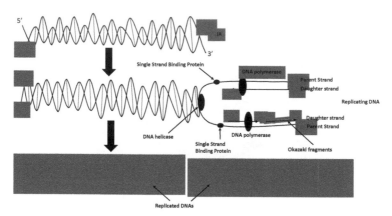

Figure 2.2: DNA replication.

- The enzyme called DNA helicase, and single-stranded binding proteins open the DNA and prevent the rewinding of the DNA helix, respectively.

- Hydrogen bonds break the A=T and G≡C base pairs.

- DNA polymerase, a DNA synthesizing enzyme, binds to one of the parent DNA strands and adds nucleotides, one by one, to the DNA strand that is starting to form.

- Simultaneously, the same process occurs with another parent DNA strand.

- ATP, the energy-carrying molecule, adds nucleotide bases to the new DNA strand.

- After the synthesis of the new DNA molecule, the DNA replication complex disassembles, and the daughter molecules are released.

2.3.2   Transcription

Transcription is the copying or transcribing process, which gives rise to a single-stranded messenger RNA (mRNA) molecule from a small portion of the DNA strand, as shown in Figure 2.3. The steps in transcription are as follows.

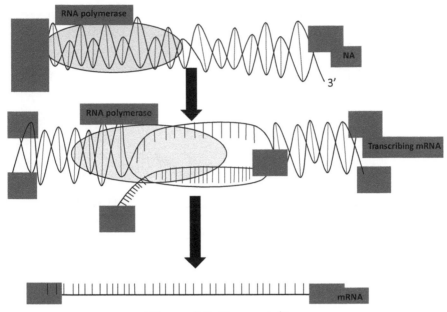

Figure 2.3: Transcription.

- RNA polymerase, the transcribing enzyme, binds to the promoter sequence.

- The DNA double helix is unwound and RNA elongation is initiated when nucleotides are added to the growing RNA chain. The transcription process is terminated after the RNA polymerase reaches the terminator sequence in the DNA molecule strand.

- The mRNA molecule is spliced to remove introns that are not required for future steps.

2.3.3 Translation

Translation is the process in which mRNA directs protein synthesis using the assistance of transfer RNA. After transcription, the mRNA is exported out of the nucleus into the cytoplasm where translation is processed in association with ribosomes, as illustrated in Figure 2.4. The steps in translation are as follows.

- Each three-base stretch of mRNA is known as a codon. mRNA passes through the ribosome subunit, where an mRNA codon pairs with anticodon transfer RNA (tRNA), as per Watson-Crick base pairing.

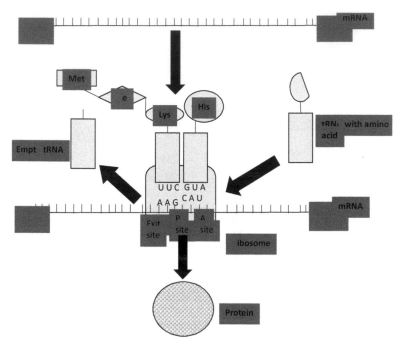

Figure 2.4: Transcription.

- After the mRNA-tRNA base pairing has occurred, the amino acid attached to the tRNA gets incorporated into the growing protein chain.

- Finally, the tRNA is expelled and the new amino acid attached tRNA is paired with the next codon of the mRNA.

2.4   Ambiguity in Central Dogma

The Central Dogma is the popularly known mechanism for DNA, RNA and protein synthesis. As initially proposed by Francis Crick, there is a sequential transfer of genetic information from DNA to RNA to protein, and it could not be transferred from protein to protein or to nucleic acid. There were many questions raised about this sequence of actions proposed by the Central Dogma of Molecular Biology, and no one had any evidence until Howard Temin formulated the DNA provirus hypothesis in 1964 [401]. This hypothesis states that RNA viruses can use viral RNA as a template for DNA synthesis in a direction that is the reverse of the one proposed by the Central Dogma. It forced Crick to re-explain the concept of Central Dogma in 1970 [93]. Crick, however, was correct in explaining that the sequences of nucleotides and amino acids are the most important factors governing the production of nucleic acids and proteins, respectively.

2.4.1   Reverse Transcription

The process of synthesizing DNA from an RNA template with the help of the reverse transcriptase enzyme is known as reverse transcription, and the resulting DNA is known as complementary DNA (cDNA). The Central Dogma held that DNA is transcribed to RNA, and this hypothesis was challenged in 1970 by Howard Temin and David Baltimore who independently identified the reverse transcriptase enzyme associated with the replication of RNA viruses called retroviruses. Reverse transcription has functional roles in biological systems, to produce cDNA to understand actively expressed genes at specific times and conditions and to build libraries of cDNAs. The reverse transcriptase enzyme has three molecules that help it in carrying out reverse transcription (See Figure 2.5):

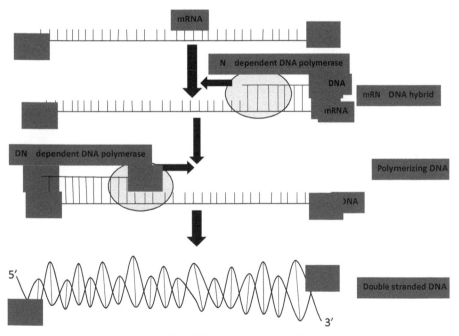

Figure 2.5: Reverse transcription.

- RNA-dependent DNA polymerase,

- Ribonuclease H (Rnase H), and

- DNA-dependent DNA polymerase.

The steps in RNA transcription are given below.

- The RNA-dependent DNA polymerase gets attached to the RNA and synthesizes a DNA strand by adding the nucleotides one by one to the growing DNA chain using the Watson-Crick base pairing rule.

- After synthesizing a DNA strand complementary to the RNA strand, the RNA strand is separated from the DNA-RNA hybrid by using the Rnase H enzyme.

- Finally, the DNA-dependent DNA polymerase enzyme gets attached to the single-stranded DNA, and a new complementary DNA strand is synthesized by adding nucleotides one by one to the growing DNA chain using the Watson-Crick base pairing rule.

2.4.2   RNA Replication

RNA viruses contain RNA as their genetic material; therefore, they need to replicate the RNA to reproduce themselves. Because of limited genetic and molecular content in an RNA virus, it depends upon the host cell's molecular machinery for RNA replication, as seen in Figure 2.6.

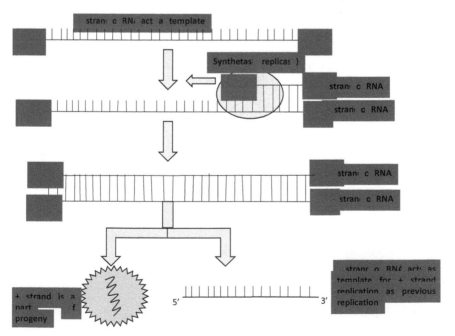

Figure 2.6: RNA replication.

The steps in RNA replication are as follows.

- An RNA (+/-) strand undergoes replication using the RNA synthetase (replicase) enzyme in the same way as in DNA replication.

- Newly formed replicated RNA strands are passed to the virus progeny.

- DNA-dependent DNA polymerase.

2.5 Discussion

2.5.1 Biological Information Flow from a Computer Science Perspective

From a computer science perspective, the process stated by the Central Dogma may be less important than the flow of information, which it represents. A computer scientist describes the process in an inherently digital manner, where a four-character alphabet containing A, T, C, and G, is used to represent nucleotides or bases. From the encoded information in the template DNA in the nucleus, sophisticated machinery reads and copies the information to messenger RNAs. The messenger RNA later translates the information to a protein. Theoretically, A binds to T with a double covalent bond and C binds to T with a triple covalent bond. However, in reality, the A=T and C≡G base pairings are not always exact. There are errors and mismatches. The rate of error is about one base pair in every million base pairs. The effect of an error might be negligible or tremendous, depending on the specific position(s) where it occurs. Even a single error may result in deleterious effects by leading to severe diseases such as cancer. An error introduced in early stages may be propagated further in multiple manners. The DNA copy error in a single base pair position may be compared to a single bit error in a computer program.

2.5.2 Future Perspective

Computer science has tremendous potential to assist in the analysis of huge amounts of biological data and solve biological problems, which are beyond the capacity of manual analysis. There is an urgency to solve biological problems with robust methods of computational algorithms since huge amounts of data are available and are being produced continuously. Unlike biologists, a computer scientist must understand the process of Central Dogma, which they can then replicate and simulate in computational programs, e.g., they can write programs that re-create in software those processes that convert DNA to RNA, RNA to DNA, RNA to protein, and from protein to RNA and DNA, compare actual sequences to expected sequences, and perform analysis to discover where things may have gone wrong.

Chapter 3

Gene Expression Data Generation

3.1 History of Gene Expression Data Generation

The nucleus of every cell in an organism contains the complete genome established in the fertilized egg. However, each cell or tissue possesses a particular identity of its own and responds differently to real situations. This is possible because of the variability in expressions of gene levels in different cells or tissues with respect to time and condition. The entire information controlling the behavior of a cell is encoded in its DNA, from which it is transcribed. Various strategies have been employed by humans to study mRNA and its behavior. We focus on the methodologies formulated for detection and study of gene expression changes until the present date.

Strategies employed to quantify mRNA have evolved over the years, and can be divided into three broad stages of development. These include (i) classical ways, (ii) low-throughput sequencing, and (iii) today's state-of-the-art technology called high-throughput sequencing. Classical methods for detecting and quantifying mRNA included Northern Blotting, ribonuclease protection assay, microarray and QRT-PCR. Although with passage of time, these procedures have been upgraded for efficiency and precision, limitations identified in them paved the way for introduction of low- and high-throughput sequencing methods like SAGE, CHIP-Seq and RNA Seq and transcriptome study, utilizing the full potential of sequencing platforms. The advent of sequencing technologies in the late 1970s revolutionized the whole perspective of studying genomes and transcriptomes, starting a new era. This method for sequencing was first developed by Sanger in 1977 and

DOI: 10.1201/9780429322655-3

was popularly called Sanger sequencing [359]. Although highly popular in its heyday, its time, cost and low throughput were major limitations. Over time, Sanger sequencing gradually paved the way for NGS, also called high-throughput sequencing or second-generation sequencing [361].

The word *transcriptome* was coined in the 1990s. A transcriptome refers to the whole set of mRNA molecules produced by a cell under a particular condition at a particular time. Various attempts have been made to decipher transcriptomes of humans. The first attempt was performed in 1991, using 609 mRNA molecules from the human brain [5]. Later in 2008, two human transcriptomes covering 16,000 genes were published [314]. By 2015, transcriptome studies were published for hundreds of individuals [260].

Classical methods for quantifying mRNA did not have a standard approach to store data for public availability and use. However, with the advent of sequencing technologies, the focus was on developing standards for storage of data so that data can be shared easily. EST libraries, which came into existence in the 1990's, provided lists of short nucleotide sequences generated from single mRNA transcripts. mRNA was first copied as cDNA by the reverse transcriptase enzyme and then the resultant was sequenced using Sanger sequencing to generate ESTs. These EST libraries provided sequence information for early microarray experiments. In the year 1995, SAGE Serial analysis of gene expression, which worked with Sanger sequencing became the earliest sequencing method for base transcriptome generation [228]. With the introduction to NGS technology, studies of transcriptomes became more feasible. RNA-Seq is such an approach, using the capabilities of NGS methods to decipher the transcriptome of a cell. It has far better and higher coverage and greater resolution for transcriptome study compared to Sanger sequencing and microarray-based methods [21]. The availability of transcriptomic data follows two main approaches, either through sequencing of individual transcripts, i.e., ESTs or RNA-Seq, or by hybridization techniques followed in microarray creation.

Classical methods could only identify and quantify a handful of mRNA molecules, and required prior knowledge of the transcripts. Hence, de novo identification and study of unknown transcripts were not possible. To overcome the limitations of wet-lab techniques, combining novel sequencing methods and advanced bioinformatics tools appear to be the only viable way for analyzing gene expression changes. We provide a brief perspective on the principles of the various methods for quantifying mRNA along with their limitations, which have encouraged researchers to recently switch to more advanced technologies.

Figure 3.1: Timeline showing the evolution of methods for studying gene expression changes.

3.2 Low-Throughput Methods

3.2.1 Northern Blotting

Northern Blotting is a classical way of analyzing the presence of a specific type of RNA sequence within a sample. This technique was developed in 1977 by James Alwine, David Kemp and George Stark [171], and is one of the oldest methods for mRNA detection. It examines relative sizes and quantities of particular types of RNA. The RNA within the sample is denatured and separated into single strands. The RNA molecules are then separated according to their size using gel electrophoresis. The RNA is transferred into a blotting membrane, to which DNA or RNA probes or genes of interest are added. The probe is typically a florescent dye. Using complementarity and hybridization, one can visualize the presence of interesting RNA [74]. Because of the relative ease of use, it had become a widely popular method for RNA detection and quantification. However, this method has several drawbacks such as efficiency of RNA transfer, membrane binding, and kinetics of probe hybridization. However, this method is considered good for mRNA quantification, specifically for those RNAs that are expressed in low amounts [71].

3.2.2 Ribonuclease Protection Assay

Ribonuclease protection assay (RPA) is a similar technique to detect and quantify mRNA molecules, but is more effective than Northern Blotting when it comes to detection of mRNAs that are expressed in low amounts. This is because it ensures perfect complementary hybridization [420]. It is a highly sensitive method and uses in vitro transcribed P32-labeled antisense RNA probes that hybridize to their complementary mRNA in a solution. RNAase is used to remove non-hybridized RNA species. From the cRNA fragments, a standard curve can be generated and used to perform the quantification. Another advantage is that it can be used to detect and quantify multiple mRNA transcripts simultaneously. One of its greatest drawbacks lies in its inability to generate a clear gel background, resulting in long smears in the lanes of gel, due to incomplete nuclease digestion [222].

3.2.3   qRT-PCR

Another widely used method to analyze gene expression changes is Quantitative RT–PCR. Polymerase chain reaction was invented by Kary B. Mullis in 1983, for which he was awarded the Nobel Prize in Chemistry in 1993. PCR is often described as a simple, sensitive and rapid technique to amplify DNA using oligo primers, dNTPs and heat-stable taq polymerase. The emergence of PCR to amplify mRNA in real time has proved to be a revolution in gene expression studies and has gained much popularity in the last 10 to 15 years. Quantitative RT–PCR follows a single tube format and uses florescent reporter dyes to combine the amplification and detection steps. The method relies on measuring increase in florescent signal, which is directly proportional to the amount of DNA produced in each cycle.

The incorporation of the reverse transcription step has made RT-PCR a powerful tool for amplifying any type of RNA. Currently, Quantitative RT-PCR is a benchmark technology for detection and comparison [71]. This process is often described as a gold standard, but has limitations in its ability to deal with problems caused by variability of RNA templates, assay designs, inappropriate data normalisation and inconsistent data analysis [71].

3.2.4   SAGE

Serial Analysis of Gene Expression was a direct product of EST sequencing technology. The term EST was coined in 1991. Following advances in sequencing technology, including the availability of EST, in 1995, Victor Velculescu published the protocol for SAGE [420]. SAGE performs simultaneous quantitative analysis of thousands of transcripts, which experiments prior to the invention of DNA/RNA sequencing could not perform. Unlike the microarray approaches, no prior knowledge of the sequence is required. It is often regarded as an open platform for identification of unknown transcripts [222]. In SAGE, short 10-11bp sequence tags are derived from a defined position within a transcript, which can further unequivocally identify this transcript. These tags are isolated, ligated and sequenced in a low- or high-throughput manner. Thousands of tags are sequenced per sample and the frequency of each tag in the sequenced concatemers corresponds directly to transcript abundance. To determine differential expression of a gene in two or more samples, the frequencies of the corresponding tags in these samples are compared. However, short tags used in this protocol often have reported false positive results for identification of respective transcripts and hence, an advanced version of SAGE called Super-SAGE was developed. In Super-SAGE, 26bp long tags allow secure tagging for gene annotation. The process was further advanced by switching from conventional Sanger sequencing to various NGS platforms to sequence the tags, and is often called as HT-Super SAGE [279].

3.3 High-Throughput Methods

3.3.1 Microarray

To overcome the biggest drawback of Northern Blot analysis and RNase protection assays, i.e., the inability to simultaneously analyze a large number of genes, microarrays emerged as a more efficient tool for quantifying mRNA. The discovery of this technique dates back to 1975, when Grunstein and Hogness applied the process of molecular hybridization to DNA released from blotted microbial colonies. The colony hybridization method was incorporated by randomly cloning *E.coli* DNA into agar petri plates covered with nitrocellulose papers. Radioactive probes were added to bind to complementary DNA within the sample. This is considered the first microarray. Adapting the afore mentioned method in 1979, Gergan et al. produced multiple plates on agar to produce arrays using a 144-pin device for sample placing. The bacterial colonies could be easily transferred to paper filters for necessary lysis for producing hybridized DNA. In the early 1990s, robotic technology was used to quickly array clones from microtiter plates onto filters, increasing efficiency with automatized placing of samples on the array [77].

DNA microarray technology is a robust tool for parallel investigation of a number of nucleic acid hybridization studies in an efficient manner. The DNA microarray chip discovery was driven by techniques for miniaturization, integration, and microelectronic fabrication, to produce the "laboratory-on-chip" tools. DNA microarrays are microscopic slides printed with thousands of tiny spots with each spot containing a known DNA sequence or gene, called a probe. The mRNA extracted from samples is converted to cDNA molecules that hybridize to their complementary probes in the micro slide and express color in different intensities to indicate their abundance. The main procedure involved is an alteration of the Southern Blot technique. Bioinformatics tools are needed to produce meaningful and reproducible inference from large amounts of results distilled from complex

DNA chip data. Access to electronic datasets to get reliable and valid annotation and productive data mining tools is crucial for comparative data analyses, and so advanced statistical methods (e.g., superparamagnetic curve) or other specialized analytical software packages are needed for analysis of microarray data.

With advances in science and technology, the microarray technique to measure gene expression had become a widely popular method, although it had some drawbacks. It was suitable for the study of genes with low or very high expressions as it could only detect two or three folds of expression levels because of background and signal overload. In addition, its accuracy for a known genome could worsen if the errors were found in the database when investigating unknown genomes. It also lacked standards for data collection and could detect only those genes that were prespecified to be present in the array [461].

3.3.2 RNA-Seq

RNA-Seq, developed in the mid-2000s, is an advanced technique that uses high-throughput sequencing (HTS) or next-generation sequencing (NGS) technologies to decode a transcriptome. A transcriptome includes the complete set of transcripts in a tissue, organism, or a specific cell for a given physiological condition. These are the protein-coding messenger RNA (mRNA) and non-coding RNA like ribosomal RNA (rRNA), transfer RNA (tRNA), and other ncRNAs. The content of a transcriptome varies with physiological conditions. RNA-seq helps look into the genome and quantify the expressions of the transcripts. RNA-seq can be used as a powerful tool for dissecting and understanding many biological phenomena like the underlying mechanisms and pathways controlling disease initiation, development, and progression [404].

The emergence of high-throughput sequencing platforms in 2005 like the Illumina or Solexa technology, Ion Torrent semiconductor sequencing technology, Single-Molecule Real-Time (also known as SMRT sequencing), and Oxford Nanopore Technologies ultimately led to a revolution in DNA sequencing [112].

The first attempt to completely sequence a gene was made by Holley in 1964, in order to look specifically at the structure of transfer RNA (tRNA) [44]. Studies of individual transcripts were also performed before any transcriptomic approach was developed. In the late 1970s, mRNA libraries from silkmoths were collected and converted to complementary DNA (cDNA) for storage using reverse transcriptase [315]. In the 1970s, Maxam and Gilbert developed a DNA sequencing technology based on chemical modification of the DNA and subsequent cleavage at specific bases. Sanger in 1977 developed a DNA sequencing method based on chain termination which proved to be highly efficient due to low radioactivity. Sanger's approach came to be known as the first-generation sequencing technology. In 1987, the first automatic sequencer that used the Sanger method appeared. It used capillary electrophoresis, and was the most accurate and efficient sequencing method until 2001 [109]. Revolution in DNA sequencing occurred due to the emergence of high-throughput sequencing platforms in 2005. Such platforms included the Illumina/Solexa technology, Ion Torrent's semiconductor sequencing technology, Single-molecule real-time (SMRT) sequencing, and the Oxford Nanopore Technologies [26].

Initial transcriptomic studies were conducted using hybridization-based microarray techniques, but such approaches failed to fully systematize and quantify the levels of the diverse RNA molecules that are expressed across a wide range of genomes [260]. RNA sequencing is a technique in which a combination of high-throughput sequencing technologies and computational methods are used to capture and quantify transcripts present in the sequenced RNA fragments in 25—400 base pairs [228]. Transcriptomic analysis first requires the isolation of RNA from the organism, construction of cDNA libraries using reverse transcriptase enzyme, and amplification of cDNA libraries by PCR to enhance fragments containing the required $5'$ and $3'$ adapter sequences. After preparation of the transcript molecules, they can be sequenced in a single direction (single-end) or both directions (paired-end). It is easier and cheaper to generate single-end sequences than paired-end sequences. Single-end sequences are sufficient for gene expression quantification study. However, paired-end sequences generate solid alignments or assemblies that help gene annotation and transcript isoform discovery. The results generated in the form of raw sequence reads need to be processed using a combination of bioinformatics tools that are suitable for the particular experimental study and its objectives. The basic pipeline of RNA-Seq data analysis includes the following four steps; quality control, alignment, quantification and differential gene expression study.

Compared to other techniques for gene expression studies, RNA-Seq provides deeper coverage and resolution of the transcriptome, facilitating RNA editing, newly transcribed region detection, analysis of alternative splicing and allele-specific expression [228]. Though microarray technology has been used for large-scale sequence analysis, it cannot be performed for many non-model organisms. However, if information about the genomes and sequences is available, the transcriptome can still be examined using RNA-Seq for many species of organisms, even when no reference genomes are available.

Integration of results from RNA-Seq data with other biological data sources can help generate a better picture of gene regulation. For example, compilation of RNA-Seq with genotyping data can help identify gene loci involved in variations of gene expression among individuals. Integrating expression data with RNA interference, transcription factor binding, DNA methylation and histone modification information can lead to better understanding of many regulatory mechanisms [86]. The use of polyadenylated messenger RNA (mRNA) enables the investigation of different populations of RNA, including pre-mRNA, total RNA and noncoding RNAs such as long ncRNA and microRNA.

However, this technique is costly and biased. It also requires extensive resources in contrast to other available techniques. The datasets in RNA-seq are larger and more complex, and the generated data cannot be interpreted easily without extensive bioinformatics intervention. RNA-Seq analysis is challenging when studying organisms without reference genomes, requiring more computational sources for de novo gene assembly [393].

3.3.3   Types of RNA-Seq

Small RNA/Non-Coding RNA Sequencing

For the past 50 years, the term 'gene' has been used to denote the regions of the genome that encode mRNAs, which are translated into proteins. However, recent genome-wide studies have found that the human genome is pervasively transcribed and produces many regulatory non-protein coding RNAs (ncRNAs) including microRNAs, small interfering RNAs, PIWI-interacting RNAs and various classes of long ncRNAs that fulfill critical roles as transcriptional and post-transcriptional regulators and also guide chromatin-modifying complexes [214], [393]. For generation of ncRNAs, the cellular RNA is isolated taking into account the desired size range and then small RNA targets, such as miRNA, are isolated through ribo-depletion libraries. As these small RNAs are lowly abundant, short in length (15—30nt) and lack polyadenylation, a separate strategy is often preferred to profile these RNA species by size selection through size exclusion gel electrophoresis, size selection magnetic beads, or with commercially developed kits such as RiboMinus (Life Technologies) or RiboZero (Epicentre). Once isolated, linkers are then added to the 3' and 5' ends and purified. Then, through reverse transcription, cDNA is generated followed by sequencing on a specific NGS platform and subsequent bioinformatic analysis [5].

Direct RNA Sequencing

cDNA-based transcriptome analysis approaches used today exhibit shortcomings that prevent us from understanding the real nature of transcriptomes and ultimately genome biology. So, a method allowing a comprehensive and unbiased view of transcriptomes using small quantities of total RNA obtained from a single or a few cells, with no pre-treatment would enable significant advances toward better characterization of complex biological processes through direct RNA sequencing (DRS). This technique allows massive parallel sequencing of RNA molecules directly without prior synthesis of cDNA or the need for ligation and amplification steps. DRS represents an extension of single-molecule DNA sequencing technology (tSMS) that relies on the stepwise synthesis and direct imaging of billions of single DNA strands on a planar surface. Sequencing-by-synthesis reactions are performed using a modified polymerase and proprietary fluorescent nucleotide analogues, called Virtual Terminator nucleotides (VT) that contain a fluorescent dye and chemically cleavable groups that allow step-wise sequencing [310], [224].

Single-Cell RNA Sequencing (scRNA-Seq)

Single-cell transcriptomics started with the development of single-cell qPCR, which along with single-molecule FISH, is still the method of choice for targeted analysis of transcripts. The first transcriptomes generated via single-cell RNA-sequencing (scRNA-seq) were published in 2009, only two years after the first applications of RNA-seq to bulk populations of cells. Historically, scRNA-seq has been applied in situations in which the amount of biological material was limited, such as cells from early embryonic development [224], [463]. Single-cell RNA-sequencing (scRNA-seq) now allows for transcriptome-wide analysis of individual cells, revealing exciting biological and medical insights. scRNA-seq requires the isolation and lysis of single cells, the conversion of their RNA into cDNA, and the amplification of cDNA to generate high-throughput sequencing libraries. As the amount of starting material is very small, this process results in substantial technical variation [463]. A prominent feature of scRNA-seq is the sparsity of data, i.e., the high proportion of zero read counts that arises for both biological reasons (e.g., the presence of subpopulations of cells or transient states where a gene is not expressed) and technical reasons (e.g., dropouts, where a gene is expressed but not detected through sequencing). In addition, technical noise in scRNA-seq is also reflected in high variability among technical replicates, even for genes with medium or high levels of expression [414]. Single-cell isolation is the first step in obtaining transcriptome information from an individual cell, which can be achieved by limiting dilution. Micromanipulation is a classical method used to retrieve cells from early embryos or uncultivated microorganisms, and microscope-guided capillary pipettes to extract single cells from a suspension. These methods are time-consuming and have low throughput. So, recently, flow-activated cell sorting (FACS) has become the most commonly used strategy for isolating highly purified single cells. Common steps required for the generation of scRNA-seq libraries include cell lysis, reverse transcription into first-strand cDNA, second-strand synthesis, and cDNA amplification. Further sequencing is performed using high-throughput sequencing platforms such as Illumina and Smarte Seq2 [187].

3.3.4   Gene Expression Data Repositories

With genome sequencing projects, gene expression data are generated daily in very large quantities worldwide. In order to use these data effectively and efficiently, a structured filing system of the data is necessary, so that the data is readily accessible to those interested. Biological databases are divided mainly into two broad categories, primary and secondary.

Primary

Primary databases are populated with experimentally derived data such as nucleotide sequences, protein sequences or macromolecular structures. Experimental results are submitted directly to the database by researchers, and the data are archived. Given a database accession number, the data in primary databases are never changed; they form part of the scientific record (REFF). Various types of primary databases are briefed below.

- **GEO**: GEO is currently the largest, fully public gene expression resource [46]. The Gene Expression Omnibus (GEO) repository was established in 2000 to host and freely disseminate high-throughput gene expression data. The database was built and is maintained by the National Center for Biotechnology Information (NCBI). The GEO at NCBI is the largest public repository of high-throughput gene expression data. The database has a flexible infrastructure that can capture fully annotated raw and processed data, enabling compliance with major community-derived scientific reporting standards such as 'Minimum Information About a Microarray Experiment' (MIAME) [90].

GEO accepts data that use a wide variety of technologies, including DNA microarrays, protein or tissue arrays, high-throughput nucleic acid sequencing, SAGE and RT-PCR. Although the majority, approximately 90%, of the data in GEO are indeed gene expression data, the applications have also expanded to include studies on genome methylation, genome binding and occupancy, protein profiling, chromosome conformation studies, and genome variations and copy numbers [90].

The GEO database is a MIAME-compliant infrastructure. The MIAME guidelines outline the minimal information that should be provided to allow unambiguous interpretation of microarray experiment data. While the submission procedures promote MIAME-compliance, ultimately it is the submitter's responsibility to ensure that their data are sufficiently well annotated [45].

- **ENA**: The European Nucleotide Archive is Europe's primary nucleotide-sequence repository. The ENA consists of databases such as the Sequence Read Archive (SRA) and the Trace Archive and EMBL-Bank. ENA operates as a public archive for nucleotide sequence data. By bringing together databases for raw sequence data, assembly information and functional annotation, the ENA provides a comprehensive and integrated resource for this fundamental source of biological information [408].

  ENA content is organized into a number of data types. Core data types, such as raw sequence data and derived data, including sequences, assemblies and functional annotation, are supplemented with accessory data, such as studies and samples, to provide experimental context. Primary data (such as reads) and several derived data types (sequence, assembly and analysis) are submitted, while the remaining derived data types (coding, non-coding, marker and environmental) are derived from submitted content as part of processing and indexing within ENA. ENA offers a comprehensive range of submission options through the Webin system. This system provides both an interactive web application that offers spreadsheet upload support and a powerful RESTful programmatic submission interface. The former is recommended for infrequent submitters and first-time users, including those setting up programmatic submission systems. The latter is recommended for those with informatics skills who wish to establish ongoing regular data flow for a project or submitting center [242].

ENA data can be browsed and retrieved in XML, HTML, fasta, fastq and flat file formats using the ENA Browser, which can be used both interactively and programmatically through REST URLs. Bulk download of data is supported through FTP [225].

- **SRA**: The Sequence Read Archive (SRA) is an international public archival resource for next-generation sequence data, established under the guidance of the International Nucleotide Sequence Database Collaboration (INSDC). INSDC's partners include the National Center for Biotechnology Information (NCBI), the European Bioinformatics Institute (EBI) and the DNA Data Bank of Japan (DDBJ) [223].

Secondary Databases

Secondary databases are those which consist of data derived from the analysis of primary data. Secondary databases often draw upon information from numerous sources. They are highly curated.

3.3.5 Standards in Gene Expression Data

High-throughput technologies generate large amounts of complicated data that have to be stored in databases, communicated to data analysis tools and interpreted by scientists. Hence, data representation and communication standards are needed [65]. Formatting and describing data using community standards enables them to be understood, compared, exchanged and reused by both collaborators and the wider community [384]. Standards help fit together many pieces to make a useful whole to enable exchange of large volumes of information when many smaller transactions are needed. Various gene expression studies also generate large amounts of data that need standardization to manage and exchange [65].

For describing microarray experiments, Minimum Information About a Microarray Experiment (MIAME), is a type of standard used. MIAME provides an informal means for modeling a microarray experiment. The consensus was achieved by establishing the Microarray Gene Expression Data (MGED) Society, which started as a grass-roots movement comprising all the main microarray players at the time, and has continued to be inclusive. [65]. Some other microarray standards include the following.

- **MAGE-ML, MicroArray Gene Expression Markup Language and MAGE-OM, MicroArray Gene Expression Object Model:** The Microarray Gene Expression Markup Language

(MAGE–ML) is based on a conceptualization of microarray experiments modeled using the unified modeling language (UML) named MAGE–OM (Microarray Gene Expression Object Model) [38]. The Gene Expression Markup Language (GEML) is a file format for storing DNA microarray and gene expression data for chip patterns and chip scans (profiles). GEML is an open-standard XML format. GEML stores which data collection methodology was used, without making assumptions about the meaning of a measurement. This enables possible normalization, integration, and comparison of data across methodologies. This data standard was designed to separate data reporting and collection from methodology used.[1] MAGE was (and continues to be) developed in a collaborative and cooperative manner by many members of the grass–roots Microarray Gene Expression Data Society (MGED). MAGE developers and adopters have contributed to a freely available software tool kit (MAGE-STK) that eases the integration of MAGE-ML into end users' systems [38].

- **MO MGED Ontology:** The MGED Ontology (MO) was developed to provide the semantics required to support the MAGE-OM and as a resource for the development of tools for microarray data acquisition and query. The MO contains concepts that are universal to other types of functional genomics experiments such as protocol and experiment design, and can also be used for annotation of the data in these domains. The major component of the ontology involves biological descriptors relating to samples or their processing [432].

  In case of RNA-Seq data, there are still no general guidelines for the minimum requirements that should be applied as a standard for the recording and reporting of next-generation sequencing-based gene expression data. This makes the exchange of RNA-Seq and other massive sequencing data a delicate issue. However, there are a number of attempts by different consortia to change this prevailing lack of a standard. As an example, the Functional Genomics Data Society has published the MINSEQE guidelines. In addition, the ENCODE (Encyclopedia of DNA Elements) consortium has released experimental data standards and processing pipelines for next-generation sequencing platform technologies [324].

- **MINISEQE:** It provides information about high-throughput nucleotide sequencing experiments. It facilitates reproducibility of the

---

[1]http://xml.coverpages.org/geml.html

results of experiments and avoids ambiguous interpretations. Its use improves integration of multiple experiments across modalities. This guideline provides recommendations for: (1) description of biological systems, samples, and experimental variables, (2) sequence read data for each assay, (3) summary data for a set of assays, (4) general information about experiment and sample-data relationships, and (5) essential experimental and data processing protocols.[2]

- **ENCODE:** ENCODE makes the following recommendations when mapping RNA-Seq data.

  (a) It is preferable to include not only the genome reference, but also a set of annotations in the mapping set, since this greatly increases the specificity of mapping to splice junctions.

  (b) The sequences for any exogenous RNA spike-ins that were added should also be included to obtain mapped spike-in reads.

3.4   Chapter Summary

Gene expression studies have become important since they help understand molecular mechanisms, diseases, their progression and associated pathways. Such studies help unfold molecular complexities across systems of organisms as well as plants through the analysis of gene expression profiles. Continuous evolution in the relevant technologies have helped quantify, measure and understand better the patterns of mRNA with much ease. This in turn, has helped researchers identify marker genes associated exclusively with particular diseases and develop targeted therapies based on the findings of whole cells as well as single cell experiments. Considering such great potential impact, this chapter provides an outline of the available techniques for gene expression studies with advantages and limitations. Additional details will help readers choose suitable techniques considering their strengths and drawbacks as they design their experiments. Knowledge of available tools and how they compare against one another may also help develop better future techniques to study gene expression by combining the strengths, and overcoming the limitations of currently available resources. As a consequence, it may also help develop cost-effective and user-friendly techniques for gene expression studies that can be easily applied by individuals with biological knowledge, but without deep technical expertise.

---

[2]https://omictools.com/minseqe-tool

Chapter 4

Statistical Foundations and Machine Learning

4.1 Introduction

This chapter provides background knowledge on statistical modeling and machine learning techniques prior to their applications in the gene expression analysis. In presents theoretical foundations supported by illustrative examples for easy understanding of the concepts and their applications.

4.2 Statistical Background

This section focuses on statistical modeling, probability distributions, hypothesis testing, various tests, and common data distributions. It introduces these concepts and illustrates with suitable examples for easy understanding.

4.2.1 Statistical Modeling

Statistical modeling plays a key role in many application domains, including bioinformatics, in support of effective decision making. It helps model real-world scenarios, summarize available information, and support predictions based on the summaries and trends. To build a statistical model, appropriate mathematical formulation is needed. To understand and analyze a real-life problem, one usually gathers the raw data, pre-processes them as necessary to generate a dataset, and then finally applies appropriate formulas for estimation of summary statistics such as variance and standard deviation, or perform linear regression to estimate trends.

DOI: 10.1201/9780429322655-4   53

4.2.2   Probability Distributions

To analyze data to draw effective inference, one has to understand the underlying distribution patterns in the data. A probability distribution describes all the values that a random variable can take, under certain assumptions. Fitting collected data to a distribution and visualization of the data help analysis of a real-world situation effectively. Some examples of distribution types are the Bernoulli distribution, uniform distribution, and normal distribution. Three commonly used axioms of probability are as follows:

A_1 The probability of an event is a real number greater than or equal to 0.

A_1 The total probability that at least one of all possible outcomes of a process will occur is 1.

A_1 If two events A and B are mutually exclusive, then the probability of either A or B occurring is equal to $P(A) + P(B)$.

4.2.3   Hypothesis Testing

A hypothesis is an unsubstantiated supposition put forward without evidence as a point of initiation for further investigation. It is a proposition or claim that needs to be tested or validated. Hypothesis testing is a systematic statistical investigation based on interpretation of data and associated results to confirm whether the proposition or claim holds. The main objective of hypothesis testing is to estimate the likelihood of or level of confidence in the hypothesis, based on interpretation of the results. Generally, hypothesis testing is tested on the analysis of a set of random samples of the data rather than the complete dataset.

Two hypotheses commonly used in statistics are *null* and *alternate* hypotheses. A *null* hypothesis, also referred to as H_0, assumes that no variation exists in the variables with reference to the *mean*. In other words, this hypothesis assumes that there is no statistical significance in the set of observations. The *alternate* hypothesis, also referred to as H_1 or H_a, refers to a hypothesis to be established by research. It refers to a new claim (or value) which needs to be validated with adequate evidence. It assumes that some differences exist between two or more variables.

It is clear from the above discussion that these hypotheses are mathematical opposites. To reject a *null* hypothesis, adequate evidence must be provided. So, we summarize that the process of hypothesis testing may lead to two possible outcomes.

a. If the original claim could not be established with substantial evidence, it will lead to a *rejection* of the *null* hypothesis.

b. If enough evidence could not be provided to establish the *alternate* hypothesis, it will lead to *acceptance* of the *null* hypothesis or *fail* to *reject* it.

To conclude with any of the above two possible outcomes, a statistician uses a measure (say, β) called the level of significance. If the probability value (also referred to as p-value) $\geq \beta$, we cannot reject H_0; otherwise, it can be rejected. Let us illustrate with an example for better understanding. Assume a chocolate manufacturing company makes chocolate bars, each of which is 5 grams in weight. After performing maintenance on the manufacturing unit, a worker claims that the new chocolate bars made are not 5 grams in weight. To test, we can create the hypotheses as follows:
H_0 = weight of a chocolate bar is 5 grams, or $w = 5$ gm.
H_a = weight of a chocolate bar is not 5 grams, or $w \neq 5$ gm.
To validate these hypotheses, one can take a random sample of 50 chocolate bars, and measure their weights. If weights of the bars are found to be 5 grams, we fail to *reject* the *null* hypothesis. If weights are found to greater than or less than 5 grams, we *reject* the *null* hypothesis.

4.2.4 Exact Tests

An exact test requires that all the assumptions upon which the distribution of the tests are based are valid. It causes a significance test with a false reject rate equal to the significance level of the test. An exact test at significance level 5% will reject the null hypothesis exactly 5% of the time. To understand better, let us take the classic example of *'Lady Tasting Tea'*. In this example, a lady claims to be able to taste a cup of tea and identify whether it was made by pouring the milk first or the tea first. The lady was given 8 cups of tea, out of which 4 were made by pouring the milk first and 4 were made by pouring the tea first. She is to taste the tea and decide (guess) correctly which was poured first in each case. Table 4.1 shows the type of tea and her guesses.

Table 4.1: Sample Data: Tea and Guesses

Items	Guess Tea	Guess Milk
Tea	4	0
Milk	0	4

We now consider the hypotheses in the context of the above example, as follows.

H_0 = There is no association between her guess and what was poured first. Let, the significance value, α = 0.05. If the p-value < 0.05, then H_0 is rejected. The total number of ways in which 4 cups of tea can be chosen from 8 cups = $\binom{8}{4}$ = 70. If the lady can decide accurately, it implies that there is only 1 combination of cups for the correct guess, i.e., p = 1/70 = 0.014. Since, $p < 0.05$, and H_0 is rejected.

4.2.5   Common Data Distributions

In this section, we introduce some common data distributions used in a statistical study.

Uniform Distribution

A uniform distribution is defined as a probability distribution of events each with constant probability occurrence. In other words, in a uniform distribution, the events are equally likely. In this distribution, the probability distribution function (PDF), $f(x)$, is constant for all the values of x. An example of uniform distribution function $f(x)$ for all possible values of x is shown in Figure 4.2.5. It is equally likely for a value near c to occur as it is

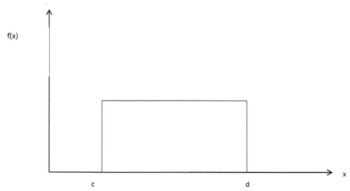

Figure 4.1: Uniform distribution.

for a value near d to occur. The area under the curve is equal to 1 since it is the sum total of the probabilities of occurrence for all events.

Area under the curve = area of the rectangle

$$\implies 1 = (d - c) \times f(x)$$

$$\implies f(x) = \frac{1}{(d - c)} \tag{4.1}$$

where, $[c, d]$ is the range of values the random variable can take. Thus, the PDF for a uniform distribution,

$$f(x) = \frac{1}{(d - c)}, for \; c \le x \le d$$
$$= 0, \qquad otherwise.$$

(4.2)

To understand better, let us take an example. Let us consider the rolling of a fair die with 6 sides. All six faces have the same probability of occurrence, $1/6$. So, we can say that the probability distribution of occurrence of any of the sides of a fair die follows a uniform distribution.

t-Distribution

t-distribution, a continuous probability distribution function, is strongly related to the normal distribution. If we select a sample of n independent observations from a normally distributed population with mean μ and variance σ^2, we know that the standard normal distribution, Z, is given as

$$Z = \frac{\bar{X} - \mu}{\sigma/\sqrt{n}}$$

where $\bar{X}$ is the mean of the sampled data. In practice, we may not know the value of σ, the population standard deviation. Instead, we use the standard deviation of our selected samples, denoted S. The value of S would change according to the sample selected. This new distribution is referred to as a *t*-distribution, with $n - 1$ degrees of freedom.

$$t = \frac{\bar{X} - \mu}{S/\sqrt{n}}$$

The PDF for a *t*-distribution with $\nu > 0$ degrees of freedom is given as

$$f(t) = \frac{\Gamma(\frac{\nu+1}{2})}{\sqrt{\nu\pi}\Gamma(\frac{\nu}{2})} \cdot \left(1 + \frac{t^2}{\nu}\right)^{-\frac{\nu+1}{2}}$$

for $-\infty < t < \infty$, Γ is the Gamma function and ν is the degree of freedom. In addition,
median $= 0$, $\mu = 0$ (for $\nu > 1$, otherwise undefined), and

$$\text{variance, } \sigma^2 = \begin{cases} \nu/(\nu - 2), & \text{for } \nu > 2 \\ \infty, & \text{for } 1 < \nu \le 2 \\ \text{undefined}, & \text{otherwise} \end{cases}$$

For example, let us consider a sample of 8 values from a normally distributed population with $\mu = 0$. The values are
$-4, -2, -2, 0, 2, 2, 3, 3$.
To find the t-distribution for the sample, we need the mean and the standard deviation of the sample, which are $\bar{X} = 0.25$ and $S = 2.65$, respectively. Therefore,
$t = \frac{0.\bar{2}5-0}{2.65/\sqrt{8}} = 0.266$.
As the degrees of freedom increases, t-distribution tends to exhibit the behavior of a standard normal distribution, which can be seen in Figure 4.2.5.

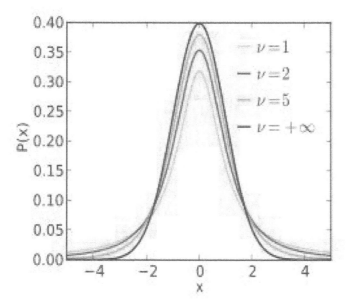

Figure 4.2: Behavior of t-distribution

Exponential Distribution

It is a continuous probability distribution used to model the time elapsed between successive events within a process in which events occur continuously and independently at a constant average rate. It is also concerned with the amount of time until a specific event occurs. Many examples can be cited such as the time until an earthquake occurs, the length of telephone conversations, and the running time of a battery before it dies. Given that events occur with a rate of change of λ, the distribution of waiting times between successive events is

$$D(x) = P(X \leq x) = 1 - P(X > x) = 1 - e^{-\lambda x}.$$

The PDF is given as

$$P(x) = D'(x) = \begin{cases} \lambda e^{-\lambda x}, & \text{for } x < 0 . \\ 0, & \text{otherwise.} \end{cases}$$

$$\int_0^\infty P(x)dx = \int_0^\infty \lambda e^{-\lambda x} = \lambda \int_0^\infty e^{-\lambda x} = 1$$

Let us consider an example to better explain the distribution's behavior. Let us assume that any telephone conversation follows an exponential distribution such that $f(x) = \frac{1}{5}e^{\frac{-x}{5}}$, for $x \geq 0$, where x is the length of the conversation. Then, what is the probability that the conversation exceeds 5 minutes?

We are given that $\lambda = 1/5$.

The probability of a conversation exceeding 5 minutes can be obtained as

$$P(x > 5) = \int_5^\infty f(x)dx = \int_5^\infty \frac{1}{5}e^{\frac{-x}{5}} = \frac{1}{5}\left[\frac{e^{-x/5}}{-1/5}\right]_5^\infty$$

$$\Rightarrow P(x > 5) = \frac{1}{e}$$

Figure 4.2.5 shows the PDFs for an exponential distribution for several values of λ.

Chi-Squared Distribution

The chi-squared distribution represents the distribution of the sum of squared standard normal deviates. The degree of freedom (DF) of the distribution is equal to the number of standard normal deviates being summed. For the summation of k standard normal deviates, the DF is $(k - 1)$.

A chi-squared test checks the independence of the variables in a given distribution. It attempts to find whether any form of association exists among the variables based on the chi-squared statistic. It calculates the statistic using the observed values (say, 'O') and expected values (say, 'E'), and

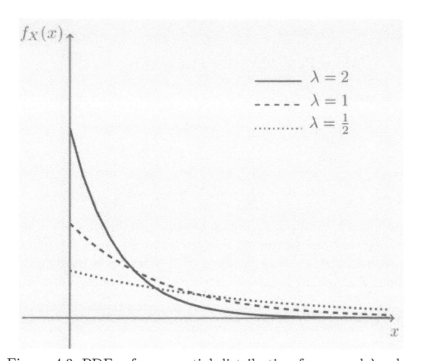

Figure 4.3: PDFs of exponential distribution for several λ values

comparing it to the chi-square distribution value. We can calculate the Chi-squared statistic, χ^2, as follows.

$$\chi^2 = \sum \frac{(O - E)^2}{E}$$

Mathematically, a chi-squared distribution can be stated as follows.

$$Y = Y_0 \times (\chi^2)^{(\frac{n-1}{2})-1} \times e^{\frac{-\chi^2}{2}}$$

where, Y_0 is a constant that depends on the number of degrees of freedom, and it is defined such that the area under the chi-squared curve $= 1$. e is the base of natural $log \approx 2.71828$. χ^2 is positively skewed, with the degree of skew decreasing with increase in the degree of freedom. For example, let us consider the temperatures of 6 random days in a month. Table 4.2 shows the expected temperatures and observed temperatures for the 6 days.

Table 4.2: χ^2 Distribution: Sample Data

Types	1	2	3	4	5	6
Expected	20	20	30	40	60	30
Observed	30	14	34	45	57	20

In this case, the two hypotheses are

$H_0 =$ Expected values are correct, $H_a =$ Expected values are not correct.

$\chi^2 = \frac{(30-20)^2}{20} + \frac{(14-20)^2}{20} + \frac{(34-30)^2}{30} + \frac{(45-40)^2}{40} + \frac{(57-60)^2}{60} + \frac{(20-30)^2}{30}$

$\Rightarrow \chi^2 = 11.44$

Using $DF = 5$, $\chi^2 = 11.44$ and the significance value, $\alpha = 0.05$, the chi-squared statistic is found to be 11.07. Thus, the p-value for 11.44 will be less than the given significance value, i.e., $p < 0.05$. Therefore, H_0 will be rejected. With an increase in the degrees of freedom, the χ^2 distribution approaches the normal distribution. It is shown in Figure 4.2.5.

Poisson Distribution

The Poisson distribution refers to a discrete probability distribution for a given number of events occurring in a given interval of time, where the average number of times an event occurs in the same interval of time is also given. The four conditions that need to be valid for Poisson distribution are: (i) an event may occur any number of times during the interval, (ii) events occur independently of each other, not influenced by the probability of occurrence of other events during the same time interval, (iii) the rate of occurrence of an event is constant, and (iv) with the increase in the length of the time period, the probability of occurrence of an event also increases. In other words, the probability of occurrence of an event is proportional to the length of the time period.

Let X be the discrete random variable that represents the number of events observed over a given time period. Let λ be the expected value or the average of X. If X follows a Poisson distribution, the probability of observing k events over the time period is

$$P(X = k) = \frac{\lambda^k e^{-\lambda}}{k!} \tag{4.3}$$

where e is *Euler's constant*. For example, assume that in a soccer match, an average of 2.5 goals is scored in each game. Using Poisson distribution, what is (i) the probability that 2 goals are scored? (ii) the probability that 4 goals are scored?

(a) Here, $\lambda = 2.5$, k $= 2$; therefore,

$$P(X = k) = \frac{2.5^2 e^{-2.5}}{2!} = 0.2562. \tag{4.4}$$

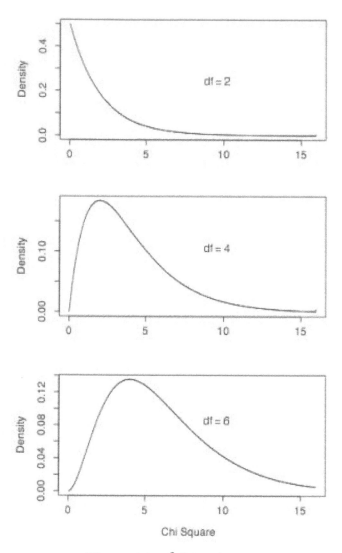

Figure 4.4: χ^2 Distribution

(b) Here, $\lambda = 2.5$, k $= 4$; therefore,

$$P(X = k) = \frac{2.5^4 e^{-2.5}}{4!} = 0.1334. \tag{4.5}$$

Bernoulli Distribution

The Bernoulli distribution represents a discrete probability distribution, having two possible outcomes labeled $n = 0$ and $n = 1$, where $n = 1$ represents the occurrence of success with probability p, and $n = 0$ represents the occurrence of failure with probability $q = 1 - p$, where $0 < p < 1$. The PDF is given as –

$$P(n) = \begin{cases} p, & for\ n = 1 \\ 1 - p, & for\ n = 0. \end{cases} \tag{4.6}$$

This can also be written as

$$P(n) = p^n.(1 - p)^{(1-n)}. \tag{4.7}$$

Performing a fixed number of trials with a fixed probability of success for each trial is known as a *Bernoulli trial*. For example, tossing a fair coin with an equal distribution of heads and tails is a Bernoulli distribution. Here, the probability of heads $= p = \frac{1}{2}$ and the probability of tails $= q = 1 - p = \frac{1}{2}$.

Binomial Distribution

The Binomial distribution is another example of a discrete probability distribution for obtaining n successes out of N *Bernoulli trials*. The Binomial distribution is given as –

$$
\begin{aligned}
P(n) &= \binom{N}{n} p^n.q^{(N-n)} \\
&= \frac{N!}{n!(N-n)!} \cdot p^n.(1-p)^{(N-n)}
\end{aligned} \tag{4.8}
$$

For example, if a fair coin is tossed 10 times, what is the probability that (i) exactly 4 tosses include a head? (ii) the probability of either 2 heads or 3 heads? Here, $N = 10$, the probability of *heads* $= p = \frac{1}{2}$, and the probability of *tails* $= q = 1 - p = \frac{1}{2}$.

(a) With $n = 4$,

$$P(4) = \binom{10}{4}(\frac{1}{2})^4.(\frac{1}{2})^{(10-4)} = 210 * (\frac{1}{2})^4 * (\frac{1}{2})^6 = 0.205. \tag{4.9}$$

(b) With $n = 2$ and $n=3$,

$$
\begin{aligned}
P(2) + P(3) &= \binom{10}{2}(\frac{1}{2})^2.(\frac{1}{2})^{(10-2)} + \binom{10}{3}(\frac{1}{2})^3.(\frac{1}{2})^{(10-3)} \\
&= 45 * (\frac{1}{2})^2 * (\frac{1}{2})^8 + 120 * (\frac{1}{2})^3 * (\frac{1}{2})^7 \\
&= 0.043 + 0.117 \\
&= 0.16.
\end{aligned} \tag{4.10}
$$

4.2.6   Multiple Testing

For an experiment, it is possible that one can perform 10000 separate hypothesis tests. If we use a p-value cut off of 0.05, we would expect 500 tests to be significant. Therefore, if we perform m hypothesis tests, we need to find the probability of 1 false positive (rejecting the *null* hypothesis, also called *type 1* error).

P(making an error) $= \alpha$ [significance value $= \alpha$]
P(not making an error) $= 1\text{-}\alpha$
P(not making an error in m tests) $= (1 - \alpha)m$
P(making atleast 1 error in m tests) $= 1 - (1 - \alpha)m$

Adjusting the p-value controls the false positives or type 1 error. Different approaches to compute the number of false positives are

(a) Per Comparison Error Rate (PCER): The expected value of the number of *type 1* errors over the number of hypotheses. Mathematically,

$$PCER = E(V)/m \qquad (4.11)$$

(b) Per Family Error Rate (PFER): The expected number of *type 1* errors. Mathematically,

$$PFER = E(V) \qquad (4.12)$$

(c) Family-wise Error Rate (FWER): The probability of at least one *type 1* error. Mathematically,

$$FWER = P(V \geq 1) \qquad (4.13)$$

(d) False Discovery Rate (FDR): The expected proportion of *type 1* errors among the rejected hypotheses. Mathematically,

$$FDR = E(V/R|R > 0)P(R > 0), \qquad (4.14)$$

where R is the number of rejected hypotheses.

4.2.7   False Discovery Rate

It is used to control the proportion of false positives among the rejected hypotheses (R). It gives the number of false positives in all of the rejected hypotheses:

$$FDR = E(V/R|R > 0)P(R > 0) \qquad (4.15)$$

where V is the number of false positives (or type 1 errors) and R is the number of rejected hypotheses. The information after the 'given' ('||') symbol states the following:

(a) We have at least 1 rejected hypothesis.

(b) The probability of getting at least one rejected hypothesis is greater than 0.

For example, in medical testing, FDR quantifies situations when we get a positive test result, but actually do not have the disease. It is the complement of Positive Predictive Value (PPV), which tells us the probability of a positive test result being accurate. If PPV is 60%, FDR is 40%. Out of 10,000 people given a test, if there are 640 positive results, out of which 450 are true positives and 190 are false positives, then the FDR is $\left(\frac{190}{640}\right) \times 100$ = 29.6%.

4.2.8 Maximum Likelihood Estimation

It is a method that can determine the values of the parameters for a distribution model. The parameter values are found such that they maximize the likelihood of the occurrence of the data of a given observation. If θ is the set of parameters that we need to estimate and x_i is a set of observations of the probability distribution, where $i = 1, 2, ..., n$, then, the likelihood of θ for observations x_i is given as

$$L(\theta; x_i) = f(x_i|\theta)$$

where f denotes a probability distribution function.
Assuming all observations are independent of each other, the likelihood can be given as

$$
\begin{aligned}
L(\theta; x_i) &= & f(x_1, x_2, ..., x_n|\theta) \\
&= & f(x_1|\theta)f(x_2|\theta)...f(x_n|\theta)
\end{aligned}
$$

To get the maximum likelihood of the data, we have to take the derivative of $L(\theta; x_i)$ and equate it to 0. Since the terms are in the product, we have to use the chain rule of derivation. To simplify this process, we take the log likelihood of $L(\theta; x_i)$ and denote it as l.

$$
\begin{aligned}
l &= & \log L(\theta; x_i) \\
&= & \log(f(x_1|\theta).f(x_2|\theta)...f(x_n|\theta)) \\
&= & \log(f(x_1|\theta)) + \log(f(x_2|\theta)) + ... + \log(f(x_n|\theta))
\end{aligned}
$$

$$\Rightarrow l = \sum_{i=1}^{n} \log(f(x_i|\theta))$$

To get the maximum likelihood of the parameter θ, we can calculate the derivative of l with respect to θ and equate it to 0.

$$\frac{dl}{d\theta} = \frac{d}{d\theta} \sum_{i=1}^{n} \log(f(x_i|\theta)) = 0$$

Let us consider the *Bernoulli distribution*, where the parameter is the probability of success of an event, denoted by p.

$$
\begin{aligned}
f(x_1, x_2, ..., x_n|p) \quad &= P(X_1 = x_1, X_2 = x_2, ..., X_n = x_n|p) \\
&= p^{x_1}(1-p)^{1-x_1}.p^{x_2}(1-p)^{1-x_2}...p^{x_n}(1-p)^{1-x_n} \\
&= p^{\sum x_i}(1-p)^{\sum(1-x_i)} \\
&= p^{\sum x_i}(1-p)^{n-\sum x_i}
\end{aligned}
$$

where $x = 0$ or 1 and $i = 1, 2, ...n$

$$\log f = \sum x_i \log p + \left(n - \sum x_i\right)\log(1-p)$$

$$\frac{d}{dp}\log f = \frac{\sum x_i}{p} - \frac{n - \sum x_i}{1-p} = 0$$

$$\Rightarrow \frac{\sum x_i}{p} = \frac{n - \sum x_i}{1-p}$$

$$\Rightarrow \sum x_i - p\sum x_i = np - p\sum x_i$$

$$\Rightarrow \hat{p} = \frac{\sum x_i}{n}$$

Thus, the maximum likelihood estimation, $\hat{p}$, is $\frac{\sum_{i=1}^{n} x_i}{n}$, for a Bernoulli distribution.

Let us consider the *normal distribution*, where the parameters are mean (μ) and variance (σ^2). The probability distribution for normal distribution is given as

$$f(x; \mu, \sigma^2) = (2\pi\sigma^2)^{-1/2}exp\left(-\frac{1}{2}\frac{(x-\mu)^2}{\sigma^2}\right)$$

The likelihood function is

$$f(x_1, x_2, ..., x_n; \mu, \sigma^2) = (2\pi\sigma^2)^{-n/2}exp\left(-\frac{1}{2\sigma^2}\sum_{j=1}^{n}(x_j - \mu)^2\right)$$

$$\log f = -\frac{n}{2}\log 2\pi - \frac{n}{2}\log \sigma^2 - \frac{1}{2\sigma^2}\sum_{j=1}^{n}(x_j - \mu)^2$$

. Taking the first derivative of the log likelihood function, we get the maximum likelihood estimation for the sample mean, $\hat{\mu}_n$ and sample variance, $\hat{\sigma}_n^2$.

$$\hat{\mu}_n = \frac{1}{n}\sum_{j=1}^{n} x_j$$

$$\hat{\sigma}_n^2 = \frac{1}{n}\sum_{j=1}^{n}(x_j - \mu_n)^2$$

.

For example, let us take a set of weights of bags (in kgs), where values are normally distributed.
$X = (115, 122, 130, 127, 149, 160, 152, 138, 149, 180)$

We have $\sum_{j=1}^{10} X_j = 1422$.

The maximum likelihood estimation for parameters mean, $\hat{\mu}_n$ and variance, $\hat{\sigma}_n^2$, will be

$$\hat{\mu}_n = \frac{1}{n}\sum_{j=1}^{n} x_j = \frac{1422}{10} = 142.2,$$

$$\hat{\sigma}_n^2 = \frac{1}{n}\sum_{j=1}^{n}(x_j - \mu_n)^2 = \frac{3479.6}{10} = 347.96.$$

4.3 Machine Learning Background

Ever-increasing growth in application development based on machine learning (ML) has been noticed in all fields in recent times. Scientists, professionals and application developers are exploring and applying machine learning in every field to better their outputs, productivity and business. The impact of this evolving field is being felt in every aspect of life. However, the question arises: What makes ML so important and popular today? The response to this query from professionals, researchers and application developers may not be exactly the same, although there are significant commonalities. Machine learning finds intrinsic, previously unknown and interesting patterns inherent in the data. Such patterns can help identify or predict significant outcomes to solve real-life problems. Machines can be trained to imitate humans so as to provide decision support with high precision in fields like healthcare service, speech recognition, satellite data analysis, face recognition, and weather forecasting just to name a few. There are also many applications of machine learning in areas where adequate human expertise is not available. In such cases, machine learning can help computers learn how to work with minimum or little direct human intervention. Machine learning can be used to train machines to explore comprehensive possibilities with minimum human supervision, such as simply providing inputs and corresponding correct outputs. The machines meticulously map the inputs to the appropriate output, using machine learning algorithms.

Computers and computer networks have been an integral part of human lives for many decades. Every sector of society, industry and government depend extensively on computers. The rise of the Internet has played both a beneficial and crucial role in our lives, especially in healthcare support. Everything accessible in the world is now just a click away, even if it is situated far away. One important application domain where there is high growth of web-based ML applications is healthcare services using computers and computer networks. An appropriate use of ML algorithms can immensely help support healthcare in real time to save precious lives from critical diseases. The use of ML in the treatment of critical diseases in various capacities, especially in identifying prognostic and diagnostic biomarkers, is an established fact.

4.3.1   Significance of Machine Learning

Machine learning solutions can be used to solve a wide range of problems. Wherever there are problems which can be readily solved by human experts and machines alike, there are some, whose complexity requires computers and machine proficiency, surpassing human expertise. In addition, there are many problems for which there exists no or little human skill, or where human expertise is unexplainable. There are other cases where data keep on evolving from time to time, requiring solutions that can evolve as well [59].

Thus, the issue is how to make a machine learn to solve a problem by itself, or improve its performance on a task by itself. Machine learning has truly changed how the world views computer programs. It has been a boon to almost every field in the territory of technology. Machines learning algorithms feed on data and extract underlying interesting useful patterns. Learning from the existing data and improving its performance is pivotal in building a reliable machine learning model because later the model has to undertake predictive tasks on unseen data. This is the reason why existing data should be of high quality, and represent the input data distribution well.

Data for a specific problem are available in the form of datasets. A dataset can be thought of as a matrix, where each row represents represents an instance (for example, a gene expression instance), and each column depicts a feature or attribute describing the characteristics (for example a patient or a sample) of each instance with the exception of the last column. The last column of such a matrix usually serves as the knowledge (to be used during training) used to solve a specific task. In particular, the last column denotes the class label (for example, "benign" and "malignant" in case of a two-class cancer patient dataset) for each instance. The values of different features can be binary, categorical or continuous. The problem to be solved can be a two-class problem or a multi-class problem. In a two-class problem, each instance belongs to either class A or class B, depending on the characteristics it exhibits. One key requirement for machine learning is the concept of *generalization*. During training, a model learns from known data how to differentiate between instances belonging to two separate classes or groups. This learning model is put to test on unseen instances, i.e., data examples, which are from the same distribution but were not seen during training. It is expected that after training, the model performs well on the unknown or unseen data and is also able to grasp the trends underlying in the data. This quality of adaptability to previously unseen instances is called generalization. An ML model that can adapt in such situations is said to generalize well. During training, if a model only memorizes the outcomes of the data it has seen, it may not perform well when put to test. It will miss the trends in the data and consequently not generalize well. It should learn the concepts well from the training data and apply these concepts effectively to the test data.

The two types of errors that need to be considered when building a model are mentioned below.

(a) *Bias error*: It is an estimation of the number of times, on average, a prediction made by a learner deviates from the actual value. A learner with high bias error usually underperforms and can misclassify an instance easily. Such a learner can be assumed to be incapable of properly learning the trends in the data.

(b) *Variance error*: Variance is the quantification of deviation in predictions made by the learning model on the same samples. If a model has high variance, it is bound to suffer from overfitting when trained, and as a result, will not perform as desired during the testing phase.

Depending on how models learn from the existing data, machine learning approaches can be of different types, namely, (i) supervised learning, (ii) unsupervised learning, (iii) semi-supervised learning, and (iv) reinforcement learning. Unlike unsupervised learning, in supervised machine learning, the label information corresponding to each instance in the dataset is used as the guiding knowledge. However, in semi-supervised learning, the existence of prior knowledge is limited. In other words, some data instances are labeled while some or most are unlabeled.

4.3.2   Machine Learning and Its Types

This section provides a comprehensive background on various types of machine learning approaches, using practical examples.

(a) *Supervised Learning*: *Figure* 4.5 illustrates supervised learning. Suppose we have a problem to correctly identify a given shape to be either a circle or something else. As input, a learning model is provided with a set of objects along with their respective class labels (or responses). Our task is to predict the shape of an input instance as either "circle" or "not circle". In other words, the given task is a 2-class classification problem. The class "circle" includes all the samples whose shapes are actually a "circle" and the class "not circle" will include rectangles, pentagons and triangles. During the training phase, the model learns how to differentiate a circle from other shapes based on characteristics such as the number of corners (size and color do not play any role in this example). After the model is built, it is ready to be tested with a test set. This test set consists of previously unseen instances whose classes need to be predicted by the model based on the training it received. If the model receives a test instance of circular shape, the model should predict it's shape to be a circle. This means the test instance belongs to the class "circle". In this way, the test instances are mapped to their respective classes.

(b) *Unsupervised Learning*: Unlike supervised learning where prior knowledge is provided, in unsupervised learning, prior knowledge is not provided to the learner in any form. In *Figure* 4.6, the input to the learning model is a set of objects. The problem is to organize similar objects into *groups* based on basic characteristics like the number of *corners* (not

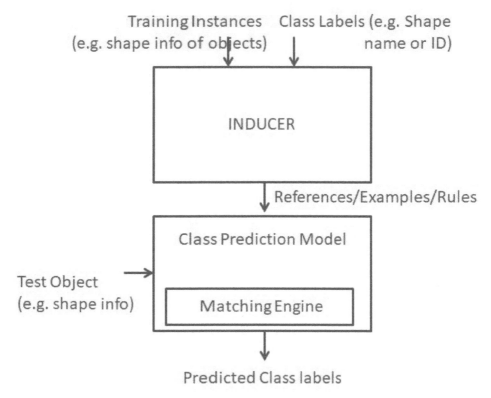

Figure 4.5: An example of supervised learning.

size or *color*). So, the groups in this case may be something like shapes with no corners, shapes with three corners, shapes with four corners or shapes with five corners. Accordingly, circles, triangles, rectangles and pentagons may be arranged into different groups.

(c) *Reinforcement Learning*: Like unsupervised learning, in reinforcement learning, an agent is not furnished with any kind of prior knowledge to carry out a task that requires one or more actions in sequence. A simple case can be the identification task discussed earlier. It attempts to accomplish the task on its own. It exploits the learning algorithm to explore and choose the best possible action(s) to carry out a task. Let us consider an illustrative example to explain reinforcement learning. We consider a game where a player is supposed to identify the geometrical shapes of objects. On successful identification of a shape, the player is "rewarded", otherwise penalized, and accordingly its state is changed. The example is illustrated in *Figure* 4.7.

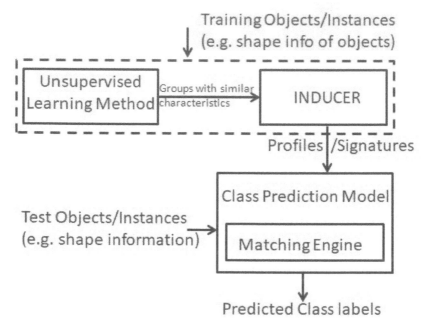

Figure 4.6: An example of unsupervised learning.

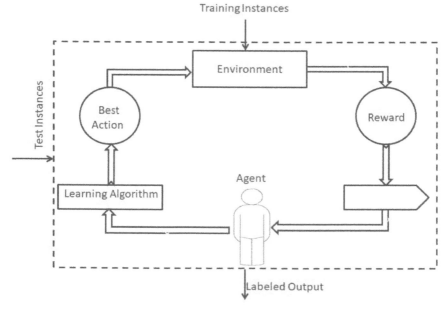

Figure 4.7: An example of reinforcement learning.

4.3.3   Supervised Learning Methods

A supervised machine learning method attempts to identify the class label of a given test object or instance based on the knowledge (references, rules or signatures) provided. The problem for such a method can be defined mathematically as follows. Given a set X of input variables to a function f that attempts to map each element $x_i \in X$ to an output variable y, using the equation $y = f(x)$, with high accuracy. Each instance x_i is characterized by a set of features and the target variable or class label, i.e. 'y' indicating to which category x_i belongs. A supervised learning method has two major steps: *training* and *testing*. In the *training* step, the method exploits the knowledge (i.e., examples of known instances tagged with class labels) to build a predictive model to help identify classes or to assign class labels to test instances. In the testing step, the prediction model is used to evaluate its performance in assigning class labels for input test instances or objects.

ID	Age	Sex	Likes Android/iOS smartphone
01	22	M	1
02	49	F	0
03	21	F	1
04	52	M	0
05	22	M	1
06	24	F	1
07	54	F	0
08	51	F	0
09	58	F	0
10	56	M	0
11	25	F	1
12	20	M	1
13	23	M	1
14	60	F	0
15	26	M	1

Figure 4.8: An example dataset.

Based on the desired outcome type or target output, the task of a supervised learning method is categorized into two types: (i) classification and (ii) regression. In classification, the outcomes or class labels are discrete or categorical, whereas in regression, it is numeric or continuous. A classic example of classification concerns data from healthy and cancer patients (clinical or gene expressions), where a supervised learning model is first trained to differentiate a malignant instance from a benign instance. Generally, it attempts to decide the class label of a given test instance based on a matching score obtained after matching against the class-specific knowledge or reference. The performance of a supervised learning method is largely dependent on the quality, quantity and relevance of training instances. Recently, a large number of sources of data have become available, including general machine learning repositories, gene expression data repositories, Kaggle [146], AWS [458], and Google Cloud [134], from which a researcher can download data to train and evaluate the performance of a machine learning method.

Once data is available, it needs to be pre-processed so that there are no inconsistencies, exceptions, or incorrect and skewed examples. Data preprocessing may involve a number of activities like finding or removing missing values, normalization, discretization, batch effect removal, feature extraction, and feature selection. The preprocessed data is split into two parts, namely the *training set* and the *test set*. The *training set* should be an appropriate representative of the unseen input that is to be received later during prediction. The *test set* is put aside for later use, and the training set is given as input to the selected supervised learning algorithm. In this step, the hyper-parameters of the learning algorithm can be specified, for example, the number of trees to grow in a random forest. The learning model is built from the training set which is then evaluated on the test set using evaluation metrics. If the performance is not as expected, there can be several possible reasons, such as (i) the presence of inconsistencies or outliers in the collected data, (ii) the presence of missing values, (iii) the presence of redundant examples, (iv) inexhaustive collection of features extracted, (v) non-optimal subset of features selected, and (vi) improper training data selection.

A large number of supervised learning methods are currently available. The next section presents some popular supervised methods under six distinct categories, (i) tree-based classifiers, (ii) statistical classifiers, (iii) classification using K nearest neighbors, (iv) classification using a Support Vector Machine, (v) classification using soft computing, and (vi) ensemble learning.

Tree-Based Classifiers: Decision Tree Algorithm

In supervised learning, decision trees represent a popular and effective method for both classification as well as regression analysis. The simple logic structure with high predictability makes this technique attractive in handling any supervised learning problem. It attempts to solve a classification problem using a divide-and-conquer approach based on a set of IF-THEN-ELSE rules. The three basic components of a decision tree are: *nodes, edges* and *leaves*. Each node helps partition the data based on an attribute.

Depending on the parent node's attribute values, the edges are defined to connect child node(s) with the parent. There may be none as well as multiple child nodes and each leaf in the tree is labeled with a distinct *Class-id* or *name*. The splitting criterion in a decision tree algorithm is usually based on the computation of the *Gini* value or *Entropy*. In *Gini* approach, it calculates a *Gini value* for each subset after a split to support finding the probability for identification of a class. If the resultant instances belong to a single class after a split, the *Gini value* becomes '0' and it suggests no more split.

To calculate the *Gini value*, it uses Equation 4.16.

$$Gini = 1 - \sum_{i=1}^{N} p(i|t)^2. \tag{4.16}$$

Here, N represents the number of classes. Since it uses IF-THEN-ELSE construct to partition the data, on fulfilment of IF part, it traverses the left branch, otherwise for the ELSE part, it traverses the right branch and continues until a leaf node is encountered. An example dataset is shown in Table 4.3. A corresponding decision tree is shown in Figure 4.9 for illustration. The three most well-known decision tree-based classifiers are *ID3*, *CART*, and *C4.5*.

Table 4.3: Sample Job-Income-Vehicle Data

Job	Income	Vehicle
Government	1,55,000	Four wheeler
Private	50,000	Two wheeler
Government	1,75,000	Four wheeler
Self-employed	90,000	Two wheeler
Private	58,000	Two wheeler

Statistical Classifiers

This section presents discussion on a popular statistical classifier called Naive Bayes (NB) [162]. NB has been widely used in most classification problems. The heart of this classifier is a Bayesian method that helps the predictions for any unseen or previously unknown test case. The method calculates the probabilities of target values for any given test instance, and assigns the target value that is associated with the maximum probability. For illustration, let us take an n-dimensional test instance, say $A(a_1, a_2, \ldots, a_n)$. In order to classify A by assigning a class label b, this method computes class probability for A using Bayes rules. The Equation 4.17 provides the basis of the

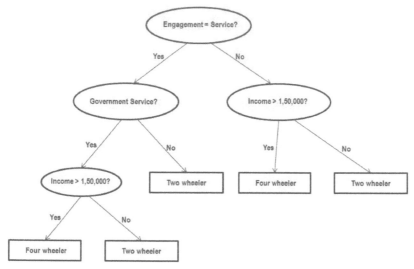

Figure 4.9: An example of decision tree.

Bayes' theorem.

$$p(b|A) = (p(A|b) \times p(b))/p(A) = p(a_1, a_2, \cdots a_n|b) \times p(b)/p(A) \qquad (4.17)$$

It assigns a class label to an instance when the associated probability is maximum.

Classification Using K Nearest Neighbor Algorithm

The K-nearest neighbors (KNN) [162] classifier is a robust, yet simple supervised learning algorithm that stores all the available cases or examples and attempts to classify new instances based on a proximity measure, e.g., a distance function. The KNN algorithm is based on the assumption that objects or instances of similar nature should be near each other or exist in close proximity.

The KNN classifier is an effective non-parametric technique in statistical estimation and pattern recognition. It attempts to classify a test instance based on a majority voting consensus function of its neighbors. It assigns the instance to the class most common (or, majority) amongst its K nearest neighbors, recognized by a distance function. In case of $K = 1$, the test instance gets simply the class label of the nearest neighbor. An important advantage of this classifier is that it can use any proximity (i.e., similarity or dissimilarity) measure to compute the closeness of a given test object or instance with the training objects or instances.

KNN Algorithm

Input : D_{train}: Training dataset; D_{test}: Test dataset; K: number of nearest neighbors;
Output: S, labeled test dataset;

(a) Read D_{train} and set K to the given number of neighbors

(b) For each test instance, $x_i \in D_{train}$.

    (a) Calculate the distance between x_i and training examples, i.e., $y_i \in D_{test}$

    (b) Insert the distance information and the index of xi to an ordered collection, L

(c) Sort L in ascending order by the distances.

(d) Pick the first K entries from the sorted collection.

(e) Get the labels of the selected K entries.

(f) If classification, return the mode of the K labels.

(g) If regression, return the mean of the K labels.

Choosing the Right Value for K

In the KNN algorithm [162], K plays a crucial role. When the value of K changes, a significant change in the outcomes of the algorithm can be observed. To select an appropriate value for K, we follow a heuristic approach. We run the KNN algorithm several times with different values of K and choose the K that gives us maximum prediction accuracy. However, it is extremely difficult to obtain a single K value for multiple datasets from different domains, if we want to ensure maximum accuracy. So, we recommend an optimal range of K values for best possible accuracy. The value of K should be odd to avoid any tie situation. Another important point to be noted is that with an increase in the value of K, the stability of predictions becomes higher due to the involvement of more numbers of samples during majority voting. Similarly, the stability of predictions decreases with the decrease in the value of K.

The major advantages of the KNN algorithm are: (i) it is simple to implement, (ii) except K, no other tuning parameters are required, (iii) it is independent of any proximity measure, (iv) it is free from any additional assumptions, and (v) it is applicable to both classification and regression problems. However, a serious issue with the KNN algorithm is that its execution time performance deteriorates with an increase in number of training instances or predictors.

Classification Using Support Vector Machine (SVM)

This classification method [162] aims to solve a classification or regression problem by constructing hyperplane(s) over a k-dimensional space where k is the number of attributes of the example or training instances. It attempts to achieve a good separation between two class instances by choosig an appropriate hyperplane, and it does so by considering the hyperplane that has the largest distance to the example instances for a given class. However, there could be a large number of hyperplanes so that task is not straightforward. The higher the margin between the instances of two given classes, the better the accuracy of classifying future test instances with high confidence.

For a given training dataset D_{train} and test dataset D_{test} of k-dimensional instances, we can define the SVM model as a representation of the instances $\in D_{train}$ in k-dimensional space so that the training instances belonging to two different classes are separated by a distinct gap as wide as possible. It is well understood that the higher the separation achieved by the hyperplane (larger functional margin), the lower the generalization error of the classifier. During testing, any test instance $\in D_{test}$ is mapped to a class separated by a gap. So, instead of computing probability of a class, this algorithm attempts to assign any new instance to one of the class labels. It uses the term 'support vector' to represent the instances closest to the hyperplane. It identifies the location and orientation of the hyperplane using the 'support vectors', maximizing the margin. Unlike a linear SVM, in case of non-linear instances, SVM attempts to classify the test instances using a kernel trick that helps learn a linear model to classify non-linear instances. Figure 4.10 illustrates examples of both linear and non-linear SVM.

Ensemble Learning

An ensemble of classifiers [296] aims to combine the predictions of multiple individual classifiers in a meaningful way to classify new instances so that an improved and consistent classification accuracy is achieved than the individual participating classifiers. In other words, an ensemble learning approach deals with a multiple and diverse set of learners, and attempts to combine them in a manner so that a best possible result is achieved. The basic philosophy behind the introduction of ensemble learning is that the decision of a team is always better than the decision of an individual.

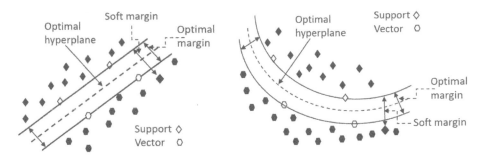

Figure 4.10: An example of support vector machine

The learning algorithms that participate in the ensemble process are called *base learners*. For quality and unbiased output generation, an ensemble learning approach should include the base learners from diverse families and independent of each other. This avoids correlated errors made by the base learners, where misclassifications by any of the base learners can be averaged out when correctly classified by other learners. The purpose of using a combination function is to aggregate the decisions of the base learners to provide an unbiased decision. One such example of a combination function is *majority voting*. For a given test instance, if three learners assign class label X and two learners assign Y, then the final class label output would be the instance classified as class X as a majority of learners voted for it.

To develop an approach based on a combination of several models for solving a classification or clustering problem is not difficult. To refer to the method of combining models using multiple learning algorithms, researchers [3] have coined terms such as 'Blending' by Elder and Pregibon [115], 'Ensemble of Classifiers' by Dietterich [106], 'Committee of Experts' by Steinberg and Colla [385], and 'Perturb and Combine (P&C)' by Breiman [67]. We can train multiple models using either the same dataset or using different samples from the same dataset and finally, combine the predictions of the individual models based on (i) voting (to solve a regression problem) or (ii) averaging the individual outputs (to solve estimation problem). In 1996, Breiman introduced [67] an effective ensemble approach, called the *'Bagging'* approach, to achieve better classification accuracy. It takes multiple decision tree models generated from bootstrap samples (with replacement) of a training dataset and combines their outputs by simple voting. In another approach, Freund and Schapire [128] introduced *'AdaBoost'*, an iterative process of weighing more heavily the wrongly classified cases based on the decision tree models, and then combining all the models generated during the process. For example, ARCing [67] is a kind of boosting approach which weighs wrongly classified cases more heavily, but unlike the formula given by Freund and Schapire (1996) [128], weighted random samples are chosen from the training data. In another approach, Wolpert (1992) [435] exploited the regression technique to combine ANN (Artificial Neural Network) models, which was later known as *Stacking*. We summarize the ensemble construction approaches as follows.

(a) *Using diverse classifier models*: An ensemble can be constructed from a diverse set of classifier families or from the same classifier with varied hyper-parameter values.

(b) *Using different feature subsets*: Characteristics or features can be chosen in a deterministic way or randomly from a training set to form several subsets of features, to build different learners.

(c) *Using different training sets*: A single learning algorithm is used, but different classifiers are built with different random sub-samples from the training data.

(d) *Using different combination schemes*: The outputs of several classifiers trained on the same training data are combined in some fashion like majority voting, algebraic combination, and weighted sum. Some combination rules used in ensemble learning are tabulated in Table 4.11.

Table 1 Summary of combination rule

Approach	Method	Formula	Effectiveness
Class label combination	Majority voting	$\sum_{t=1}^{T} d_{t,j}(x) = \max_{j=1}^{C} \sum_{t=1}^{T} d_{t,j}$	Gives average performance when majority does not give accurate prediction
	Weighted majority voting	$\sum_{t=1}^{T} w_t d_{t,j}(x) = \max_{j=1}^{C} \sum_{t=1}^{T} w_t d_{t,j}$	Performs well only if the weights of the classifiers are assigned precisely
Continuous output combination	Sum rule	$\mu_j(x) = \sum_{t=1}^{T} d_{t,j}(x)$	Gives average performance when majority does not give accurate prediction
	Weighted sum rule	$\mu_j(x) = \sum_{t=1}^{T} w_t d_{t,j}(x)$	Performs well only if the weights of the classifiers are assigned precisely
	Mean rule	$\mu_j(x) = \frac{1}{T} \sum_{t=1}^{T} d_{t,j}(x)$	Performance is significantly affected by outliers
	Product rule	$\mu_j(x) = \prod_{t=1}^{T} d_{t,j}(x)$	Sensitive to low probability value
	Maximum rule	$\mu_j(x) = \max_{t=1}^{T}\{d_{t,j}(x)\}$	Chooses the most optimistic value
	Minimum rule	$\mu_j(x) = \min_{t=1}^{T}\{d_{t,j}(x)\}$	Performance is significantly affected by outliers
	Median rule	$\mu_j(x) = \text{median}_{t=1}^{T}\{d_{t,j}(x)\}$	Performance is significantly affected by outliers
	Generalized rule	$\mu_{j,\alpha}(x) = \left[\frac{1}{T}\sum_{t=1}^{T} d_{t,j}(x)^{\alpha}\right]^{\frac{1}{\alpha}}$	Affected by outliers

Figure 4.11: Some combination rules used in ensemble learning

Next, we present some approaches commonly used in ensemble learning, followed by discussion of some well-known ensemble learning algorithms.

Bagging: In 1996, Breiman [67] introduced an ensemble method, called *Bagging*, where it uses the same learning algorithm for different training datasets to obtain different classifiers. It uses a bootstrap technique to re-sample the training dataset towards achievement of better diversity among the training instances. *Figure* 4.12 demonstrates an example of the *Bagging* approach, where each classifier is trained with a re-sample of instances. Finally, the predictions (having equal weight) of the individual classifiers are combined based on *majority voting*.

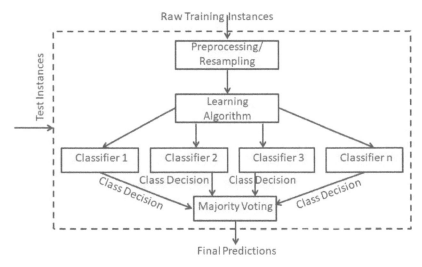

Figure 4.12: An example ensemble learning: Bagging

Boosting: In 1996, Freund and Schapire [128] introduced another effective ensemble approach, called *Boosting*, which uses a different re-sampling technique. This approach generates a new training dataset based on its sample distribution. *Figure* 4.13 shows an example boosting approach, where the first classifier is created from the original dataset considering all samples to have an equal weight. In the subsequent training datasets, it reduces the weights for those samples which have been correctly classified, otherwise (misclassification cases) it increases the weights of the samples. To combine the individual decisions, it uses a *weighted majority voting* method where the more accurate classifiers are given higher weightage than a less accurate classifier.

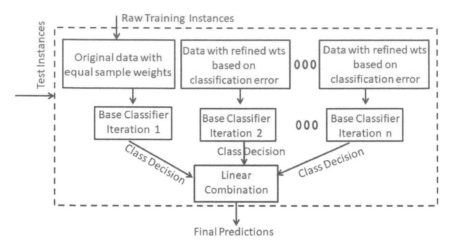

Figure 4.13: An example ensemble learning based on boosting

Stacked generalization: Stacked generalization [435] aims to generate an unbiased output using a *meta learner* concept. It is a two-tier ensemble approach, where *Tier*-0 comprises a set of classifiers from diverse backgrounds to generate predictions. These predictions are used as input for the *meta learner* in the next tier, i.e. *Tier-1*, to generate the final prediction. For better illustration, let us take sets of training and test instances, say D_{train} and D_{test}, respectively. The stacked generalization approach divides the training instances D_{train} into say, m sets, and for training, all but one of these are provided as input to all the base generalizers. It evaluates the learners on the hold-out set say, D_{test}, not seen previously. The predictions generated, say $O_1, O_2, \cdots, O_k$ by the base generalizers for the test instances on which it was not trained, along with the correct labels of those instances, constitute the training data for the second level (*Tier*-1) meta-classifier. This approach aims to learn the training data as precisely as possible. If a base generalizer repeatedly misclassifies instances belonging to a certain region of the feature space, then this trait can be learned by the meta-learner and corrected with the help of the other learners in the ensemble. Once the *meta learner* is trained, all data are pooled, and the individual base generalizers are retrained on the entire dataset, using a suitable resampling technique.

(i) *Random Forests*: The *Random Forests* [247] approach constructs a forest with the *Decision Tree* algorithm as the base component. In this ensemble, each decision tree is uniquely trained using the *Bagging* approach. As discussed previously, the bagging approach is based on multiple subsets of training instances chosen randomly from the original training set. To gather subsets of training instances, if the sampling is carried out with replacement,

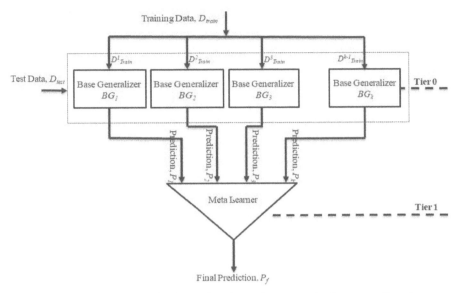

Figure 4.14: Ensemble learning based on stacked generalization

we call it *Bootstrap Aggregating* or *Bagging*; otherwise, it is called *Pasting* (without replacement). Unlike the decision tree algorithm, in this method while constructing the forest, each node is represented by a randomly chosen subset of features (sometimes called *feature bagging* [239]. It combines the outcome of each decision tree using a combination function (usually, *majority voting*) to achieve the best possible accuracy. Figure 4.15 shows an example of random forest. Random forests are able to give significantly better performance than an individual decision tree algorithm.

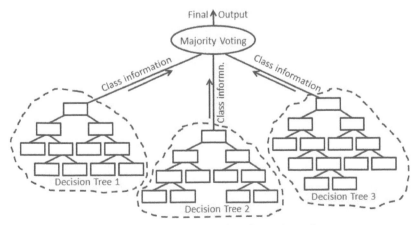

Figure 4.15: An Example of random forest

(ii) *Extra trees*: Extremely Randomized Trees [138] is another effective supervised ensemble algorithm. Although the *Extra trees* method is similar to
Random Forest, it differs from it because it does not depend on the concept of *bagging*. It uses all the instances in the training set for constructing
the individual trees and not just the bootstrap samples (to minimize the
bias). *Random forests* introduce some degree of randomness into the existing decision tree algorithm by choosing a subset of features randomly during
splitting a node. However, *Extra Trees* moves one step ahead by introducing a higher degree of randomness by randomly choosing the cut point for
each feature as it uses a criterion like *Gini* or *Information Gain*, to split the
nodes. Our experimental study shows that *Extra Trees* is relatively faster
than other tree-based classifiers. It is mostly due to the random selection of
the splitting values, which is otherwise a time-consuming process. Finally,
Extra Trees combines the individual predictions to yield a best possible prediction.

AdaBoost: Adaptive Boosting (*AdaBoost*) [128] is a robust and effective supervised ensemble algorithm which attempts to provide classification performance with high accuracy by focusing mostly on the misclassified instances.
It works sequentially and in iteration, with an objective to overcome the
misclassification issues of the previous iterations. In each iteration, it focuses more in handling the misclassified instances of the previous iteration.
The main motive of this method is to create several weak learners towards
solving a hard problem, which an individual weak learner may not be able
to solve. For better illustration, let us consider the decision tree as a base
learner and train it to make predictions on a training dataset. Adaboost initially assigns equal weights to all the instances. In subsequent iterations, it
emphasizes the previously misclassified instances by assigning more weight
while feeding to the decision tree algorithm. In other words, it boosts the
weights for the wrongly classified instances so that better accuracy with
minimum false positives is achieved.

4.3.4   Unsupervised Learning Methods

Unsupervised learning is one of the three main approaches to machine learning, along with *supervised* and *reinforcement* learning. As discussed in the
previous subsections, unlike supervised learning, an unsupervised learning
method is not provided with any form of knowledge or labeled data. This
learning approach is also referred to as self-organized Hebbian learning
[136] [333] that helps extract interesting, yet previously unknown patterns

in the data.

One major unsupervised learning approach is *Cluster analysis*. Cluster analysis aims to cluster, group, or segment instances (unlabeled), based on the proximities among the characteristics of the instances. This unsupervised learning approach identifies commonalities and differences among the instances while grouping and segmenting them. It is capable of identifying anomalous or non-conforming instances not fitting into any of the identified groups. The performance of an unsupervised learning method is dependent on the nature of input data and the ability of the proximity measure(s) used. Next, an in-depth discussion is provided on various types of input data and proximity measures.

Nature of Data

A key aspect of any anomaly detection technique is the nature of the input data. Input is generally a collection of data instances (also referred as objects, records, points, vectors, patterns, events, cases, samples, observations, entities) [394]. Each data instance can be described using a set of attributes (also referred to as variables, characteristics, features, fields, dimensions). The attributes can be of different types such as binary, categorical or continuous. Each data instance may consist of only one attribute (univariate) or multiple attributes (multivariate). In the case of multivariate data, all attributes may be of the same type or may be a mixture of different data types. The nature of attributes determine the applicability of clustering techniques. For example, different statistical models have to be used for continuous and categorical data. Similarly, for nearest-neighbor-based techniques, the nature of the attributes determines the distance measure to be used. Often, instead of the actual data, the pairwise distance between instances may be provided in the form of a distance (or similarity) matrix. In such cases, techniques that require original data instances cannot be used; these include many statistical and classification based techniques.

Similarity and Dissimilarity Measures

Distance is defined as a quantitative measure of how far apart two objects are. Similarity is a numerical quantity that reflects the strength of relationship between two objects or two features. Similarities are higher for pairs of objects that are more alike. This quantity is usually in the range of either -1 to $+1$ or may also be normalized to the range of 0 to 1. If the similarity between object x and object y is denoted by $S_{x,y}$, we can measure this quantity in several ways depending on the scale of measurement (or data type) that we have.

A distance measure is also known as a dissimilarity measure. Similarity and dissimilarity measures are often called proximity measures. Dissimilarity measures the discrepancy between the two objects, i.e., it measures the degree to which two objects are different. There are many distance and similarity measures. Each similarity or dissimilarity measure has its own strengths and weaknesses. Next, we consider several important issues concerning proximity measures.

Relationship between Similarity and Dissimilarity

Let the normalized dissimilarity between object x and object y be denoted by $d(x, y)$. Then the relationship between dissimilarity and similarity is given by- $S_{x,y} = 1 - d(x, y)$. Here, $S_{x,y}$ is a normalized similarity measure between objects x and y. Similarity is bounded by 0 and 1. When similarity is one (i.e., two objects are exactly similar), the dissimilarity is zero and when the similarity is zero (i.e., two objects are very different), dissimilarity is one. If the value of similarity has range of -1 to $+1$, and the dissimilarity is measured with range of 0 and 1, then $S_{x,y} = 1 - 2d(x, y)$. When dissimilarity is one (i.e., two objects are very different), similarity is minus one and when dissimilarity is zero (i.e., two objects are very similar), similarity is one.

In many cases, measuring dissimilarity (i.e., distance) is easier than measuring similarity. Once we measure dissimilarity, we can easily normalize it and convert it to a similarity measure. It is also common for dissimilarities to range from 0 to ∞. Frequently, proximity measures are transformed to the interval $[0, 1]$. The transformation of similarities to the interval $[0, 1]$ is given by Equation 4.18,

$$S'_{x,y} = (S_{x,y} - min_{S_{x,y}})/(max_{S_{x,y}} - min_{S_{x,y}}) \qquad (4.18)$$

where $min_{S_{x,y}}$ and $max_{S_{x,y}}$ are the minimum and maximum similarities, respectively. Similarly, dissimilarity measures with a finite range can be mapped to the interval $[0, 1]$ by using Equation 4.19,

$$d'(x, y) = (d(x, y) - min_{d(x,y)})/(max_{d(x,y)} - min_d(x, y)) \qquad (4.19)$$

where $min_{d(x,y)}$ and $max_{d(x,y)}$ are the minimum and maximum dissimilarities, respectively. If the proximity measure has values in the range $[0, \infty]$, a non-linear transformation is needed, and the values in the transformed scale do not have the same relationship to one another as the original. But, whether such a transformation is desirable or not depends on the application.

Different Similarity and Dissimilarity Measures

In this section, we discuss a large number of similarity and dissimilarity measures for numeric, categorical and mixed-type data. Table 4.4 presents a list of proximity measures for numeric data.

Table 4.4: Similarity Measures for Numeric Attributes

L_p Minkowski	
1) Euclidean, $L_2 d_{Euc} = \sqrt{\sum_{i=1}^{d} \|X_i - Y_i\|^2}$	2) City block, $L_1 d_{CB} = \sum_{i=1}^{d} \|X_i - Y_i\|$
3) Minkowski, $L_p d_{Mk} =$ $\sqrt[p]{\sum_{i=1}^{d} \|X_i - Y_i\|^p}$	4) Chebyshev, $L_\infty d_{Cheb} = \overset{max}{i} \|X_i - Y_i\|$

L_1	
1) Sorensen, $d_{sor} = \frac{\sum_{i=1}^{d}\|X_i - Y_i\|}{\sum_{i=1}^{d}(X_i + Y_i)}$	2) Gower, $d_{gow} = \frac{1}{d}\sum_{i=1}^{d}\frac{\|X_i - Y_i\|}{Z_i} = \frac{1}{d}\sum_{i=1}^{d}\|X_i - Y_i\|$
3) Soergel, $d_{sg} = \frac{\sum_{i=1}^{d}\|X_i - Y_i\|}{\sum_{i=1}^{d} max(X_i, Y_i)}$	4) Kulczynskid, $d_{kul} = \frac{\sum_{i=1}^{d}\|X_i - Y_i\|}{\sum_{i=1}^{d} min(X_i, Y_i)}$
5) Canberra, $d_{Can} = \sum_{i=1}^{d}\frac{\|X_i - Y_i\|}{X_i + Y_i}$	6) Lorentzian, $d_{Lor} = \sum_{i=1}^{d} \ln(1 + \|X_i - Y_i\|)$

Intersection	
1) Intersection, $S_{IS} = \sum_{i=1}^{d} min(X_i, Y_i)$	2) $d_{non-IS} = 1 - S_{IS} = \frac{1}{2}\sum_{i=1}^{d}\|X_i - Y_i\|$
3) Wave Hedges, $d_{WH} = \sum_{i=1}^{d}(1 - \frac{min(X_i, Y_i)}{max(X_i, Y_i)})$	4) $d_{WH} = \sum_{i=1}^{d}\frac{\|X_i - Y_i\|}{max(X_i, Y_i)}$
5) Czekanowski, $S_{Cze} = \frac{2\sum_{i=1}^{d} min(X_i, Y_i)}{\sum_{i=1}^{d}(X_i + Y_i)}$	6) $d_{Cze} = 1 - S_{Cze} = \frac{\sum_{i=1}^{d}\|X_i + Y_i\|}{\sum_{i=1}^{d}(X_i + Y_i)}$
7) Motyka, $S_{Mot} = \frac{\sum_{i=1}^{d} min(X_i, Y_i)}{\sum_{i=1}^{d}(X_i + Y_i)}$	8) Motyka, $d_{Mot} = 1 - S_{Mot} = \frac{\sum_{i=1}^{d} max(X_i, Y_i)}{\sum_{i=1}^{d}(X_i + Y_i)}$
9) Kulczynski s, $S_{Kul} = \frac{1}{d_{Kul}} = \frac{\sum_{i=1}^{d}(min(X_i, Y_i)}{\sum_{i=1}^{d}\|X_i - Y_i\|}$	10) Ruzicka, $S_{Ruz} = \frac{\sum_{i=1}^{d} min(X_i, Y_i)}{\sum_{i=1}^{d} max(X_i, Y_i)}$
11) Tanimoto, $d_{Tani} =$ $\frac{\sum_{i=1}^{d} X_i + \sum_{i=1}^{d} Y_i - 2\sum_{i=1}^{d} min(X_i, Y_i)}{\sum_{i=1}^{d} X_i + \sum_{i=1}^{d} Y_i - \sum_{i=1}^{d} min(X_i, Y_i)}$	12) $d_{Tani} =$ $\frac{\sum_{i=1}^{d} max(X_i, Y_i) - min(X_i, Y_i)}{\sum_{i=1}^{d} max(X_i, Y_i)}$

Inner Product	
1) Inner Product, $S_{IP} = X.Y = \sum_{i=1}^{d} X_i Y_i$	2) Harmonic Mean, $S_{HM} = 2\sum_{i=1}^{d}\frac{X_i Y_i}{X_i + Y_i}$
3) Cosine, $S_{Cos} =$ Kumar Hasserbrook, $S_{Jac} = \frac{\sum_{i=1}^{d} X_i Y_i}{\sum_{i=1}^{d} X_i^2 + \sum_{i=1}^{d} Y_i^2 - \sum_{i=1}^{d} X_i Y_i}$	4) Jaccard, $S_{Jac} =$ $\frac{\sum_{i=1}^{d} X_i Y_i}{\sum_{i=1}^{d} X_i^2 + \sum_{i=1}^{d} Y_i^2 - \sum_{i=1}^{d} X_i Y_i}$

5) $d_{Jac} = 1 - S_{Jac} =$ 6) $Dice, S_{Dice} = \frac{2\sum_{i=1}^{d} X_i Y_i}{\sum_{i=1}^{d} X_i^2 + \sum_{i=1}^{d} Y_i^2}$

$$\frac{\sum_{i=1}^{d}(X_i - Y_i)^2}{\sum_{i=1}^{d} X_i^2 + \sum_{i=1}^{d} Y_i^2 - \sum_{i=1}^{d} X_i Y_i}$$

7) $d_{Dice} = 1 - S_{Dice} = \frac{\sum_{i=1}^{d}(X_i - Y_i)^2}{\sum_{i=1}^{d} X_i^2 + \sum_{i=1}^{d} Y_i^2}$

Fidelity

1) $Fidelity, \; S_{Fid} = \sum_{i=1}^{d} \sqrt{(X_i Y_i)}$ 2) $Bhattacharyya,$ $d_B =$
$-\ln \sum_{i=1}^{d} \sqrt{(X_i Y_i)}$

3) $Hellinger,$ $d_H =$ 4) $d_H = 2\sqrt{1 - \sum_{i=1}^{d} X_i Y_i}$
$\sqrt{2 \sum_{i=1}^{d} \left(\sqrt{X_i} - \sqrt{Y_i}\right)^2}$

5) $Matusia, \; d_M = \sqrt{\sum_{i=1}^{d} \left(\sqrt{X_i} - \sqrt{Y_i}\right)^2}$ 6) $d_M = \sqrt{2 - 2\sum_{i=1}^{2} \left(\sqrt{X_i} - \sqrt{Y_i}\right)^2}$

7) $Squared$ $-$ $chord,$ $d_{sqc} =$ 8) $S_{sqc} = 1 - d_{sqc} = 2\sum_{i=1}^{d} \sqrt{X_i Y_i} - 1$
$\sum_{i=1}^{d} \left(\sqrt{X_i} - \sqrt{Y_i}\right)^2$

Squared L_2 or χ^2

1) $Squared \; Euclidean, \; d_{sqe} = \sum_{i=1}^{d}(X_i - Y_i)^2$ 2) $Pearson$ $\chi^2,$ $d_p(X, Y) =$
$\sum_{i=1}^{d} \frac{(X_i - Y_i)^2}{Y_i}$

3) $Neyman$ $\chi^2,$ $d_N(X, Y) =$ 4) $Squared \; \chi^2, \; d_{SqChi} = \sum_{i=1}^{d} \frac{(X_i - Y_i)^2}{X_i + Y_i}$
$\sum_{i=1}^{d} \frac{(X_i - Y_i)^2}{X_i}$

5) $Probabilistic \; Symmetric \; \chi^2, d_{PChi} =$ 6) $Divergence, \; d_{Div} = 2\sum_{i=1}^{d} \frac{(X_i - Y_i)^2}{(X_i + Y_i)^2}$
$2\sum_{i=1}^{d} \frac{(X_i - Y_i)^2}{(X_i + Y_i)^2}$

7) $Clark, \; d_{Clk} = \sqrt{\sum_{i=1}^{d} \left(\frac{|X_i - Y_i|}{X_i + Y_i}\right)^2}$ 8) $AdditiveSymmetric\chi^2,$ $d_{AdChi} =$
$\sum_{i=1}^{d} \frac{(X_i - Y_i)^2 (X_i + Y_i)}{X_i Y_i}$

Shannon's entropy

1) $Kullback \; Leibler \; d_{KL} = \sum_{i=1}^{d} X_i \ln \frac{X_i}{Y_i}$ 2) $Jeffreys \; d_J = \sum_{i=1}^{d}(X_i - Y_i) \ln \frac{X_i}{Y_i}$

3) K $divergence$ $d_{Kdiv} =$ 4) $Topsd_{Top} =$
$\sum_{i=1}^{d} X_i \ln \frac{2X_i}{X_i + X_i}$ $\sum_{i=1}^{d} \left(X_i \ln(\frac{2X_i}{X_i + Y_i}) + Y_i \ln(\frac{2Y_i}{X_i + Y_i})\right)$

5) $JensenShannon, d_{JS} =$ 6) $Jensendifference, d_{JD} =$
$\frac{1}{2}\left[\sum_{i=1}^{d} X_i \ln(\frac{2X_i}{X_i + Y_i}) + \sum_{i=1}^{d} Y_i \ln(\frac{2Y_i}{X_i + Y_i})\right]$ $\sum_{i=1}^{d}\left[\frac{X_i \ln X_i + Y_i \ln Y_i}{2} - (\frac{X_i + Y_i}{2})\ln(\frac{X_i + Y_i}{2})\right]$

Combinations

1) $Taneja$ $d_{TJ} =$ 2) $Kumar$ $Johnson$ $d_{KJ} =$
$\sum_{i=1}^{d} \left(\frac{X_i + Y_i}{2}\right) \ln \left(\frac{X_i + Y_i}{2\sqrt{X_i Y_i}}\right)$ $\sum_{i=1}^{d} \frac{(X_i^2 - Y_i^2)^2}{2(X_i Y_i)^{\frac{3}{2}}}$

3) $Avg(L_1, L_\infty)$ $d_{ACC} =$
$\frac{\sum_{i=1}^{d} |X_i - Y_i| + max|X_i - Y_i|}{2}$

Proximity Measure for Categorical Data

Computing similarity between categorical data instances is not straight-forward owing to the fact that there is usually no explicit notion of ordering between categorical values. The simplest way to find similarity between two categorical attributes is to assign a similarity of 1 if the values are identical and a similarity of 0 if the values are not identical. Several data-driven similarity measures have been proposed [61] for categorical data. The behavior of such measures directly depends on the data. Table 4.5 presents the mathematical formulas for the measures.

Table 4.5: Similarity Measures for Categorical Attributes

No.	$S_k(X_k, Y_k)$	$w_k, k=1 \ldots d$
1.	$Overlap = \begin{cases} 1 & \text{if } X_k = Y_k \\ 0 & \text{otherwise} \end{cases}$	$\dfrac{1}{2}$
2.	$Eskin = \begin{cases} 1 & \text{if } X_k = Y_k \\ \dfrac{n_k^2}{n_k^2+2} & \text{otherwise} \end{cases}$	$\dfrac{1}{d}$
3.	$IOF = \begin{cases} 1 & \text{if } X_k = Y_k \\ \dfrac{1}{1+\log f_k(X_k)x\log f_k(Y_k)} & \text{otherwise} \end{cases}$	$\dfrac{1}{d}$
4.	$OF = \begin{cases} 1 & \text{if } X_k = Y_k \\ \dfrac{1}{1+\log \frac{N}{f_k(X_k)}x\log \frac{N}{f_k(Y_k)}} & \text{otherwise} \end{cases}$	$\dfrac{1}{d}$
5.	$Lin = \begin{cases} 2\log \hat{p}_k(X_k) & \text{if } X_k = Y_k \\ 2\log(\hat{p}_k(X_k) + \hat{p}_k(Y_k)) & \text{otherwise} \end{cases}$	$\dfrac{1}{\sum_{i=1}^{d}\log \hat{p}_i(X_k) + \log \hat{p}_i(Y_i)}$
6.	$Lin1 = \begin{cases} \displaystyle\sum_{q\in Q}\log \hat{p}_k(q) & \text{if } X_k = Y_k \\ 2\log \displaystyle\sum_{q\in Q}\hat{p}_k(q) & \text{otherwise} \end{cases}$	$\dfrac{1}{\sum_{i=1}^{d}\sum_{q\in Q}\log \hat{p}_i(q)}$
7.	$Goodall1 = \begin{cases} 1 - \displaystyle\sum_{q\in Q}p_k^2(q) & \text{if } X_k = Y_k \\ 0 & \text{otherwise} \end{cases}$	$\dfrac{1}{d}$

Table 4.5

No.	Measures	Measures
8.	$$Goodall3 = \begin{cases} 1 - p_k^2(X_k) & \text{if } X_k = Y_k \\ 0 & \text{otherwise} \end{cases}$$	$\dfrac{1}{d}$
9.	$$Goodall4 = \begin{cases} p_k^2(X_k) & \text{if } X_k = Y_k \\ 0 & \text{otherwise} \end{cases}$$	$\dfrac{1}{d}$
10.	$$Smirnov = \begin{cases} 2 + \dfrac{N-f_k(X_k)}{f_k(X_k)} + \displaystyle\sum_{q\in\{A_k\setminus X_k\}} \dfrac{f_k(q)}{N-f_k(q)} & \text{if } X_k = Y_k \\ \displaystyle\sum_{q\in\{A_k\setminus\{X_k,Y_k\}\}} \dfrac{f_k(q)}{N-f_k(q)} & \text{otherwise} \end{cases}$$	$\dfrac{1}{\sum_{k=1}^{d} n_k}$
11.	$$Gambaryan = \begin{cases} -[\hat{p}_k(X_k)\log_2 \hat{p}_k(X_k) + \\ (1-\hat{p}_k(X_k))\log_2(1-\hat{p}_k(X_k))] & \text{if } X_k = Y_k \\ 0 & \text{otherwise} \end{cases}$$	$\dfrac{1}{\sum_{k=1}^{d} n_k}$
12.	$$Burnaby = \begin{cases} 1 & \text{if } X_k = Y_k \\ \dfrac{\sum_{q\in A_k} 2\log(1-\hat{p}_k(q))}{\log \frac{\hat{p}_k(X_k)\hat{p}_k(Y_k)}{(1-\hat{p}_k(X_k))(1-\hat{p}_k(Y_k))} + \sum_{q\in A_k} 2\log(1-\hat{p}_k(q))} & \text{otherwise} \end{cases}$$	$\dfrac{1}{d}$

The various measures described in Table 4.5 compute the per-attribute similarity $S_k(X_k, Y_k)$ as shown in *column 2* and compute the attribute weight w_k as shown in *column 3*. For the sake of notation, consider a categorical data set D containing N objects, defined over a set of d categorical attributes where A_k denotes the kth attribute. $S_k(X_k, Y_k)$ is the per-attribute similarity between two values for the categorical attribute A_k. Note that for $X_k, Y_k \in A_k$, IOF is *Inverse Occurrence Frequency* and *OF* is *Occurrence Frequency*.

Proximity Measure for Mixed Data

A mixed dataset contains attributes some of which are categorical and others are numeric. A common practice to cluster a mixed dataset is to transform categorical values into numeric values and then proceed to use a numeric clustering algorithm. Another approach is to compare the categorical values directly, in which two distinct values result in distance 1 while identical values result in distance 0. Nevertheless, these two methods do not take into account the similarity information embedded between categorical values. Consequently, the clustering results do not faithfully reveal the similarity structure of the dataset. Huang [184] proposed a new incremental clustering algorithm for mixed datasets, in which the similarity information embedded in categorical attributes is considered during clustering. Each attribute of the data is associated with a distance hierarchy, which is an extension of the concept hierarchy [378] with link weights representing the distance between concepts. The distance between two mixed data patterns is then calculated according to distance hierarchies.

It is worth mentioning that the representation scheme used for a distance hierarchy can generalize some conventional distance computation schemes including simple matching and binary encoding for categorical values, and the subtraction method for continuous values and ordinal values.

Cluster Analysis

Cluster analysis is an unsupervised learning approach that constitutes a cornerstone of an intelligent data analysis process. It plays a major role in the overall knowledge discovery process, and groups objects in a way that intra-similarity among group members is maximized, whereas the inter-similarity among objects across groups is minimized. It is called unsupervised learning because, unlike classification (also known as supervised learning), no a priori labeling of some patterns is available to use in categorizing others and inferring the cluster structure of the whole data. Similarity measurement is a subjective concept since it varies with the variation of proximity measures such as Manhattan distance, Euclidean distance, Cosine Distance or Pearson correlation coefficient, which are generally used to quantify the degree of proximity (dissimilarity or similarity) between a pair of objects. Intra-connectivity is a measure of the density of connections between the instances of a single cluster. A high intra-connectivity indicates a good clustering arrangement because the instances grouped within the same cluster are highly dependent on each other. Inter-connectivity is a measure of the connectivity between distinct clusters. A low degree of interconnectivity is desirable because it indicates that individual clusters are largely independent of each other. Every instance in the data set is represented using the same set of attributes. The attributes are continuous, categorical or binary. To induce a hypothesis from a given data set, a learning system needs to make assumptions about the hypothesis to be learned. These assumptions are called biases. Since every learning algorithm uses some biases, it behaves well in some domains where its biases are appropriate while it performs poorly in other domains. A problem with the clustering methods is that the interpretation of the clusters may be difficult. In addition, the algorithms always assign the data to clusters even if there were no natural clusters in the data.

When using this unsupervised learning technique to solve a real-life problem from any application domain, the researcher has to ensure that the employed clustering approach deals with the *intrinsic properties* of the data such as the type of correlation or exclusivity of clusters. In addition to these properties, the researcher also is to ensure whether the clustering approach is able to handle *extrinsic properties* such as the number of features or attributes, i.e., dimensionality. A good clustering algorithm is able to handle both types of properties in an effective manner. The problem of clustering can be defined in the context of gene expression data analysis as follows.

Problem formulation: Given a gene expression dataset D with expression values of a set of genes $G = g_1; g_2; : : : ; g_m$ over the set of features $F = f_1; f_2; : : : ; f_n$, a clustering technique needs to group genes into clusters based on their expression values over features such that the overlap between the generated clusters and biological groups corresponding to biological functions can be maximized.

Challenges of Data Clustering

Cluster analysis is a difficult problem because of following factors [162]:

- Scalability

- Ability to deal with different types of attributes

- Discovery of clusters with arbitrary shapes

- Minimal requirement of domain knowledge

- Ability to deal with noisy data

- Insensitivity to the order of input parameters

- High dimensionality

- Interpretability and Usability of clustering results

Classification of Variables

A comprehensive categorization of the different types of variables met in most data sets provides a helpful means for identifying the differences among data elements. We can broadly classify [162] these variables as follows:

(i) A continuous variable has an uncountably infinite range. Typically, such a variable may assume any value in an interval (say 1.5 to 6.7) or a collection of such intervals.

(ii) A discrete variable has a finite, or at most countably infinite range.

(iii) A binary or dichotomous variable is a discrete variable which may take only two values.

A variable can also be classified based on the scale of measurement. It is convenient to illustrate this scheme with a variable X and two objects, say A and B, whose values of X are xA and xB, respectively.

(a) A *nominal scale* helps distinguish two categories of instances. This means that we can only say if $xA = yB$ or $xA \neq yB$. Examples include balls of various colors, and different types of favorite food.

(b) An *ordinal scale* involves nominal-scaled attributes with the additional feature that their values can be totally ordered, i.e., $xA > yB$ or $xA < yB$. Examples include the number of medals won by several athletes.

(c) An *interval scale* assigns a measure of the difference between two objects. One may say not only that xA¿yB but also that A is xA- yB units different from B. Examples include the number of a book in series or TV channel numbers.

(d) A *ratio scale* is an interval scale with a meaningful zero point. If $xA > yB$, then one may say A is xA/yB times superior to B. Examples include weight, height and the number of children in a family.

(e) *Mixed type* variables are combinations of variables described above. In general, a real database can contain all of the above types.

[A] Partitioning Approaches

Given a database of objects, a partitional clustering algorithm constructs partitions of the data, where each cluster optimizes a clustering criterion, such as the minimization of the sum of squared distances from the mean within each cluster. One of the issues with such algorithms is their high complexity, as the simplest algorithm would exhaustively enumerate all possible groupings and try to find a global optimum. Even for a small number of objects, the number of partitions is huge. This is why common solutions start with an initial, usually random partition, and proceed with its refinement. A better practice would be to run the partitional algorithm for different sets of initial points (considered as representatives) and investigate whether all solutions lead to the same final partition. Partitional Clustering algorithms try to locally improve a certain criterion. They compute the values of the similarity or distance, they order the results, and pick the one that optimizes the criterion. Hence, the majority of them can be considered greedy algorithms. Some well-known partitioning techniques are discussed in brief next.

A.1 k-means   [162], [406]: The k-means algorithm is a typical partitioning-based clustering method. Given a pre-specified number of partitions, say k, the algorithm partitions the data set into k disjoint subsets so that the similarity between a pair of objects within a partition or cluster is high, whereas the similarity between a pair of objects from two distinct partitions or clusters is the minimum. The objective function for the k-means algorithm attempts to minimize the sum of the squared distances of objects from their respective cluster centers. k-means is a simple, yet fast clustering algorithm. The time complexity of k-means is $O(l \times k \times n)$ where l is the number of iterations, k is the number of clusters, and n is the number of objects in the dataset. k-means typically converges in a small number of iterations. However, it also has several limitations, especially in gene-based cluster analysis. First, the number of gene clusters in a gene expression data set is usually unknown in advance. To detect the optimal number of clusters, users usually run the algorithm repeatedly with different values of k and compare the clustering results. For a large gene expression data set, which contains thousands of genes, this extensive parameter fine-tuning process may not be practical. Second, gene expression data typically contain a large amount of noise. The k-means algorithm forces each gene into a cluster, which may cause the algorithm to be sensitive to noise. Further, it is incapable of detecting concave-shaped clusters, and cannot detect overlapped (non-exclusive) clusters. Furthermore, k-means often converges to a locally optimum solution.

A.2. k-Modes   [162], [58]: The k-modes algorithm is an extension of the k-means algorithm to handle categorical data. The k-modes algorithm is also a simplified version of the k-protypes algorithm. It differs from the k-means algorithm in three ways: (i) it uses a different similarity measure, (ii) it replaces the use of means with modes, and (iii) it uses a frequency-based approach to update its modes. These three updates allow it to directly cluster categorical data without the need for data conversion. Another advantage of the k-modes algorithm is that the modes give characteristic descriptions of clusters. These descriptions are important to the user in interpreting clustering results. The k-modes algorithm is scalable to very large and complex data sets in terms of both the numbers of records and the number of clusters. The k-modes algorithm is faster than the k-means algorithm.

A.3. Fuzzy c-means [56]: Bezdek and Dunn introduced the fuzzy c-means (FCM) algorithm in 1973. The three basic steps involved in Fuzzy c-means are: (i) compute new prototypes as weighted averages of all Q feature vectors, (ii) compute new weights and standardize, and (iii) check the stopping criterion. To better understand the algorithm, let us consider an example. Let $x(q) : q{=}1\cdots, Q$ be a feature vector or a vector of samples. During cluster expansion, it assigns Q feature vectors to k clusters according to their proximities with the prototypes of the initial clusters. Feature vectors or samples that are far away from the prototypical vector for a cluster should not be larger than those which are in the near proximity, or are centrally and densely located. These weights are called *fuzzy weights*. The clusters are assumed to be of approximately the same size. Each cluster is represented by its center. This representation of a cluster is also called a prototype, since it is often regarded as a representative of all data assigned to the cluster. As a measure of estimation of proximity, the Euclidean distance between a data sample vector and a prototype is used. FCM first draws initial weights w_{qk} randomly and then standardizes them so that for each fixed q, we have: $w_{q1} + w_{q2}, \cdots + w_{qk} = 1$.

A.4 PAM [162], [58]: Partition Around Medoids (PAM) is a robust clustering algorithm that works based on the k-medoid approach. PAM initially selects k objects from the given data arbitrarily as medoids. These k objects are considered representatives of k clusters. Other objects in the database are classified based on their distances to these k medoids. The algorithm iteratively attempts to improve its medoid selection using a *cost* function. In each step, a swap between a selected object O_i and a non-selected object O_h is made as long as it results in an improvement of the quality (with

reference to the *cost* function) of clustering. To calculate the effect of such a
swap between O_i and O_h, it computes the cost C_h^i (i.e., average dissimilarity), which is related to the quality of partitioning the non-selected objects
around k medoids towards generation of an optimal set of clusters.

A.5 CLARA [162], [58]: A major disadvantage of the PAM algorithm is
its significant computational requirement to determine k medoids through
an iterative optimization. To address this scalability issue, CLARA (Clustering LARge Applications) was introduced. Though CLARA follows the same
principle, it attempts to reduce the computational efforts. Instead of finding
representative objects for the entire data set, CLARA draws a sample of the
data set, and applies PAM on this sample. It then classifies the remaining
objects using partitioning. If the samples were drawn in a sufficiently random way, the medoids of the sample would approximately be the medoids
of the entire data set.

A.6 CLARANS [301]: Clustering Large Applications based on RANdomized Search (CLARANS), is a k-medoid algorithm. It is a variant of PAM
and CLARA. It uses a graph abstraction of the data and relies on a randomized search over the graph to identify the best possible set of medoids which
represent the clusters. The algorithm is dependent on two user-defined input
variables, i.e. *maxneighbor* and *numlocal*. *Maxneighbor* represents the maximum number of neighbors of a node that are to be examined and *numlocal*
is the maximum number of local minima that can be collected. CLARANS
initiates its cluster expansion process by selecting a random node. It then
checks a sample of the neighbors of the node, and if a better neighbor is
found based on the "cost differential of the two nodes", it moves to the
neighbor and continues processing until the maxneighbor criterion is met.
Otherwise, it declares the current node a *local minimum* and starts a new
pass to search for other *local minima*. After a specified number of *local minima* (*numlocal*) are collected, the algorithm returns the best of these local
values as the *medoids* of the cluster.

[B] Hierarchical Approaches:

As the name suggests, a hierarchical approach attempts to form clusters
by following a sequence of partitioning operations. These can be either (i)
bottom-up, performing repeated amalgamations of smaller groups of data
until some pre-defined threshold is reached, or (ii) *top-down*, recursively

dividing the data until some pre-defined threshold is reached. The effectiveness of a hierarchical clustering method is largely dependent on the use of an appropriate measure to help evaluate the goodness of a 'split' or 'merge' operation during the cluster formation process. It is widely used in document and text analysis. We discuss a few hierarchical clustering algorithms next.

B.1. CURE [150]: CURE follows a bottom-up clustering approach. Unlike a centroid-based approach, this algorithm initiates cluster formation through an iterative merging process by choosing a constant number of well-scattered points for each candidate cluster. CURE uses these points to identify the shape and size of the cluster. CURE shrinks the selected points toward the centroid of the cluster using a pre-determined fraction. To facilitate effective cluster formation, this algorithm uses a spatial index structure called $k - d$ tree, to store the representative points for the clusters.

B.2. ROCK [151]: Unlike most other numeric data clustering algorithms, this bottom-up or agglomerative hierarchical clustering algorithm was developed to handle Boolean and categorical data. ROCK (Robust hierarchical Clustering with Link) introduces an effective similarity measure called 'link' to support the merging process between a pair or groups of data objects. The number of 'links' between two k-dimensional objects is the number of common neighbors they have in the data set. ROCK introduces the concept of non-metric similarity between a pair of objects, to effectively handle Boolean or categorical data.

B.3. CHAMELEON [211]: CHAMELEON is a hybridized version of graph partitioning and hierarchical clustering approaches to support the creation of clusters dynamically. Initially, it partitions the data using a multi-level graph partitioning approach based on the k-nearest-neighbor approach. It recognizes the density of a region in the graph as the weight of the connecting edge(s). The data is partitioned into a large number of smaller sub-clusters. An important advantage of CHAMELEON is that the partitioning algorithm it uses ensures high-quality partitions with a minimum number of edge cuts. In the next step, it uses an agglomerative, or bottom-up hierarchical clustering algorithm to combine the sub-clusters and find the real clusters. For n items and m clusters, the overall complexity of CHAMELEON is $O(n \times m + nlogn + m \times 2logm)$. This algorithm is capable of finding clusters of various shapes, including nested and concave shaped clusters.

B.4. CACTUS **[133]:** This is a categorical data clustering technique based on subspace identification. CACTUS (CAtegorical data ClusTering Using Summaries) splits the dataset vertically and attempts to cluster the set of projections to only a pair of attributes. Initially, it identifies the cluster projections for all pairs of attributes by fixing one attribute. Then, it generates an interesting set to represent the cluster projections on this attribute for n clusters, involving all the attributes. Once all the cluster projections with reference to fixed individual attributes are generated, the next task is to synthesize them to obtain the final clusters for the given dataset.

B.5. BIRCH **[460]:** BIRCH (Balanced Iterative Reducing and Clustering using Hierarchies) is a cost-effective and scalable hierarchical clustering method. It exploits two basic concepts, i.e. cluster features (CF) and cluster feature trees, to summarize cluster representations. These two concepts help BIRCH achieve better speed and ability to handle large amounts of data. In addition, BIRCH also supports incremental and dynamic clustering for effective handling of incoming objects. BIRCH yields good clustering performance with a single scan of the dataset. However, one or more additional (optional) scan ensure further improvement in quality of clustering.

B.6. AMOEBA **[119]:** AMOEBA is another effective hierarchical clustering method for spatial data. It uses the concept of the Delaunay triangle as the source of analysis for identifying clusters. The algorithm consists of two major steps. Initially, it constructs the Delaunay diagram. A connected plane-embedded graph is passed to the algorithm to act as the diagram. Subsequently, clusters are formed in a recursive fashion using the points in a connected component and the points in the clusters. Every edge is matched against the criteria for selection, and the passive edges and noises are discarded. Passive edges are Delaunay edges greater than or equal to some criterion function $F(p)$, whereas active edges are Delaunay edges that are less than $F(p)$. Noise consists of points that do not have active edges meeting them. Active edges and their points form proximity sub-graphs at each level of the hierarchy. AMOEBA recursively eliminates the passive edges and noise, and it continues until no new connected components are created.

[C] Density-based Clustering Approaches

A density-based clustering attempts to form clusters by growing continuously as long as the density (i.e., the number of objects or data points) in the "*neighborhood*" exceeds some threshold. In other words, for each data point within a given cluster, the neighborhood (defined with reference to a given radius, say ϵ) has to contain at least a minimum number of points, say *MinPts*. A density-based clustering method can discover clusters of any shape. Further, it can handle noise effectively. Some well-known density-based clustering algorithms are discussed next.

C.1. DBSCAN [118]: DBSCAN is a popular density-based clustering algorithm that operates on spatial data and is able to discover clusters of any shape. During the expansion of clusters, it recognizes the data points to be any of the three following types.

(a) *Core points*: A data point is considered a *core* point, if there are enough ($\geq$ *MinPts*) points in its ϵ-neighborhood. *Core* points are interior points of clusters and these are potential candidate points for further expansion towards the formation of clusters.

(b) *Border points*: A data point is considered a *border* point, if there are not enough ($<$ *MinPts*) points in its ϵ-neighborhood. Although a border point is not a core point, it falls within the neighborhood of a core point.

(c) *Noise points*: A data point is considered a *noise* point, if it is neither a *core* point nor a *border* point.

To discover a cluster, DBSCAN starts with an arbitrary instance, say p in a given dataset, D, and retrieves all instances of D with parameters ϵ and a *MinPts*. To minimize the cost of ϵ-neighborhood finding, DBSCAN uses a spatial index structure such as R*-tree to locate relevant points within ϵ distance from the core points of the clusters easily.

C.2. GDBSCAN [360]: GDBSCAN is an effective variant of the DB-SCAN algorithm. It is more generalized than DBSCAN and is able to handle both numeric and categorical attributes. Within a density-based framework, it offers generic features in two major dimensions. First, it allows a flexible notion of neighborhood, instead of only ϵ-neighborhoods. It introduces a novel binary predicate called *NPred*, which is symmetric and reflexive. Second, instead of simply counting the objects in the neighborhood of an object, it allows the use of other measures to define equivalence, considering the "cardinality" of that neighborhood. GDBSCAN is capable of discovering clusters of any shape over scattered as well as spatial data.

C.3. OPTICS **[24]:** Several other variants of DBSCAN have also been proposed. OPTICS (Ordering Points to Identify the Clustering Structure) is one such variant that addresses two major issues with DBSCAN regarding (i) the strict requirement of two input parameters and (ii) handling of non-uniformly distributed data. OPTICS determines an augmented cluster ordering based on density to automatically and interactively discover clusters of any shape. The density-based ordering of the data represents information that is equivalent to density-based clustering obtained by a range of parameter settings. OPTICS considers a minimum radius that makes a neighborhood legitimate for the algorithm, i.e., with the minimum number of objects, and extends it to a maximum value. OPTICS is able to handle data with both uniform as well as non-uniform density with the same computational complexity of DBSCAN.

C.4. DBCLASD **[443]:** DBCLASD (Distribution-Based Clustering of Large Spatial Databases) is another effective locality-based clustering algorithm. Unlike DBSCAN, this algorithm assumes a uniform distribution of points inside each cluster. Each cluster is assumed to follow a probability distribution, considering all the points in it. Thus, it defines a cluster as a probability set. DBCLASD uses a grid-based representation to approximate the clusters as part of the probability calculation. For a given pair of objects $(O_i, O_j) \in C$, where C is a cluster, O_i, O_j are connected, or, there is a path of occupied grid cells connecting O_i and O_j. Further, the cluster C is maximal, i.e., any extension of C does not fulfill the expected probability distribution condition.

C.5. DENCLUE **[178]:** DENCLUE (DENsity-based CLUstEring) is a generalization of partitioning, locality-based, hierarchical, and grid-based clustering approach. The algorithm models the overall point density analytically using the sum of the influence functions of the points. Determining the density-attractors causes the clusters to be identified. DENCLUE can handle clusters of arbitrary shape using an equation based on the overall density function. The authors claim three major advantages for this method of higher-dimensional clustering: (a) firm mathematical base for finding arbitrary shaped clusters in high-dimensional data sets, (b) good clustering properties in data sets with large amounts of noise, and (c) significantly faster than existing algorithms. DENCLUE first performs a pre-clustering step, which creates a map of the active portion of the data set, used to speed up the density function calculations. The second step is the clustering, including the identification of density-attractors and their corresponding points.

C.6. TURN* **[127]:** TURN* consists of an overall algorithm and two component algorithms: (i) an efficient resolution-dependent clustering algorithm, TURN-RES, which returns both a clustering result and certain global statistics (cluster features) from that result, and (ii) *TurnCut*, an automatic method for finding the important or "optimum" resolutions from a set of resolution results from TURN-RES. To date, clustering algorithms have returned clustering results for a given set of parameters, as TURN-RES does, and some have presented graphs or dendrograms of cluster features from which the user may be able to adjust or select the parameters to optimize the clustering. TURN* takes clustering into new territory by automating this process removing the need for input parameters. Clustering Validation is a sub-area where attempts have been made to find rules for quantifying the quality of a clustering result. Though developed independently, Turn-Cut can be seen as an advance in this sub-area, which has been integrated into the clustering process.

[D] Grid-based Clustering Approach

For grid-based clustering approaches [191], it is assumed that the object space has been quantized into a finite number of cells that form a grid structure. All of the clustering operations are performed on the grid structure. The main advantage of this approach is its fast processing time, which is typically independent of the number of data objects and dependent only on the number of cells in each dimension in the quantized space.

D.1. WaveCluster **[370]:** WaveCluster takes a grid-based approach to clustering. It maps the data onto a multidimensional grid and applies a wavelet transformation to the feature space instead of to the objects themselves. In the first step of the algorithm, the data are assigned to units based on their feature values. The number or size of these units affects the time required for clustering and the quality of the output. The algorithm then identifies the dense areas in the transformed domain. It does this by searching for connected components. If the feature space is examined from a signal processing perspective, then a group of objects in the feature space forms an n-dimensional signal. Rapid change in the distribution of objects, i.e., the borders of clusters, corresponds to the high-frequency parts, which are used to find areas of low and high frequencies, and thus identify the clusters. Wavelet transformation breaks a signal into its frequency sub-bands, creating a representation that shows multiple resolutions, and therefore provides for efficient identification of clusters. Areas with low frequency and low amplitude are outside the clusters. With a high number of objects, that is if the set is large, signal processing techniques can be used to find areas of low and high frequencies, and thus identify the clusters.

WaveCluster has several significant positive contributions. It is not affected by outliers, and is not sensitive to the order of input. The main advantage of *WaveCluster*, apart from its speed at handling large datasets, is its ability to find clusters of arbitrary and complex shapes, including concave and nested clusters. It has been found that clustering results are quite sensitive to parameters settings. Moreover, according to [20], prior knowledge of the number of clusters helpful in choosing the parameter values for *WaveCluster*.

D.2. STING [428]: STING (Statistical Information Grid-based method) exploits the clustering properties of index structures. It divides the spatial area into rectangular grid cells using, for example, longitude and latitude, This makes *STING* order-independent, i.e., the clusters created in the next step of the algorithm are not dependent on the order in which the values are placed in the grid. A hierarchical structure is used to manipulate the grid. Each cell at level i is partitioned into a fixed number k of cells at the next level. This is similar to spatial index structures. Since the default value chosen for k is 4, the hierarchical structure used in *STING* is similar to a quad-tree structure. The algorithm stores the parameter values in each cell to help answer certain types of statistically based spatial queries. In addition to storing the number of objects or points in the cell, *STING* also stores some attribute-dependent values. Since *STING* uses density-based methods to form its clusters, its ability to handle noise is similar to *DBSCAN*. Although it can handle large amounts of data, and is not sensitive to noise, it cannot handle high-dimensional data without a serious degradation of performance.

D.5. STING+ [427]: STING+ builds on *STING* to create a system that supports user-defined triggers, and works on dynamically evolving spatial data. Spatial data mining triggers have several advantages: (a) From the user's perspective, queries do not need to be submitted repeatedly in order to monitor changes in the data. Users only need to specify a trigger and the pattern desired as the trigger condition. In addition, interesting patterns are detected immediately if trigger conditions are used. (b) From the system's perspective, it is more efficient to check a trigger on an incremental basis than to process identical queries on the data multiple times. (c) Data mining tasks on spatial data cannot be supported by traditional database triggers since spatial data consist of both non-spatial and spatial attributes, and both may be used to make up the trigger condition. In STING+, users can specify the type of trigger, the duration of the trigger, the region on which the trigger is defined, and the action to be performed by the trigger.

[E] Model-based Clustering Approaches

Model-based clustering methods attempt to optimize the fit between the given data and a mathematical model. Such methods are often based on the assumption that the data are generated by a mixture of underlying probability distributions. Model-based clustering methods follow two major approaches: statistical approach or neural network.

E.1. COBWEB [27]: COBWEB is a popular and simple method for incremental conceptual clustering. Its input objects are described by categorical attribute-value pairs. COBWEB creates hierarchical clustering in the form of a classification tree. A classification tree differs from a decision tree. COBWEB uses a heuristic evaluation measure called *category utility* to guide the construction of the tree. Category utility is the increase in the expected number of attribute values that can be correctly guessed given a partition over the expected number of correct guesses with no such knowledge. Category utility rewards intra-class similarity and inter-class dissimilarity. COBWEB has two additional operators' called *merging* and *splitting*, which help COBWEB to become less sensitive to input order and also allow it to perform bi-directional search.

E.2. SOM [191]: The SOM (Self-Organizing feature Map) net can be thought of as a two-layer neural network. Each neuron is represented by an n-dimensional weight vector, m (i.e., $m_1, m_2, \cdots, m_n$), where n is equal to the dimension of the input vectors. The neurons of the SOM are themselves cluster centers; but to accommodate interpretation, the map units can be combined to form bigger clusters. The SOM is trained iteratively. In each training step, one sample vector x from the input data set is chosen randomly, and the distances between it and all the weight vectors of the SOM are calculated using a distance measure, e.g., Euclidean distance. After finding the Best-Matching Unit (the neuron whose weight vector is closest to the input vector), the weight vectors of the SOM are updated so that the Best-Matching Unit is moved closer to the input vector in the input space. The topological neighbors of the BMU are also treated in a similar way. An important property of the SOM is that it is robust. An outlier can be easily detected from the map, since the distance in the input space from other units are large. The SOM can deal with missing data values, too.

E.3. AutoClass [78]: AutoClass uses the Bayesian approach, starting from a random initialization of the parameters, incrementally adjusting them in an attempt to find their maximum likelihood estimates. Moreover, it is assumed that, in addition to the observed or predictive attributes, there is a hidden variable. This unobserved variable reflects the cluster membership for every sample in the data set. Therefore, the data-clustering problem is also an example of supervised learning from incomplete data due to the existence of such a hidden variable. Their approach for learning has been called RBMNs (Recursive Bayesian Multinets).

[F] Graph-Based Clustering Approaches

This category of clustering algorithms is still in the evolution stage. Several good algorithms have been proposed, so far. In this section we discuss a select few.

F.1. Autoclust [120]: Autoclust, automatically extracts boundaries of clusters based on Voronoi modeling and Delaunay diagrams. Parameters are not specified by users in Autoclust. Rather, values for parameters are obtained from the proximity structures of the Voronoi model. Autoclust calculates them from the Delaunay diagram. This not only removes human-generated bias, but also reduces exploration time. The effectiveness of the approach allows detection, not only of clusters of different densities, but sparse clusters near the high-density clusters. Multiple bridges linking clusters are identified and removed. All this is performed within $O(nlogn)$ expected time, where n is the number of data points.

[G] Ensembles of Clustering Algorithms [296]: The theoretical foundation for combining multiple clustering algorithms is still in early stages of development. Combining multiple clustering algorithms is more challenging than combining multiple classifiers. The reason that impedes the study of clustering combination is that various clustering algorithms produce largely different results due to different clustering criteria. Therefore, combining the clustering results directly with integration rules, such as sum, product, median and majority vote cannot generate a meaningful result. Cluster ensembles can be formed in a number of different ways:

1. using a number of different clustering techniques (either deliberately or arbitrarily selected),

2. using a single technique many times with different initial conditions,

3. using different partial subsets of features or patterns.

In recent work under this category, a split-and-merge strategy also has been adopted. According to this strategy, the first step is to decompose complex data into small, compact clusters. The k-means algorithm serves this purpose. An ensemble of clustering algorithms is produced by random initialization of the cluster centroids. Data partitions present in these clusterings are mapped into a new similarity matrix between patterns, based on a voting mechanism. This matrix, which is independent of data sparseness, is then used to extract the natural clusters using the single link algorithm. More recently, the idea of combining several different clustering algorithms for a set of data patterns based on a Weighted Shared nearest-neighbor Graph WSnnG [32] has been introduced. Due to the increasing size of current datasets, constructing efficient distributed clustering algorithms has attracted considerable attention. Distributed clustering assumes that the objects to be clustered reside on different sites. Instead of transmitting all objects to a central site (also called a server) where we can apply standard clustering algorithms to analyze the data, the data are clustered independently on the different local sites, also called clients). In a subsequent step, the central site establishes a global clustering based on the local models, i.e., the representatives. There are different scenarios for distributed clustering: (a) *Feature-Distributed Clustering* (FDC) combines a set of clusterings obtained by clustering algorithms having partial views of the data features,

(b) *Object-Distributed Clustering* (ODC) combines clusterings obtained by clustering algorithms that have access to the entire set of data features and to a limited number of objects, and (c) *Feature/Object-Distributed Clustering* (FODC) combines clusterings obtained by clustering algorithms having access to a limited number of objects and/or features of the data.

[H] Projected Clustering Approaches

To overcome information loss that occurs in feature selection or dimension reduction methods, projected clustering techniques have been introduced. It was proposed that *projected clusters* [8] give more meaningful information about the underlying clustering structure. A projected cluster C_p can be defined [8] [162] as a set of orthogonal vectors together with a set C of data points such that the points in C are closely clustered in the subspace defined by the vectors C_p. The subspace defined by the vectors C_p may have lower dimensionality than the full dimensional space. Some projected clustering techniques are discussed below.

H.1. PROCLUS [8]: PROCLUS (PROjected CLUStering) works like the k-means algorithm, extended with the idea of projected clustering. First, a greedy approach is applied to select a set of potential medoids (the centers of the clusters). With the set of medoids, it estimates the correlated subspace for each cluster by examining its locality. Locality is defined as the set of data objects in a neighborhood in the full-dimensional space. The projections of these data objects on different single dimensions are examined to find those dimensions that have closer average distances to the corresponding medoids. These are chosen as the dimensions of the correlated subspace. After the estimation of subspaces, a data object is assigned to its closest medoid with the distance measured with respect to the corresponding subspace. The quality of clustering, which is the sum of intra-cluster distances, is evaluated. Medoids are replaced using a hill-climbing approach, targeting improvement in the clustering quality. One problem we see is that the full dimensionality is used in forming the locality. This may not include the real neighbors in the correlated subspace and may actually include unrelated points. Therefore, it makes little sense to look for neighbors in the high-dimensional space. In addition, the parameters of k and l, the dimensionality of clusters, also may greatly affect the result quality.

H.2. ORCLUS [9]: ORCLUS makes use of Singular Value Decomposition (SVD), which is a well-known technique for dimension reduction, with the least loss of information. SVD transforms the data to a new coordinate system (defined by a set of eigenvectors) in which the correlations in the data are minimized. In contrast, ORCLUS chooses the eigenvectors with minimum spread (eigenvalue) to do the projection, so that the greatest amount of similarity among the data points in the clusters can be detected. OR-CLUS works as a hierarchical merging method, starting with a group of initial seeds. During the merging process, the dimensionalities of the subspaces associated with each clusters are gradually reduced, by using the SVD technique mentioned above. The merging is terminated when the number of clusters and the dimensionalities of subspaces reach the user input parameters.

[C] Subspace Clustering [317]: Although projected clustering offers more flexibility than other dimensionality reduction techniques, it often leads to information loss. Due to clustering of the objects over varied projected attributes or feature subspaces, some interesting information contained in the objects may be lost. To address this issue, the subspace clustering approach has been introduced. This approach aims to identify appropriate subspaces from the original feature space, over which the optimal clusters are extracted. A few subspace clustering techniques are described next.

C.1. CLIQUE [191] [307]: CLIQUE is a hybrid version of grid-based and density-based clustering approaches. This scalable method handles high-dimensional data using automatic feature subspace identification. It considers the problem of cluster extraction as a subspace-finding problem in high-dimensional feature space. During cluster formation, it uses a density-based approach. CLIQUE partitions each dimension into equal-length intervals to estimate the density of a region of data objects following a bottom-up approach. It exploits density information to identify the subspaces. It forms the clusters by grouping connected high-density partitions within a subspace. The clusters can also be thought of as a set of axis parallel hyper-rectangles with overlap. CLIQUE is dependent on two input parameters: (i) the number of equal-length intervals and (ii) a density threshold.

C.2. MAFIA **[145]:** MAFIA (Merging of Adaptive Finite Intervals (and more than a CLIQUE)) is an effective variant of CLIQUE. Its effectiveness can be observed in terms of both cluster quality and speed. Two distinguishing features of MAFIA are: (i) elimination of the pruning technique that restricts the number of subspaces to be examined and (ii) incorporation of adaptive interval size. For each dimension, it calculates the minimum number of bins and accumulates them in a histogram. Next, it merges the bins based on two criteria: (i) contiguity and (ii) histogram values. Unlike CLIQUE, the boundaries of the bins are not static, to help detect any random cluster shape precisely. MAFIA is scalable and is significantly faster than the original CLIQUE.

C.3. Cell-based Clustering Method (CBF) **[76]:** A common problem faced by bottom-up algorithms is scalability. With increase in dimensionality, the number of bins also increases drastically. CBF is able to address this issue successfully. It creates optimal partitions by generating a minimum number of possible bins (cells) by an iterative examination of minimum and maximum values for a given dimension. To handle high dimensionality, CBF uses an efficient filter-based index structure to store the bins to facilitate improved retrieval performance. CBF is dependent on two parameters: (i) A *'section threshold'* is used to determine the bin frequency for a dimension. With an increase in the threshold value, the retrieval time comes down because the number of records accessed is reduced. (ii) The *'cell threshold'* is the another parameter that helps determine the minimum density of data points in a bin. It considers the bins with density above this threshold as potential candidates for cluster membership. CBF can detect clusters of all shapes and sizes. However, it is slightly inferior to CLIQUE in terms of precision, although is faster in retrieval and cluster extraction.

C.4. FINDIT **[436]:** FINDIT (Fast and Intelligent Subspace Clustering Algorithm using Dimension Voting) is a fast subspace clustering algorithm. A distinguishing feature of FINDIT is the use of a unique dissimilarity measure called DOD (Dimension Oriented Distance). It follows a top-down approach like PROCLUS. It assumes that two high-dimensional instances can be considered similar when the instances are found to be close over several dimensions rather than a few. So, it checks the number of dimensions over which two instances are found similar, within a user-defined distance threshold. It maps the dimension-tallying problem into a voting problem. FINDIT comprises three phases: (i) sampling, (ii) cluster formation, and (iii) data assignment. In *phase 1*, it adopts a random sampling of the data to select two small sets. These sets are used as the initial representative medoids of the clusters. In *phase 2*, it finds the correlated dimensions using the DOD measure with reference to each medoid. FINDIT then repeats this step until the cluster quality stabilizes. In phase 3, all the instances are assigned to medoids based on the subspaces found. FINDIT is dependent on two input parameters: (i) the minimum number of objects in a cluster and (ii) the minimum distance between two clusters. A major advantage of FINDIT is that it can identify clusters over varied subspaces.

C.5. δ-Clusters [449]: The δ-Clusters [] algorithm aims to extract clusters defined by a subset of instances and a subset of features that show high coherence. When forming a cluster, it searches for instances that show similar trend based on PearsonR correlation [16] estimation, rather than checking for closeness over a number of dimensions. It starts cluster formation with an initial seed selection and attempts to improve the cluster quality by random swapping of features and instances. The cluster formation process continues until no further improvement in the quality is noticed. δ-Clusters is dependent on two parameters: (i) the individual cluster size and (ii) the number of clusters. Its execution time performance is largely influenced by the cluster size parameter. The selection of inappropriate (non-optimal) value may cause significantly high execution time to terminate. A major distinguishing feature of δ-Clusters is the use of a coherence measure during cluster formation. It makes the algorithm more suitable for gene expression analysis.

C.6. COSA [130]: COSA (Clustering On Subsets of Attributes) is based on the nearest-neighbor approach. It starts with equally wighted dimensions and follows an iterative approach to assign weights to each of the k nearest-neighbor dimensions. It assigns higher weights to those dimensions that have less dispersion within the k-nearest-neighbor dimensions. COSA uses the weights to calculate weights of the dimensions for any given pair of instances. It is subsequently used to update the dissimilarities used during the k-nearest-neighbor calculation. It repeats the process until the weights are stabilized. This process helps enrichment of the neighborhoods for each instance with member instances of a cluster. COSA generates a distance matrix as output based on a weighted inverse exponential distance. This matrix can be used as input to any distance-based clustering method. Once clusters are formed, the dimension weights of the cluster members are compared for computation of an overall importance value for each dimension for each cluster.

[D] Biclustering or Two-way Clustering Approaches

In cluster analysis of gene expression data, genes are usually grouped based on distances or correlations considering the full feature/sample space. However, Gibbons et al. [142] observed that biologically significant groups of genes have high correlation over subsets of samples, rather than the full sample space, leading to the development of a new category of clustering called biclustering or two-way clustering.

A biclustering algorithm attempts to group genes considering a subset of rows/genes (may not be contiguous) and a subset of columns/samples (may not be contiguous) that exhibit high correlation. The correlation types that exist between a pair of genes are normally of four types: absolute, shifting, scaling, and shifting-and-scaling. We define each of these types of correlations as follows.

If $g_1=[\ x_1,x_2,x_3,\ldots x_n\ ]$ and $g_2=[y_1,y_2,y_3,\ldots y_n]$ are two gene expression vectors, then (g_1,g_2) said to exhibit an

- *absolute pattern:* if their corresponding expression values are very close or very similar,i.e., y_i is similar to x_i, for $i=1,2,\cdots,n$,

- *shifting pattern:* if $x_i=k_{sh}\pm y_i$, $i=1,2,3,\ldots n$, where k_{sh} is an additive constant,

- *scaling pattern:* if $x_i=k_{sc}\times y_i$, $i=1,2,3,\ldots n$, where k_{sc} is a multiplicative constant, and

- *shifting-and-scaling pattern* if $x_i=k_{sc}\times y_i\pm k_{sh}$, $i=1,2,3,\ldots n$ where k_{sh} is an additive constant and k_{sc} is a multiplicative constant.

A given pair of genes, say (g_i,g_j), is said to exhibit shifting, scaling, or shifting-and-scaling correlation, if the expression values of any of the gene of the pair can be obtained from the other by addition/subtraction or multiplication operation with constants. Figures 4.16, 4.17, and 4.18 present a visualization of these three types of correlations.

A proximity measure used in cluster analysis of gene expression data for extraction of biologically significant groups of genes is expected to handle all the above types of correlations for any given pair of gene expressions. An appropriate shifting-and-scaling correlation measure should help identify quality clusters based on functional similarity and co-expression analysis rather than pure shifting and scaling correlation measures which operate in

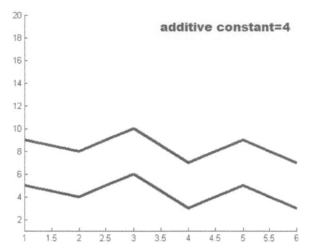

Figure 4.16: Shifting correlation: An illustration

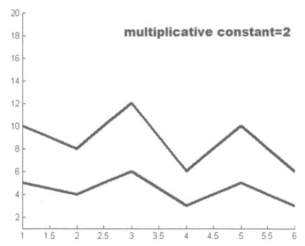

Figure 4.17: Scaling correlation: An illustration

isolation. Most correlation measures which are used in gene expression data
analysis can perform well in detecting shifting and scaling correlations in isolation only. These measures fail to detect mixed correlation correctly, and as
a consequence, some interesting co-expressed patterns with high biological
significance are missed. Although there are some techniques which claim to
extract shifting-and-scaling cluster patterns, most do not consider gene expressions pair-wise. One such example measure is a *residue score*-based [16]
measure, which attempts to handle such mixed correlation patterns by operating both row-wise and column-wise, similar to the mean square residue
[82].

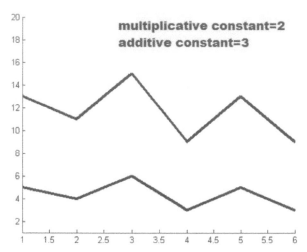

Figure 4.18: Shifting and scaling correlation: An illustration

' Residue score [16] estimates the amount of shifting-and-scaling correlation among the genes over the samples as an aggregate score (not normalized). There is another measure called *transposed virtual error* [327], used to assess the quality of a group of genes by handling shifting-and-scaling correlation, but this measure cannot be used for the estimation of pair-wise gene-gene correlation. So, although several measures have been introduced to handle mixed correlation types, these are inadequate because (i) the measures are column-biased, or (ii) like Pearson correlation, the measures are sensitive to noise due to the use of the global mean. To address these issues and to enable the handling of shifting-and-scaling correlation, another mutual similarity measure has been introduced, called *SSSim* (shifting and scaling similarity). Unlike its close competitors, including Pearson correlation, SSSim is least affected by the presence of noise. This is due to the use of *local means* during computation of gene-gene correlation instead of the global mean. It uses a 3-neighborhood (i.e., previous, current and next values) concept to compute the local mean of a feature/sample of a given gene. In other words, the local mean of a given gene g_i for a given condition c_j is computed based on the values of c_{j-1}, c_j, and c_{j+1}. When $j = 1$ and $j = n$, it considers only two neighbors, (c_j, c_{j+1}) and (c_{j-1}, c_j), respectively. Further, it uses a pair of conditions as a base line or reference condition pair. The score for any pair of gene expressions ranges always from 0 to 1, where 0 indicates that the genes are totally different, whereas 1 indicates that they exhibit perfectly similar shifting and scaling correlation.

A number of biclustering techniques have been introduced to help identify biologically significant groups of genes with high similarity, taking into

Table 4.6: Proximity Measures and their Abilities to Capture Correlation Types

MEASURE	TYPE/MODE	FORMULA	CORRELATIONS AB	SH	SC	SS
Euclidean	D/M	$\sqrt{\sum_{i=1}^{n}(a_i - b_i)^2}$	Y	N	Y	Y
Cosine	S/M	$\dfrac{\sum_{i=1}^{n} a_i \times b_i}{\sum_{i=1}^{n} a_i \times \sum_{i=1}^{n} b_i}$	Y	N	Y	N
NMRS	S/M	$1 - \dfrac{\sum_{i=1}^{n}\left|a_i - \overline{A} - b_i + \overline{B}\right|}{2 \times max\left(\sum_{i=1}^{n}\left|a_i - \overline{A}\right|, \sum_{i=1}^{n}\left|b_i - \overline{B}\right|\right)}$	Y	Y	N	N
Pearson	S/M	$\dfrac{\sum_{i=1}^{n}(a_i - \overline{A}) \times (b_i - \overline{B})}{\sqrt{\sum_{i=1}^{n}(a_i - \overline{A})^2 \times \sum_{i=1}^{n}(b_i - \overline{B})^2}}$	Y	Y	Y	Y
Spearman	S/M	$\dfrac{\sum_{i=1}^{n}(rank(a_i) - \overline{rank(A)}) \times (rank(b_i) - \overline{rank(B)})}{\sqrt{\sum_{i=1}^{n}(rank(a_i) - \overline{rank(A)})^2 \times \sum_{i=1}^{n}(rank(b_i) - \overline{rank(B)})^2}}$	N	N	N	N
MSR	D/M	$\dfrac{1}{|I||J|} \sum_{i \in I, j \in J}(a_{ij} - a_{iJ} - a_{Ij} + a_{IJ})^2$	Y	Y	N	N
Virtual error	D/M	$\dfrac{1}{|I||J|} \sum_{i \in I, j \in J}(a_{ij} - \hat{a_{iJ}})$	Y	Y	Y	N

MEASURE	TYPE/MODE	FORMULA	CORRELATIONS			
			AB	SH	SC	SS
Transposed virtual error	D/M	$\dfrac{1}{|I||J|}\displaystyle\sum_{i\epsilon I,j\epsilon J}(\hat{a}_{ij}-\hat{a}_{Ij})$	Y	Y	Y	Y
Residue error	D/M	$min(\overset{min}{_{j'\epsilon J}}A_{Ij'},\overset{min}{_{j'\epsilon I}}A_{i'J})$	Y	Y	Y	Y
SSSim	S/M	$1-\dfrac{1}{n-2}*\displaystyle\sum_{i=2}^{n-1}2\times max\left(\left|lmean_i-\dfrac{a_{i+1}-a_i}{a_2-a_1}-\dfrac{b_{i+1}-b_i}{b_2-b_1}\right|,\left|lmean_i-\dfrac{a_{i+1}-a_i}{a_2-a_1}\right|,\left|lmean_i-\dfrac{b_{i+1}-b_i}{b_2-b_1}\right|\right)$	Y	Y	Y	Y
LPCM	S/M	$\dfrac{\sum_{i=1}^{n}(X_i-LM(X_i))(Y_i-LM(Y_i))}{\sqrt{\sum_{i=1}^{n}(X_i-LM(X_i))^2}\sqrt{\sum_{i=1}^{n}(Y_i-LM(Y_i))^2}}$	Y	Y	Y	Y

account a subset of genes and subsets of samples. In biclustering, similarity indicates the coherence of the genes and conditions (samples), and it is not a function of pairs of genes/conditions. Normally, this measure is a symmetric function of genes/conditions involved and during biclustering, it helps group subsets of genes over subsets of samples simultaneously, obtaining high similarity scores.

The term *biclustering* was first coined by Mirkin [82] to introduce the concept of simultaneous clustering for a data matrix considering both rows and columns. Hartigan [82] had referred to a similar concept using the term *direct clustering*. *Box clustering* is another term that describes a node addition approach to form a *maximal constant bicluster*. Biclustering as a new paradigm was first introduced and popularized by Cheng and Church [82] for gene expression analysis. Cheng and Church considered a bicluster as a weighted bipartite graph, and referred to it as a natural generalization of the concept of a *biclique* in graph theory. Cheng and Church introduced an algorithm based on node deletion to extract biclusters (with low mean squared residue score) from a gene expression matrix. Based on this measure, they estimated a similarity score applicable to biclusters generated from a gene expression matrix transformed and augmented by logarithmically and an additive inverse, respectively. The measure, referred as *Mean Squared Residue*, is based on the definition of a residue as

$$a_{ij} - a_{iJ} - a_{Ij} + a_{IJ}. \tag{4.20}$$

For any bicluster, the residue of an element a_{ij} is represented by the subsets I and J. Here, a_{iJ} and a_{Ij} represent the means of the ith row and j^{th} column of the bicluster, respectively. Similarly, a_{IJ} represents te mean of all the elements belonging to the bicluster. The objective of Cheng and Church's [82] approach was to find large and maximal biclusters with the least mean squared residue (MSR). If for a bicluster, the MSR score is 0, it is a single-valued constant bicluster. However, for a bicluster with a non-zero score, it is possible to eliminate row(s) or column(s) to minimize the score. According to this approach, one has to delete those rows and columns iteratively which do not contribute in generating biclusters with low *mean squared residue*. The goal of finding a largest (complete) balanced bipartite graph or biclique is an NP-hard problem.

A number of biclustering techniques have been proposed in the literature for gene expression data analysis. Cheng and Church [82] proposed an algorithm that uses a node deletion approach. The algorithm uses a measure called mean squared residue that guides the discovery of subspace clusters.

The rows and columns are iteratively deleted to generate sub-matrices that have a low mean squared residue score.

OPSM (Order-Preserving Sub-matrices) [50] uses a probabilistic approach to extract order-preserving biologically relevant sub-matrices from a gene expression matrix. Contrary to the trend-preserving approach, the authors claimed that biclusters extracted based on an order-preserving approach obtain more significant co-expressed gene patterns. The use of a probabilistic approach was shown by the authors to be effective in solving the NP-hard problem of all possible order-preserving sub-matrix extraction from a matrix. BIMAX [291] is another effective biclustering technique that simplifies the submatrix extraction process by transforming the gene matrix into a binarized matrix. It uses a *cutoff* when converting the gene matrix into a binarized matrix and the cutoff is decided carefully so that the numbers of zeros and ones in the binarized matrix are almost equal. During conversion, it reassigns an expression value greater than or equal to the cutoff to be 1 and otherwise 0. Once the binary matrix is constructed, BIMAX opetates on it recursively to extract the submatrices with all zero entries.

An evolutionary biclustering technique called iBBiG [155] was introduced for bicluster generation to operate on a binary gene expression matrix. The technique extracts all biologically relevant modules. It uses a fitness function to reduce the chance of genes already in a module to become members of another module. Another good contribution in this field is the spectral biclustering technique [220]. It locates checkerboard patterns in a gene expression data matrix by exploiting eigenvectors corresponding to characteristic expression patterns across genes and conditions to detect checker-board structures. Spectral biclustering uses Singular Value Decomposition (SVD) to identify the eigenvectors.

Another significant contribution in biclustering research is ISA (Iterative Signature Algorithm) [53], which uses randomization through the selection of random seeds. It requires a very large number of initial seeds, which may not be possible to obtain practically. Another effective biclustering technique is FLOC (FLexible Overlapped biClustering) [82]. This algorithm starts cluster formation with a set of seeds (initial biclusters) and continues its expansion iteratively until no improvement in the overall quality of the subspace clusters is observed. During the extraction of a bicluster, FLOC moves rows and columns iteratively among the partial biclusters to obtain improved biclusters with lower mean squared residue. It considers the biclusters obtained in each iteration as the initial biclusters for the next iteration.

SAMBA (Statistical Algorithmic Method for Bicluster Analysis) [397], [398] is another probabilistic biclustering technique that represents a gene expression matrix as a *weighted bipartite graph* where the genes are represented as the nodes on one side of the graph and the properties of the genes are represented by the nodes on the other side. Each edge is associated with a *weight*. The weight w of an edge between a gene node g and a property node v represents an assertion that gene g has property v with a probability proportional to w. SAMBA extracts heavy subgraphs (as biclusters) in the weighted bipartite graph using combinatorial principles.

CLIQUE [191] is an efficient grid-density-based subspace clustering algorithm. It initially identifies dense clusterable subspaces and sorts them based on their subspace coverages. It discards subspaces with low coverage. Subsequently, it grows an arbitrarily dense unit into a maximal region using a depth-first search. xMOTIF [212] initially extracts a set of rows along with a set of columns to obtain a conserved gene expression motif (xMotif). An xMotif is defined as a subset of rows (genes) that is simultaneously conserved across a subset of columns (conditions) and hence, an xMotif can be considered a bicluster. For a given set of conserved rows, a column that matches a given motif, the algorithm computes the remaining conditions by checking each column for a larger xMotif. The xMOTIF algorithm can be used with any $N \times M$ data matrix to identify local patterns.

Another two-way clustering approach called CTWC (Coupled Two-Way Clustering) [137] was introduced to extract stable clusters with a set of instances and a set of samples. This algorithm uses heuristics to avoid brute force enumeration of all possible combinations. CTWC considers only stable subsets of objects or genes for use in the next iteration. Using a Self-Organizing Map (SOM), DCC (Double Conjugated Clustering) [72] forms clusters over rows and columns space of the data matrix. DCC is dependent on the angle metric as the similarity measure.

Xu et al. [398] proposed a technique to extract groups of genes that exhibit shifting-and-scaling correlation (with positive or negative factor) satisfying regulation similarity. Also, they proposed a way to measure shifting-and-scaling correlation between a pair of genes in subspaces of samples. The subspace is considered only when determining correlation between a pair of genes. A set of genes included in a cluster may be mutually correlated in different subspaces of samples.

FABIA [179] uses factor analysis in a multiplicative model to extract biclusters in a gene expression dataset. It represents linear dependency between a subset of rows and columns as an outer product of a prototype column vector and a factor vector with which the prototype vector is scaled over samples. It is able to select parameters of the multiplicative model when extracting a bicluster in gene expression data. The Additive MSBE (Maximum Similarity Biclustering Extended) [254] technique extracts additive biclusters using a greedy approach. Using a gene as the reference gene, it transforms the expression matrix using one of the columns as the reference column such that the additive constant for a pair of gene expressions becomes zero. Then, it applies a greedy deletion approach followed by greedy addition.

Teng and Chan [402] proposed a biclustering technique that sorts genes and samples using a dominant set approach. A weighted correlation measure is used to determine the correlation between two gene expressions or two samples. From the sorted gene expression data, which has highly correlated genes and samples in a corner, the biclusters are extracted removing genes or rows with low correlation values with the adjacent genes or rows. Virtual error [326] is a measure that can detect shifting or scaling correlations present in a bicluster. Transposed virtual error [273] is an extension of virtual error that can detect shifting-and-scaling correlation. The term "combined pattern" refers to a bicluster with shifting-and-scaling correlation.

Mitra and Banka [285] used a multi-objective evolutionary algorithm to extract biclusters from gene expression data using two objectives, considering size and homogeneity of biclusters. Homogeneity is measured using mean squared residue. Das et al. [96] used a multi-objective evolutionary biclustering to extract gene interaction networks. A rank-correlation-based multi-objective evolutionary biclustering method is used to extract strongly correlated gene expression pairs which correspond to edges in the gene interaction network.

The ICS biclustering algorithm [15] accepts two input parameters, SSSim threshold (τ) and bicluster association threshold (α), along with the gene expression matrix and produces a set of biclusters. The technique operates on a pair of gene expressions such that none of these are part of any bicluster that has been generated so far. All the correlated subspaces for the pair of genes are extracted. A correlated subspace for a pair of genes is a subset of samples for which corresponding expression values of the gene pair exhibit shifting-and-scaling correlation with similar values of co-efficients (p and q, as used in ICS). The correlated subspace for a pair of gene expressions corresponds to a maximal set of samples for which the gene expressions are correlated by more than the user-defined threshold τ in terms of SSSim.

Although a good number of biclustering algorithms have been introduced in the past two decades, a major issue is that if some algorithms generate a few larger biclusters, others generate too many smaller biclusters in most gene expression datasets. A consistently performing biclustering algorithm is still lacking to ensure an optimum number of biclusters with high biological significance for any gene expression data matrix. Further, most biclustering algorithms demand high computational resources for large gene expression matrices.

Clustering and biclustering techniques discussed so far to handle gene expressions over a number of cell samples where rows represent genes and columns indicate samples. However, the revolutionary growth of biological data generation technologies enables the incorporation of spatial and temporal aspects simultaneously on gene expression data to present expression levels over multiple time points for multiple samples. Such data are called 3-dimensional Gene-Sample-Time gene expression data or GST data. Not many GST datasets are available due to (i) high preparation time and (ii) a high resource requirement. Such content-rich GST data can aid tremendously in generating biologically significant results. To support analysis of GST data, existing clustering and biclustering techniques are inadequate, and hence it has resulted in the development of a new variant of clustering techniques, referred to as *triclustering*. A triclustering technique analyzes a gene expression dataset over three dimensions, gene, sample and time, simultaneously. We present tricluster analysis approaches in general, and then discuss a few popular triclustering techniques next.

Tricluster Analysis

As we know, clustering is the process of grouping data objects into a set of disjoint classes, called clusters, so that objects within a class have high similarity to each other, while objects in separate classes are highly dissimilar. Cluster analysis of gene expression data can be done in three ways. In *gene-based clustering*, the genes are treated as the objects, while the samples are the features, where as in *sample-based clustering*, the samples are treated as the objects, while the genes are the features. The samples are partitioned into homogeneous groups. Each group may correspond to some particular macroscopic phenotype, such as clinical syndromes or cancer types. Both gene-based and sample-based clustering approaches search for exclusive and exhaustive partitions of objects that share the same feature space. In *subspace clustering*, genes and samples are treated symmetrically, so that either genes or samples can be regarded as objects or features. The current thinking in molecular biology holds that only a small subset of genes participate in any cellular process of interest and that a cellular process takes place

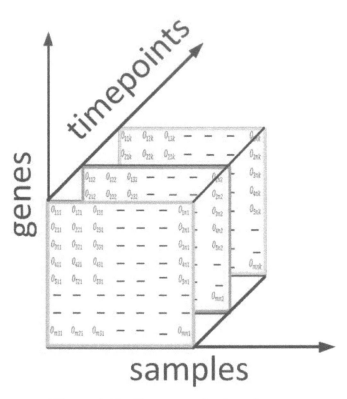

Figure 4.19: Gene-sample-time data

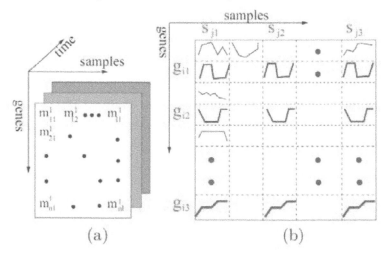

Figure 4.20: Structure of GST MIcroarray data

only in a subset of the samples. This belief calls for subspace clustering to capture clusters formed by a subset of genes across a subset of samples. Furthermore, clusters generated through such algorithms may have different feature spaces. One faces several challenges in mining microarray data using a subspace clustering technique (a) Subspace clustering is known to be an NP-hard problem and therefore, many proposed algorithms for mining subspace clusters use heuristic methods or probabilistic approximations. This decreases the accuracy of the clustering results. (b) Due to varying experimental conditions, microarray data is inherently susceptible to noise. Thus, it is essential that subspace clustering methods be robust to noise. (c) Subspace clustering methods allow overlapping clusters that share subsets of genes, samples, or time intervals, or spatial regions. (d) Subspace clustering methods should be flexible enough to mine several (interesting) types of clusters. (e) The methods should not be very sensitive to input parameters.

Some existing triclustering algorithms for mining coherent clusters in three-dimensional (3D) gene expression datasets are: (a) Mining Coherent gene clusters from GST microarray data [94], (b) TRICLUSTER [230], (c) gTRICLUSTER [158], and (d) ICSM [15].

An algorithm called Mining Coherent Gene Clusters from GST Microarray Data finds coherent genes under a subset of samples across the entire time range of a GST dataset. It accepts three input parameters, coherence threshold, minimum number of genes (min_g), and minimum number of samples (min_s), and outputs the complete set of coherent gene clusters. The steps of the algorithm are as follows: (i) For each gene, find a maximal coherent subset of samples. Coherence between two samples is measured using the Pearson correlation coefficient across the whole time range. (ii) To find the complete set of coherent triclusters from the maximal coherent samples of genes computed in the previous step, there are two alternative algorithms: (a) sample gene search and (b) gene sample search. The steps of the both algorithms can be summarized as follows: (i) Enumerate all possible subsets of samples/genes. The algorithm uses a set enumeration tree along with pruning to find the subsets efficiently. (ii) For each of the subsets of samples S/genes G, find maximal subsets of genes G/samples S such that $G \times S$ is a set of coherent gene clusters. This is done using an inverted list of maximal coherent sample genes, prepared earlier. Some positive features of this algorithm are as follows: (i) The algorithm takes care of inter-temporal coherence in the extracted triclusters. (ii) Due to the use of the Pearson correlation coefficient it can detect both shifted and scaled forms of inter-temporal coherence. (iii) It is a deterministic algorithm and it allows overlapping of triclusters. Limitations of this algorithm include the following: (i) The algorithm detects triclusters that include the entire time span. (ii) The second approach to find a complete set of coherent gene clusters, *Gene Sample Search*, is computationally very costly. (iii) Though the algorithm considers inter-temporal coherence, it does not consider coherence among genes in the triclusters. (iv) None of the proposed approaches is able to find all possible triclusters, and it is practically infeasible to use both at the same time.

TRICLUSTER [230] is an efficient and deterministic triclustering algorithm that requires four input parameters, the maximum ratio threshold (ϵ), the minimum gene threshold (mx), the minimum sample threshold (my) and the minimum time threshold (mz), and outputs coherent clusters along the gene-sample-time dimension. The steps of the algorithm are the following. (a) For each $G \times S$ time slice matrix, find the valid ratio ranges for all pairs of samples, and construct a range multigraph. (b) Mine maximal biclusters by performing depth-first search on the range multigraph. (c) Construct a graph based on the mined biclusters and obtain the maximal triclusters. (d) Delete or merge clusters if certain overlapping criteria are met. TRICLUSTER can mine only maximal triclusters satisfying certain homogeneity criteria. The clusters can be arbitrarily positioned anywhere in the input data matrix and they can have arbitrary overlapping regions. It is a deterministic and complete algorithm, which leverages the inherent unbalanced property. However, it is not free from limitations. Some major limitations are as follows. (i) The algorithm depends on four input parameters. (ii) Computation to find genes to include in terms of two samples is costly. (iii) Due to the aggregate nature of comparison, a single value of ϵ is not enough to detect relevant biclusters that may need different thresholding for their recognition. (iv) It is very difficult to control the required level of coherence using parameter ϵ. The practically recommended value of ϵ is less than 0.01. But such a value may not always be useful in detecting biologically relevant patterns. (v) The algorithm can detect clusters with scaling patterns, but cannot detect the ones with shifting patterns.

gTRICLUTER [158] is a triclustering algorithm that takes four input parameters, the minimum similarity threshold (∂), the minimum sample threshold (m_s), the minimum gene threshold (m_g) and the minimum time threshold (m_t), and outputs coherent clusters along the gene-sample-time dimension. The steps of the algorithm are as follows. (a) Identify a maximal coherent sample subset for each gene during the time series segment. (b) Perform a depth-first search in the sample space to enumerate all possible maximal cliques. (c) Find the maximal coherent gene set for a subset of samples by generating an inverted list. (d) Find the intersection of the inverted list, check whether the result is a maximal coherent gene cluster, and combine with time segment information.

An important feature of gTRICLUSTER is that it considers inter-temporal coherence while generating the triclusters. Some of the limitations of gTRICLUSTER are as follows: (i) It depends on too many input parameters. (ii) Though the algorithm overcomes one of the limitations of TRICLUSTER by considering inter-temporal coherence, it does not consider inter-gene coherence, which is even more significant than inter-temporal coherence. (iii) If there is a time latency between two similar patterns, gTRI-CLUSTER cannot detect such patterns, since two similar patterns may appear in different time ranges. (iv) Though the Spearman correlation coefficient is applied to capture the coherence across time dimension, the measure may not always be effective.

ICSM [15] attempts to address the major issue of gTRICLUSTER by considering both inter-temporal as well as inter-gene coherence by exploiting the concept of order-preserving sub-matrices. It is able to find a set of triclusters of high biological significance from GST data. Initially, the method extracts a set of modules in unordered pairs of GS plane. These modules are then extended towards formation of the final triclusters by using a planar similarity measure, referred to PMRS (Planar Mean Residue Similarity).

Another algorithm named δ-Trimax [57] extracts perfectly shifting triclusters with hubgenes from breast cancer cells. δ-Trimax uses MSR (Mean Square Residue) to mine clusters from a GST dataset. Some of the other triclustering algorithms, namely TriGen [157] and OPTricluster [400], mine clusters with coherent genes across sample and time spaces. Few more triclustering algorithms have also been proposed [205]. EOMA-δ-Trimax is a multiobjective triclustering method to mine the overlapping clusters. FTC [255] is a fuzzy triclustering approach to generate fuzzy triclusters from a GST dataset. Another triclustering algorithm, namely TWIGS [22], finds core modules from three-dimensional data using Hierarchical Bayesian and Gibbs sampling. Again, a tool known as TriClust [175] was developed based on triclustering algorithms for analysis of multiple organisms under a subset of experimental conditions. The concept of triclustering is also used as an application known as CATSeeker [25] to mine profitable stocks and in discovery of biologically significant protein residues through a unique combination of SVD (singular value decomposition), numerical optimization, and 3D frequent itemset.

THD-Tricluster [205] is a triclustering algorithm which mines $G \times S \times T$ data with absolute, shifting, scaling and shifting-and-scaling patterns, and obtain results with high biological significance. It uses the Shifting-and-Scaling Similarity (SSSim) measure [15] to identify co-expressed patterns. The range of SSSim score for any gene expression pair is $[0, 1]$ and there is no need for normalization. If the value of the score is 1, the gene expression perfectly exhibits shifting-and-scaling correlation. The measure introduces the *local mean* of a gene for a condition instead of the mean of all expressions in order to make it robust to noisy values. THD-Tricluster is composed of two algorithms, one that "generates biclusters"" and the other that "generates triclusters". The first algorithm produces a set of biclusters with high biological significance, and the second algorithm produces a set of triclusters with high biological significance and inter-temporal coherence. THD-Tricluster uses the SSSim measure to compute the similarity between any two GS planes. SSSim is used to compute the inter-temporal coherence. The technique operates on a pair of genes and mines the cluster of genes which exhibit gene expression patterns in both sample and time space.

4.3.5 Outlier Mining

Outlier mining finds instances or patterns in data that are significantly different from or non-conforming with the rest of the instances. Identification of such patterns or instances is highly influenced by the proximity measure or the metric used during the process of mining. An outlying instance or group of instances often carry interesting information that may be useful in analyzing the associated system. Outliers are also referred to as exceptions, anomalies, faults, defects, errors, surprises, peculiarities or novelties depending on the nature of the application domain. Outlier detection has received a good attention in the machine learning research community. In computational biology research, especially in gene expression data analysis, outlier mining has an important role to play. An appropriate outlier mining method can help identify a rare or non-conforming samples (or groups of samples) or a gene (or groups of genes), which may be of significant interest among domain experts during downstream analysis. It may lead to novel findings regarding the characteristics or behavior of a system, which may be useful. It is, therefore, essential to identify such non-conforming patterns or instances prior to modeling and subsequent downstream analysis.

An outlier detection method can be categorized as *supervised* or *unsupervised* depending on the availability and usage of class labels. An unsupervised outlier mining technique does not depend on prior knowledge or labeled instances or makes no assumption regarding training data whereas a supervised method assumes the existence of labeled instances for training both classes i.e., normal and outliers'. Unlike unsupervised methods, a supervised method builds predictive models based on (i) the labeled instances and (ii) a learning algorithm. For any given unseen test instance, it matches against the trained model to determine its class label. Unsupervised methods are capable of identifying novel pattern(s) or cases, and hence are more widely applicable in the domain of biology, whereas supervised outlier detection methods can identify cases with high accuracy. An unsupervised method is generally dependent on assumption(s) such as the frequency of occurrence of normal instances is higher than that of the outlier instances. A clustering, biclustering, or triclustering technique discards those gene expressions which are differently expressed across the samples or do not conform to the co-expressed patterns. However, such patterns can be of great interest for the biologists in uncovering the peculiarities or rarity of genes for identification of disease biomarkers. An outlier detection method can help significantly in identifying and analyzing such patterns. Based on the underlying approach used, an outlier detection method can be classified as follows.

Distance-Based Outlier Detection

A distance-based outlier detection method attempts to identify an object O_i as an outlier based on the distance of the object from the rest of the objects [143]. For any object O_i, an outlier factor can be defined as a function $F:O_i \longrightarrow R$. The value of function F depends on the distance of the object O_i from the rest of the objects in an input dataset. It recognizes an object O_i as an outlier if the distance between the object and other objects is significantly above or deviates from a pre-defined user threshold or does not conform to the distributions of distances for other objects. As stated by Hawkins [170], outliers are observations that deviate significantly from the other observations, as if one can suspect them of being generated by a different mechanism.

To define it formally [221], let us consider an example. Let O_i, O_j, and O_k be three objects in a dataset D and $d(O_i, O_j)$ represents the minimum distance between the objects O_i and O_j. Let C_i denotes a group of objects in D and $d(O_i, C_i)$ represents the minimum distance between the object O_i and any single object $O_j \in C_i$. We can express mathematically as

$$d(O_i, C_i) = min\{d(O_i, O_j)|O_j \in C_i\}. \tag{4.21}$$

According to one definition of an outlier, [143], for any given dataset D, an object p can be defined as a $DB(\alpha, Mindist)$ outlier, if at least $\alpha\%$ of the objects $\in D$ are located at a distance greater than *Mindist* from O_i. In other words, the cardinality of the group $\{O_j \in D | d(O_i, O_j) \leq Mindist\}$ is $\leq (100 - \alpha)\%$ of the total distance in D [143].

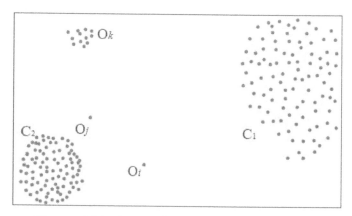

Figure 4.21: A sample 2-D synthetic data set

Thus, using the distance-based outlier definition, outliers are data objects that are located away from the rest (majority) objects based on a proximity measure and with reference to a user-defined distance threshold. The threshold can be static or can be dynamically changed. Two advantages of distance-based outlier finding methods are as follows: (i) They are simple to implement and (ii) they can achieve high outlier detection accuracy depending on appropriate selection of a proximity measure. There are some limitations such as (i) in some data distributions, distance-based methods do not perform well, and (ii) in the case of high-dimensional data, selection of an inappropriate proximity measure may lead to deteriorated performance. It is due to (a) sparsity of outliers along some dimensions and (b) the curse of dimensionality that afflicts some measures.

To address these issues, density-based outlier detection methods have been introduced.

Density-Based Outlier Detection

This method estimates the distribution of densities in a given input dataset and recognizes those objects as outliers which are located in the regions of lower densities [143]. In other words, this approach assumes that the regions with low densities are the sources of outliers. The degree of outlierness of an object O_i increases with the decrease in the density of the region, R_i, where O_i is located. Outlier factor is typically local, since ouitlierness is measured considering the local or neighborhood of the object O_i. In [69], there is only one parameter, referred to as *Minpts*, to define the local neighborhood of any object O_i. *Minpts* refers to the minimum number of nearest-neighbor objects for any given object. The LOF (local outlier factor) of any object O_i is higher if its (i) local reachability density (lrd) is lower and (ii) *lrd* of O_i's *Minpts nearest neighbors* are higher. If an object O_i is located far from another object O_j, the reachability distance between O_i and O_j is their actual distance. On the other hand, if the objects are close enough, the distance between them is substituted by k-distance of O_j. k-distance of any object O_j for any positive integer k is defined as the distance $d(O_i, O_j)$ where $O_i, O_j \in D$ such that

a. there are at least k data objects $O_k \in D|\{O_i\}$, and it is true that $d(O_i, O_k) \leq d(O_i, O_j)$, and

b. there are a maximum of $(k\text{-}1)$ data objects $O_k \in D|\{O_i\}$, and it is true that $d(O_i, O_k) < d(O_i, O_j)$

In other words, the k-distance neighborhood of O_i includes all data objects for which the distance from O_i is not more than k-distance. k can be used as a tuning parameter to minimize statistical fluctuation. With a higher value of k, a more stable reachability distance for objects in the same neighborhood is observed. In general, the performance of a density-based outlier detection method is more dependent on the optimization of the value(s) of the parameters.

If the input dataset is rife with uncertainties and inconsistencies, the distance-based and density-based methods may not perform well. In addition, if the dataset is incomplete, such methods may not work well. To handle such datasets, it is necessary to apply soft computing approaches for outlier detection. For classification, if the class objects are not noticeably different or indistinguishable, one can use approaches based on fuzzy or rough set theory to resolve such uncertainties.

In a fuzzy set theoretic approach, the membership of a data object is not crisp or in other words, it allows a data object to be in a group or set partially. Fuzzy set theoretic approaches can help deal with uncertainties or vagueness among overlapping sets. In most gene expression data, the occurrence of overlapping clusters is common. Using a fuzzy set theoretic approach, one can estimate the degree of membership of any data object for any given class and can identify the degree of outlierness of an object in a dataset with uncertainties or vagueness. There are several efforts [143], [286] to handle the outlier detection using a fuzzy set theoretic approaches or in combination with other soft computing technique(s).

Rough sets have also been used successfully for outlier detection in datasets with vagueness. Soft computing approaches generally use indiscernible or inseparable data objects to build a knowledge base about the real world. It builds and uses rough sets to handle such uncertainties in the effective detection of outliers. A number of such methods have been introduced [18], [143] to address outlier mining. The roughness appears mostly due to the presence of indiscernibility in the input pattern set, whereas fuzziness is caused by the vagueness in the output classes or clusters. For effective modeling, one can also use the concept of fuzzy-rough sets [110] to identify all the outlying objects with high accuracy.

4.3.6   Association Rule Mining

Association rule mining (ARM), introduced by Agrawal et al. [13] is an important tool in data mining and knowledge discovery. It is an effective technique that supports decision making in many real-world applications, including medical diagnosis. ARM helps uncover interesting association patterns among items (attributes or samples) or itemsets in a dataset. The core of this technique is its *support-confidence* framework, which is used to generate a set of IF-THEN rules from a binary dataset (or market basket dataset), where each attribute value indicates the presence (indicated by 1) or absence (indicated by 0) of an item in a basket. It generates rules following two steps: (i) frequent itemset generation and (ii) rule generation based on the outcome of step (i) Of these two steps, extraction of all frequent itemsets is more resource intensive. With increase in dimensionality of the data, the cost of finding frequent itemsets also becomes higher. To address this issue, a large number of algorithms have been introduced [11], [12]. However, most such algorithms fail to fulfill the two major requirements, generation of (i) correct results (*proof of correctness*) and (ii) complete set of frequent itemsets (*proof of completeness*), in a cost-effective manner. There are algorithms which fulfill both requirements, but their resource requirements are significantly high.

Next, we discuss the basics of support-confidence framework and present some popular frequent itemset mining algorithms, along with their strengths and weaknesses. We discuss these algorithms under two categories: (i) frequent itemset mining with candidate generation and (ii) frequent itemset mining without candidate generation. The reason behind this categorization is that generating too many candidates during frequent itemset mining is considered the major bottleneck with most frequent itemset mining algorithms.

The Support-Confidence Framework

A frequent itemset mining algorithm extracts an itemset (with single or multiple co-occurring items) as frequent, based on a user-given parameter, called *Minimum Support*. If the dataset contains n binary-valued items or attributes, then it is possible to generate $(2^n - -1)$ non-empty candidate itemsets. A frequent itemset mining algorithm attempts to extract itemsets, which have their *support count* not less than the user-specified *minimum support* threshold, referred to as *frequent itemsets*. We define the *Support count* of an itemset as the number of transactions or records that contain the itemset. In the support-confidence framework, it is referred to as *Support* when it is expressed in % form relative to the total number of transactions in the dataset. Though the problem looks very simple, practically it is not feasible to maintain the information for all itemsets in a computer's memory. There is a need for reduction in the search space. To reduce the search space, the non-monotonicity property of sets, called the *downward closure property* or *Apriori property*, is used [13]. According to this property, all non-empty subsets of a frequent itemset must also be frequent. Conversely, for any non-frequent itemset, there cannot be any superset of it that is frequent. Formal definitions of *Support* and *Frequent Itemsets* are given below.

Definition 1: Support: For a given market-basket dataset D of k items (or attributes), say $I_0, I_1, I_2, \cdots, I_{k-1}$, we define the *Support* of an itemset X based on its popularity or the population of transactions in which the itemset X appears. In the example dataset shown in Table 4.7, the support of the item $\{Mango\}$ is 5 out of 10 or 50% and the support of $\{Berry\}$ is 7 out of 10, i.e., 70%. For an itemset with more than one item, say $\{Guava, Berry\}$, the support is 5 out of 10, i.e., 50%. Mathematically, we define *Support* as $|X|/|D|$, where $|X|$ denotes the number of occurrences of the itemset X and $|D|$ denotes the total number transactions in D. To maintain and process a large number of candidate itemsets, a potential solution is to exploit the *downward closure property*. This property says that for any frequent itemset, all its non-empty subsets must be frequent. For example, if $\{a, b, c\}$ is frequent, all its subsets i.e., $\{a\}$, $\{b\}$, $\{c\}$, $\{a, b\}$, $\{a, c\}$, $\{b, c\}$ must be frequent.

Definition 2: Frequent Itemsets: For a given dataset D of k items, let C be all possible candidate itemsets and *Minsupp* be a user-defined threshold for an itemset to be frequent. We define an itemset L to be frequent if $Support(L) \geq Minsupp$.

Table 4.7: Market Basket Data: An Example

TID	Mango	Apple	Guava	Grapes	Berry	Pomegranate
01	1	0	1	0	1	0
02	0	0	1	0	1	1
03	1	1	1	0	1	0
04	1	1	0	0	0	1
05	0	1	0	0	1	0
06	0	0	1	1	1	1
07	0	1	1	0	0	1
08	1	0	0	1	0	0
09	0	0	1	0	1	0
10	1	1	0	1	1	0

Frequent Itemset Mining with Candidate Generation

Algorithms for frequent itemset mining with candidate generation initially generate a list of candidates or probable itemsets, which are subsequently filtered based on the fulfillment of a minimum support condition to identify the list of frequent itemsets. To find all the frequent itemsets, such algorithms carry out an exhaustive search on all possible candidate itemsets. To minimize the search complexity, these algorithms introduce or adopt various strategies. Some popular frequent itemset generation algorithms are discussed next.

Apriori is a robust frequent itemset mining technique introduced by Agrawal et al. [13]. Two major attractions of this algorithm are that it satisfies the *Proof of Completeness* (i.e., the list of frequent itemsets identified is exhaustive and complete) and *Proof of Correctness* (i.e. all frequent itemsets identified by it satisfy the *Minsupp* threshold condition). *Apriori* exploits the *downward closure property* of itemsets for cost-effective extraction of frequent itemsets. It comprises two steps: (i) candidate generation and (ii) pruning. Once all possible candidate itemsets are generated, it applies a systematic pruning technique to discard itemsets that do not fulfill the downward closure property. It discards an itemset from further consideration if any of its subsets is found to be infrequent based on its support count (precomputed and known in advance) and with reference to the threshold *Minsupp*. A large number of extensions of Apriori [191], [58] have been proposed. All these algorithms focus mainly on efficient handling of voluminous market basket data to address the generation of huge candidate itemsets and elimination of multiple database scans.

Partitioning Approach: The partitioning approach has proved to be useful in handling voluminous data with limited memory. It helps execute a frequent itemset mining algorithm faster by partitioning the data (horizontally, vertically or a combination of the two) into manageable chunks that are easy to load into the memory. Such a partitioning approach is generally dependent on an existing frequent itemset mining technique like *Apriori*, which it applies on memory-resident chunks and finally returns the results by combining the outputs. A set of itemsets found to be frequent for one partition may not be necessarily frequent for another partition. After receiving the frequent itemset information for the individual partitions, the merge routine plays an important role in finalizing the frequent itemsets for the whole dataset. This approach attempts to restrict the number of whole database scans to two. However, may generate too many candidates. Most partitioning methods take significantly high computation time due to the generation of a large number of frequent itemsets in the final stage as a consequence of merging. Both horizontal and vertical partitioning methods [191] are able to handle a large number of instances with high dimensionality.

Dynamic Itemset Counting (DIC) : Dynamic itemset counting [70] is another effective variant of Apriori. An advantage of DIC is that it requires fewer database scans on average. In addition, it generates the candidate itemsets before a complete dataset scan is over. DIC defines a set of *Stop Points* in the dataset, and at every *Stop Point* it checks whether a candidate itemset has been found to be frequent or not. If found to be frequent, it generates more candidate itemsets based on the newly found frequent itemset as well as previously found to be frequent. DIC counts support all the candidate sets from the *Stop Point* onward for the whole dataset. In the worst case, DIC requires the same number of dataset scans as *Apriori*.

Direct Hashing and Pruning (DHP): Direct hashing and Pruning (DHP) [309] is a cost-effective variant of Apriori. It addresses the issue of generating too many candidates using an effective hashing technique. It eliminates candidate itemsets which do not contribute to the generation of the next level of candidate itemsets. When counting the support of a candidate k-itemset, it accumulates information about candidate $(k + 1)$-itemsets in such a way that after pruning, all possible $(k + 1)$-itemsets of each transaction are hashed to a hash table. Each bucket in the hash table contains the number of itemsets hashed to this bucket so far. From the resulting hash table, a bit vector is obtained, where the value of a bit is set to one when the number in the corresponding entry of the hash table is $\geq$ *Minsupp*. *DHP* uses this bit vector to reduce the number of possible candidate itemsets. Although *DHP* cannot reduce the number of dataset scans compared to *Apriori*, it reduces the number of candidate itemsets. The use of the hash table makes *DHP* efficient, especially in counting the support values for the itemsets.

Pincer Search: Apriori follows a bottom-up approach that starts by finding the frequent itemsets of cardinality 1, continuing towards maximal frequent itemset finding. With increase in the cardinality of the frequent itemsets, the number of dataset scans also increases, consuming high computational resources. The Pincer search algorithm [248] addresses this issue by following both bottom-up as well as top-down approaches to make frequent itemset mining cost-effective. It identifies all single item frequent itemsets in the first pass. Next, along with the 2-item candidates, it also considers the itemset containing all the frequent items as a candidate itemset. Following the *downward closure property*, it attempts to establish that a maximal frequent itemset cannot contain any infrequent subset. Pincer Search considers both directions to generate candidate itemsets and in every subsequent pass, candidates from the top-down process are generated by removing one item at a time, and also pruned if any subset of them is known to be infrequent. This process continues untill at least one candidate is found in this side. The search process from the bottom-up side proceeds slowly, one level per pass. However, the top-down process proceeds fast to reach the maximal frequent itemset. If the size of a maximal frequent itemset is large, then the algorithm gives the results in less time due to the fewer number of dataset scans. However, with a smaller sized maximal frequent itemset, Pincer Search requires almost the same number of dataset scans as Apriori. In addition, it requires significantly more time to check the candidates using the divisive approach.

A large number of improvements to Apriori [300], [299], [455] have been introduced. These extensions are based on sampling, hashing, partitioning, incremental mining, integrative mining with relational databases, and parallel and distributed mining. Although a number of Apriori variants have been proposed, the issues of generating too many candidate itemsets and multiple dataset scans still remain the major performance bottlenecks.

Frequent Itemset Mining without Candidate Generation

Although a large number of solutions have been put forward to address two issues with Apriori, i.e. (i) too many candidate itemset generation and (ii) multiple dataset scans, the problem remains unsolved. In this subsection, we introduce a category of frequent itemset mining techniques that attempt to mine all frequent itemsets without candidate generation, and that too with a minimum number of dataset scans.

FP-Tree Growth [191][163] is a robust and cost-effective frequent itemset mining technique. Two major attractions of this technique are: (i) it requires a single scan of the original dataset to generate all frequencies of occurrence of the itemsets, and (ii) it does not generate candidate itemsets. It sorts the frequently occurring itemsets in a descending order based on their frequency values, and then filters out the infrequent itemsets. FP-Tree Growth constructs a *Frequent Pattern (FP)* tree, which not only preserves the frequencies of occurrence, but also the associations among the frequent itemsets. It then constructs a conditional pattern base (including the set of prefix paths in the FP-tree co-occurring with the suffix patterns), for each item with the lowest support. This is called the *conditional FP-tree*. FP-Tree Growth grows the patterns by concatenating the suffix patterns with the frequent patterns obtained from the conditional FP-tree.

Once the conditional FP-trees are constructed for each frequent itemset, searching becomes straightforward and significantly faster. Although FP-Tree Growth is an efficient algorithm, one major limitation is that if the tree becomes large due to high dimensionality and density (of 1s), it may not be possible to accommodate it in the memory.

There are several extensions to the FP-Tree Growth algorithm to handle the scalability issue. These include depth-first generation of frequent itemsets by Agrawal et al. [7], hyper-structure mining of frequent patterns by Pei et al. [322], top-down and bottom-up traversal in pattern growth mining by Liu et al. [252] and array-based implementation of prefix-tree structure by Grahne and Zhu [148].

Rule Generation

Rule generation creates a set of comprehensive and relevant rules based on the frequent itemsets to support decision making. It leverages the frequent itemsets along with their support counts to generate the association rules that satisfy the *confidence* threshold. The *confidence* of an association rule is the percentage that indicates how frequently the rule is found valid or true. Formally, if X and Y are two frequent itemsets and $X \cap Y = \phi$, then the confidence of an association rule $X \longrightarrow Y$ is defined as *Support*$(X \cup Y)/$*Support*(X). A rule $X \longrightarrow Y$ is considered true if its confidence, i.e., *Confidence*$(X \longrightarrow Y) \geq$ *minconf*, a user-defined threshold.

There are a few popular and effective rule generation techniques developed in the support-confidence framework.

Agrawal's Algorithm: This is a simple and straightforward rule generation technique introduced by Agrawal et al. [11], [12]. It generates rules of a simplified form with only a single item in the consequent part. For example, for a frequent itemset of size n, this algorithm considers a maximum of n candidate rules, although there may be $(2^n - 2)$ possible rules. In other words, this algorithm misses a large portion of candidate association rules from consideration, and it is a serious limitation. Srikant's Simple Algorithm [382] (discussed next), addresses this issue to a great extent.

Srikant's Simple Algorithm: Srikant's Simple Algorithm [382] is a significant contribution to association mining. It overcomes the limitations of Agrawal's algorithm of having single items in the consequent. This algorithm generates all possible rules of any dimensionality in the consequent. It initially generates all possible non-empty subsets for a given frequent itemset, P. Next, it generates a rule of the form $p_i \longrightarrow (P - p_i)$ for any subset p_i of P, if the rule satisfies the minimum confidence threshold, i.e., $support(p)/support(p_i) \geq$ *minconf*. It uses a hashing technique to maintain the frequent itemsets, allowing for faster search of the support counts of the subset of items. However, its rule generation is not cost effective due to waste of time on too many redundant checkings.

Consider an example to demonstrate how Srikant's Simple Algorithm wastes time. Let $pqrs$ be a frequent itemset used for rule generation. For the subset pqr, then for pq and then for p, Srikant's Simple Algorithm considers $pqr \longrightarrow s$, $pq \longrightarrow rs$, and $p \longrightarrow qrs$ as possible rules. However, if $pqr \longrightarrow s$ has low confidence ($<$ *minconf*), then $pq \longrightarrow rs$ cannot have confidence $\geq$ *minconf*. A similar situation occurs for $p \longrightarrow qrs$ also. However, this algorithm considers these rules when checking confidence, and hence wastes a significant amount of time. To overcome this issue, Srikant's Faster Algorithm was proposed.

[C] Srikant's Faster Algorithm: Srikant's Faster Algorithm [382] successfully uses the downward closure property to avoid unnecessary checking of candidate rules to make the rule generation process significantly faster. According to this property, if a rule with consequent r is found to be frequent, then the rules with consequents that are subsets of r are also frequent. Let us take an example. If $\bar{p} \subset p$, then the support of $\bar{p}$ is greater than p. Let $prst \longrightarrow q$ and $pqrt \longrightarrow s$ be two rules with single items in the consequent with minimum confidence, derived from the itemset $pqrst$. The previous algorithm tests whether the rules with two items in the consequent have minimum confidence, or whether the rules $prs \longrightarrow qt$, $pst \longrightarrow qr$, $rst \longrightarrow pq$, and $prt \longrightarrow qs$ hold or not. The first rule does not satisfy because $t \subset qt$ and the rule $pqrs \longrightarrow t$ also does not hold. For the same reason, the second and third rules also do not hold. Unlike others, Srikant's Faster Algorithm saves computation time significantly. It finds that the only 2-item rule that can possibly hold is $prt \longrightarrow qs$, where q and s are the consequents in the valid 1-item consequent rules. Srikant's Faster Algorithm tests only this rule.

Although this Algorithm overcomes issues with the previous two rule generation algorithms, its candidate consequent generation for rule discovery requires considerabe memory. Further, it wastes a significantly high amount of time to generate the same consequent several times for different antecedents. To overcome this issue, another effective rule generation technique called NBG was introduced.

[D] Faster Rule Generation Algorithm, NBG: NBG [300] can generate all possible rules with reference to the user-defined confidence threshold, *minconf*. It operates in a cost-effective manner to generate rules by avoiding unnecessary checking of candidate rules. It leverages the concept given in Srikant's Faster Algorithm to generate rules reducing a significant amount of time. Like Srikant's Faster Algorithm, NBG is capable of generating all rules. Since, it uses frequent itmesets stored in the memory and does not generate subsets of a given frequent itemset, it is cost effective (memory requirement is far less) in comparison to the previous algorithms. NBG can also generate all possible rules with a fixed antecedent but consequents with different dimensionalities from any given set of frequent itemsets. To generate a rule, it checks only those frequent itemsets that satisfy the minimum confidence criterion. NBG initially discovers all rules with a fixed antecedent, starting with a single item antecedent, and then it goes to the next antecedent. For the same antecedent this algorithm finds the rules with equal size consequents, starting with a 1-item consequent. Then it goes to the next level. NBG has been established to be complete.

Rare Itemset Mining

Association mining generates rules using the support-confidence framework. However, all rules derived with high confidence based on frequent itemsets may not necessarily be interesting. Some rules with high confidence, but generated based on infrequent itemsets, may also be interesting. Rare association mining attempts to identify such rules. A rare association rule is a representation of an unexpected, yet interesting association based on some infrequent itemset, but has high confidence [180]. Such rules with low support but high confidence have many real-life applications such as rare disease diagnosis, unknown attack identification in a network, and unseen object identification in a satellite image. In predicting unknown or previously unseen behavior, rare association rules can play an important role. Formally, a rare association rule can be defined as an association rule X generated using the support-confidence framework where $support(X) < minsupp$ but $conf(X) \geq minconf$, where $minsupp$ and $minconf$ are two user-defined thresholds.

An important issue related to rare association mining is that it is usually not possible to generate all rare rules with a single $minsup$ value. If we set the $minsup$ value high, we may miss frequent itemsets involving rare itemsets because rare items fail to satisfy a high $minsup$. To support the identification of both frequent and rare itemsets, we have to set $minsup$ low. However, doing so may lead to combinatorial explosion [250] and produce too many frequent itemsets. To overcome this issue, many researchers [217] suggest the use of multiple $minsup$ values, although this may drop some rare rules. Several rare itemset mining techniques have been developed. Some are discussed next.

Apriori-Rare [390] is an Apriori-based algorithm, which generates frequent as well as rare itemsets using the support-confidence framework. It divides the task of rare association mining into two sub-tasks, (i) frequent itemset traversal and (ii) testing of rare itemsets. There are two algorithms (one naive and the other optimized) for sub-task (i), and a separate algorithm for sub-task (ii). It hypothesizes that all minimal rare itemsets are generators. A major advantage of *Apriori-Rare* is that it can restore all minimal rare itemsets. However, it fails to identify all rare itemsets.

Another significant addition to rare association mining research is *Apriori-Inverse* [446]. It attempts to extract all sporadic association rules that fail to satisfy maximum support but satisfy the minimum confidence threshold. It can fast identify such sporadic itemsets. However, it is incapable of finding all rare itemsets.

MSapriori (Multiple Support Apriori) [217] is another variant of Apriori. It obtains rare itemsets by redefining minimum support. In this technique, each item is associated with a minimum item support (MIS) specified by the user. The user can specify different support requirements for different rules. *MSapriori* has several limitations. The generation of rare itemsets depends on the value of a user-defined threshold β rather than the support count. Although *MSapriori* can find rare itemsets, but high dependency on β limits the flexibility of the algorithm.

Another significant approach is the *RSAA* (Relative Support Apriori Algorithm) algorithm [454]. When generating rules, this algorithm increases the support threshold for less frequently occurring items and decreases the support threshold for frequently occurring items. RSAA spends a significant amount of time in checking for rules which are not rare. If the minimum permissible relative support count approaches zero, this algorithm behaves like Apriori in generating low support-count rules. *RSAA* can find all rules by spending a significant amount of time, and is highly sensitive to the thresholds, i.e., *minsup* and *maxsup*.

Another a priori-based rare itemset mining algorithm is *ARIMA* (Another Rare Itemset Miner Algorithm) [389]. This algorithm is dependent on *Apriori-Rare* in generating the rare itemsets based on the already identified minimal rare itemsets (MRIs). It uses the concept of a zero generator to reduce the search space. The dependency on two user-defined thresholds, *minsup* and *maxsup*, limits the flexibility of *ARIMA*.

FRIMA (Frequent Rare Itemset Mining Algorithm) [181] is another addition to rare itemset mining research. FRIMA is a cost-effective variant of Apriori. Like Apriori, this algorithm uses a bottom-up approach, and generates both frequent as well as rare itemsets (zero, frequent and rare itemsets) by leveraging the downward closure property. FRIMA performs correctly for datasets with high dimensionality. However, the major limitation of Apriori, i.e., generation of too many candidate itemsets is still a problem with this algorithm.

Incremental Association Mining

Association mining techniques discussed so far handle static market basket datasets only. Such market basket data keep on growing over time, and hence, there is need for efficient association mining techniques to extract association rules from historical as well as recent data cost effectively. Incremental association mining techniques have been developed to address this issue. An incremental technique uses the already computed information to deliver the final outcome for the whole dataset i.e., it performs union of both old and newly arrived data, rather than initiating the process from the scratch every time. An incremental algorithm operates on the newly introduced incremented data and uses the already derived frequent itemsets to minimize the overall effort. Several effective algorithms for incremental association rule mining have been introduced. Some of them are discussed next.

Fast Update (*FUP*): This algorithm [166] initially focuses on the newly arrived data to detect (i) the looser itemsets, i.e., those itemsets which become infrequent after considering newly arrived data, and (ii) winner itemsets, i.e. earlier infrequent itemsets that have become frequent after considering newly arrived data. It is straightforward to handle case (i) as the frequent itemset information for old data is already available. But case (ii) is not that simple. *FUP* attempts to generate a smaller candidate itemset (based on the previously discovered rules) to be verified, taking into account the newly arrived data. *FUP* requires multiple scanning of the whole dataset. It scans the dataset k times for the largest maximal frequent set of size k. However, this algorithm is cost-effective, compared to using a straightforward approach of re-running the itemset generation algorithm for the whole dataset from scratch. FUP can handle only insertion cases. To make *FUP* more generic, an enhanced version of *FUP* was introduced, which is discussed next.

FUP2 : *FUP*$_2$ [84] handles dynamic datasets, allowing insertions and deletions simultaneously. It generates rules from the final dataset considering both the deleted instances and the newly arrived instances. *FUP*$_2$ considers any modification of existing instances as a sequence of deletions followed by insertions. However, if the dataset is updated by only insertions, the algorithm behaves almost like *FUP*. The performance of *FUP*$_2$ is significantly better than rerunning the traditional algorithms from scratch. However, *FUP*$_2$ does not perform well if the difference between the original and newly arrived transactions is high.

IFP-Growth [440]: The Incremental Frequent Pattern Tree Growth algorithm is an extension of the FP-tree Growth algorithm, which derives frequent itemsets from a static dataset. It introduces IFP-trees to handle dynamic datasets. An IFP-tree is constructed to handle both insertions and deletions of instances from the dataset. In case of insertion, it constructs the IFP-tree based on the existing FP-tree. For each instance of newly arrived data, it updates the support count of the relevant nodes of the tree (as in the construction of FP-tree). However, if a new node results, IFP-growth keeps it as a new branch of the root node.

Once the tree is constructed, frequent itemsets are generated by comparing the old and the new trees. When a new node is identified in the updated tree, it generates frequent itemsets. Because a new node forms a separate branch in the tree, and this algorithm considers only the new branches to generate new frequent itemsets. IFP-growth avoids unproductive candidate generation while forming FP-trees and IFP-trees. Like FP-Growth, IFP-Growth may also suffer from performance degradation with increased dimensionality and the number of transactions.

[D] **Borders** [29]: Like the previously discussed algorithms, Borders also exploits the already computed frequent itemsets for old dataset during generation of frequent itemsets for newly arrived data. It uses a new set of items called the *Border set* to support extraction of frequent itemsets in a cost-effective manner for any dynamic market basket datasets. A border set is an infrequent itemset, of which all the non-empty proper subsets are frequent. With any insertion of instances, some border sets may become frequent, when they are called *Promoted Border Set*. The Borders algorithm maintains all the border sets identified for the old dataset along with the generated frequent itemsets. It generates a new set of candidate itemsets for each promoted border set and checks them for frequent sets by scanning the dataset. It then goes to the next level of candidate generation.

Some additional dynamic association rule mining algorithms are UWEP [33], ICAP [31] and MAAP [121]. UWEP generates frequent itemsets by exploiting the anti-monotone property of support. It uses the property to prune unnecessary candidates at an early stage, improving performance significatly. UWEP is known as the most economic algorithm due to its ability to work with the small number of possible candidate itemsets. However, to generate and maintain TID (Transaction ID) lists, there is significant overhead. ICAP solves this problem with only a single scan. With the newly arrived data, if there is possibility of expansion of the constrained negative borders, ICAP gives excellent performance. Another approach that solves this problem with minimum dataset scans is MAAP. This algorithm initially finds those itemsets that are frequent in the old dataset and are still frequent in the newly arrived data. It checks for possible new frequent itemsets, and if found, generates new candidates. It exploits the downward closure property of frequent itemsets to accomplish the task in a cost-effective manner. It finds new frequent itemsets and generates new candidate itemsets. MAAP can handle both update (insertions and deletions) cases successfully; however, the number of dataset passes may not be optimal. *Modified Borders* [95] is another attempt to handle this issue effectively.

4.4 Chapter Summary

Before we end, we summarize the many topics that were discussed in detail in this chapter. This summary will help refresh the main points as well as allow the readers to go back and re-read portions of the chapter when needed.

4.4.1 Statistical Modeling

- The purpose of a statistical model is not only to represent relationships among variables (say, genes), but also to help disover variabilities and uncertainties in data to gain insights.

- A statistical distribution tells us how the probabilities of measurements are distributed. Five commonly used distributions are Bernoulli, Binomial, Poisson, uniform, and normal.

- Inferential statistical analysis infers properties of a population with a probability distribution by estimating and hypothesis testing.

- A parametric statistical analysis makes assumptions about a dataset, namely, that the data are drawn from a population with a specific distribution. A non-parametric statistical analysis method does not make such assumptions.

4.4.2 Supervised Learning: Classification and Regression Analysis

- A supervised learning method has two steps: *learning* and *testing*. During learning, the method uses example instances tagged with *class labels* as prior knowledge to build a prediction model. In testing, the trained model assigns class labels to input test instances.

- Supervised learning is of two types: classification and regression. In classification, the outcome (class label) is discrete or categorical, whereas in regression, it is numeric or continuous.

- The performance of a classification or regression method is dependent on (i) the availability of an adequate number of training instances, (ii) the generality of the prediction model, and (iii) the appropriateness of decision thresholds, if any.

- For gene expression analysis, supervised learning is an effective tool for (i) gene selection, (ii) feature or sample selection, and (iii) patient data classification.

4.4.3 Proximity Measures

- Proximity measures are central to clustering or classification.

- A large number of proximity measures have been introduced for numeric, categorical and mixed-type data. A discussion on such measures is included in this chapter. A cost-effective, yet highly expressive proximity measure for gene expression data analysis is still elusive.

- In gene expression data analysis, it is not helpful to simply compute the proximity between a pair of objects or genes, rather it is necessary to capture trends present in the gene expressions. Therefore, it is essential to use an effective measure that can handle all possible correlations that exist between a pair of gene expressions.

- A proximity measure is expected to be (i) domain independent, (ii) noise insensitive, (iii) able to handle high-dimensional data, (iv) expressive enough to handle all possible relationships, and (v) also a metric that obeys non-negativity, symmetricity and transitivity properties.

4.4.4 Unsupervised Learning: Clustering

- A clustering technique groups objects, instances, or genes into clusters so that between a pair of members of a cluster, the *intra-member similarity* is *high*, but the *inter-member similarity* between a pair of objects from two different clusters is *low*.

- Clustering is an unsupervised learning technique since it does not require prior knowledge to group objects.

- A clustering technique is expected to be (i) scalable, (ii) able to handle any type of data, (iii) able to detect all shapes of clusters, (iv) able to handle noise and border objects, (v) insensitive to order of input data, (vi) dependent on a small number of input parameters, and (vii) able to generate results that are interpretable and useful.

- Clustering approaches include partitioning, hierarchical, density based, model based, grid based, and graph based. Additional categories include projected, subspace and ensemble approaches.

- In gene expression analysis, all these approaches have been applied. Each approach has its own advantages and limitations, considering the extraction of biologically significant clusters.

4.4.5   Unsupervised Learning: Biclustering

- Biologically significant groups of genes exhibit high correlation over subsets of genes and subsets of samples. Most traditional clustering approaches operating over the full sample space fail to extract interesting patterns from content-rich gene expression data.

- Biclustering or two-way clustering addresses this issue. It extracts biologically significant groups of genes called *biclusters* by operating on the rows (genes) and columns (samples) simultaneously. The subsets of genes and samples in a bicluster (i) need not necessarily be contiguously placed in a gene matrix, but (ii) they should exhibit high coherence.

- Bicluster analysis requires a robust correlation measure that can handle the four types of correlations, absolute, shifting, scaling and shifting-and-scaling.

- Mean squared residue is an effective measure introduced by Cheng and Church [82] to help the extraction of quality biclusters with high coherence.

4.4.6   Unsupervised Learning: Triclustering

- Triclustering was introduced to handle Gene-Sample-Time (GST) data to extract biologically significant triclusters with high inter-temporal coherence.

- Due to the lack of available GST datasets, research on tricluster analysis is in its infancy.

- Tricluster analysis follows three approaches: *gene-based*, *sample-based* and *subspace*. The subspace approach has been found to be most effective.

- A research issue related to triclustering is the consumption of significantly high computational resources. Although efforts have been made to optimize using distributed and parallel computing, it remains a research challenge.

4.4.7 Outlier Mining

- Outliers are instances that do not follow normal rules or satisfy expectations imposed on the majority of the data instances. Outliers may exist in isolation or in a group. Outliers impact findings in almost all domains, including gene expression mining.

- A distance-based outlier is an object that is located far away from a majority of the objects. A parametric or non-parametric distance-based outlier detection method identifies an outlying object based on distances calculated from the object to a majority of the remaining objects, considering fulfillment of a distance or conditional threshold criterion.

- A density-based outlier is an object or instance situated at a region of lower density. A density-based outlier detection method is a parametric approach to identify regions with low densities. It recognizes an object or a group of objects as outlier, if located in a region with lower density neighborhoods.

- To discover outlier(s) in a dataset with uncertainties or inconsistencies, soft computing approaches are also used.

4.4.8 Unsupervised Learning: Association Mining

- Association mining generates IF-THEN rules based on a support-confidence framework from market basket transaction data (0-1 or binary matrix).

- Rule generation is a two-step process. In step 1, it generates frequent itemsets, based on frequencies of occurrence individually as well as in association with other items. Step 2 uses the frequent itemset information to generate a set of rules in the *antecedent-consequent* or IF-THEN form.

- Association mining research has mostly focused on step 1 or development of algorithms for cost-effective frequent itemset generation. Once the frequent itemsets have been generated, rule generation becomes trivial.

- Two issues related to frequent itemset mining are: (i) the generation of too many candidate itemsets and (ii) multiple scans of the dataset.

- In gene expression analysis geared toward identify diagnostic and prognostic biomarkers for a disease, association mining can play an important role.

Chapter 5

Co-Expression Analysis

5.1 Introduction

Gene expression data are generated by high-throughput microarray and sequencing technologies and they are often presented as matrices of expression levels of genes for different conditions or samples. A major objective of gene expression data analysis is to identify groups of genes that have similar expression patterns over the full sample space or over subsets of samples. This may reveal natural structures and identify interesting patterns in the underlying data. A group (cluster or module) of genes can be defined as a set of biologically relevant genes, which are similar based on a proximity measure. The goal of clustering or network module extraction is to partition the genes into coherent groups, so that two criteria are satisfied: (i) *homogeneity*, i.e., the elements in the same cluster are highly similar to each other and (ii) *separability*, i.e., the elements from different clusters have low similarity to each other.

One can analyze gene expression data using three machine learning approaches: supervised, unsupervised, and a hybrid of these two. In supervised methods, the gene expression dataset is partitioned into disjoint classes using a class attribute. A classifier model is built based on training data that includes data samples and the corresponding correct class labels. The trained model is used to predict the class label of an unknown test sample. The goal of classification is to analyze the training set and to develop an accurate description or model for each class using the attributes present in the data. Many classification models have been developed, including decision trees, support vector machines, neural networks, evolutionary and other soft computing models.

DOI: 10.1201/9780429322655-5   145

In unsupervised methods, the dataset (or its network representation) is partitioned into disjoint groups or modules with high intra-group similarity and low inter-group similarity. In unsupervised analysis, the proximity (similarity or dissimilarity) measure plays an important role in deciding the quality of the clusters or modules. Among other things, cluster quality is dependent on the appropriateness of the proximity measure used for the expression data. A number of approaches have been developed to support gene co-expression analysis. Among clustering methods, (i) *1-way* clustering, (ii) *2-way* clustering, and (iii) *3-way* clustering are common. In turn, each of these clustering approaches can be typed as partitioning, hierarchical, density-based, grid-based, model based, and hybrid. co-expression network analysis of gene expression data can be performed by constructing various network types such as (a) unsigned, (b) signed, (c) weighted, and (d) unweighted co-expression networks. In comparison to supervised analysis, gene expression data analysis requires sophisticated unsupervised analysis method(s) that can help extract interesting gene(s), characteristics or behaviors useful for diagnosis of critical diseases.

In unsupervised learning, gene expression data is usually analyzed to find co-expressed groups of genes associated with the same biological functions. As discussed in the previous chapter, clustering is a data mining technique that groups genes based on their similarity [191]. Various traditional 1-way clustering techniques such as k-means [191] and CLINK [63] have been used to identify groups of co-expressed genes, which share similar functions. Some specialized clustering techniques such as CLICK [367] and DHC [196] have also been proposed for gene expression data analysis. But genes are often not found to be highly similar when the whole set of samples is considered because a biological phenomenon may take place over only a subset of samples in gene expression data [82]. This observation led to emergence of a variant of traditional 1-way clustering that attempts to find groups of clusters that maximize similarity across subsets of genes corresponding to subsets of samples. This variant of clustering is called 2-way or biclustering [82]. The emergence of three-dimensional gene-sample-time (GST) gene expression data presented a challenge, resulting in the development of another variant of clustering called 3-way or triclustering [269].

In addition to clustering, gene co-expression network (CEN) analysis is popular among computational biologists. In this approach, a grouping of genes is discovered with the construction of a network called co-expression network (CEN) [387], [125]. The CEN is further analyzed to extract modules or components where each module includes a group of highly connected genes in the network. A gene co-expression network extracts associations among individual genes in terms of expression similarity, providing a network-level view of the similarity among genes. A gene co-expression network represents a good approximation to the complicated web of functional associations among genes. In CENs, two genes are connected by an undirected edge if their activities have significant association computed using measurements such as Pearson correlation, Spearman correlation, or mutual information. A co-expression network may be weighted or unweighted or signed or unsigned [87]. One of the primary objectives in designing a co-expression network is to extract highly connected regions from the network. These regions are often called network modules [87], [270]. A network module corresponds to a biologically significant group of genes, which may be associated with similar biological functions or may be controlled by the same regulator.

5.2 Gene Co-Expression Analysis

Gene function prediction and disease diagnosis are two important problems in gene expression data analysis that can be addressed by clustering. In gene function prediction, one is interested in the clusters that have high overlap with known biologically curated groups of genes. Genes which are included in such clusters, but do not belong to the biological groups with which the clusters have high overlap, are genes that have not yet been found to be associated with biological phenomena corresponding to the biological groups. Such predictions can be further confirmed by wet-lab experiments by biologists. In disease diagnosis, genes associated with a disease can also be discovered in a similar way. Then, the disease can be diagnosed based on expression patterns of these genes, in addition to the already known genes. Thus, clustering in gene expression data analysis can be formulated as follows.

Problem formulation: *Given a gene expression dataset D with expression values of a set of genes $G = g_1; g_2; \cdots; g_m$ for a given set of samples $F = f_1; f_2; \cdots; f_n$, a co-expression analysis technique extracts subsets of genes based on their expression similarity over a set or subsets of samples such that the overlap between the extracted groups and biological benchmark groups corresponding to biological functions is high.*

5.2.1   Types of Gene Co-Expression

Gene co-expression means that a pair of genes exhibit similar expression patterns, given a group of samples. When two genes are functionally associated, they generally show a similar expression pattern, i.e., their expressions are regulated by each other, considering a group of samples through up- or down-regulation, resulting in some biological functions. Figure 5.1 shows an example of four co-expressed patterns, g_i, g_j, g_k, and g_l. Co-expression analysis can identify such functionally enriched co-expressed genes. Its applications include associating functions to an unknown gene through guilt by association analysis, prioritizing candidate disease genes or gene groups and identifying disease sub-types based on the functions of genes associated with a co-expressed group of genes.

Patterns of Gene Co-Expression

The computational biologists can associate biological significance between a pair of genes with absolute, shifting and scaling correlations. A pair of genes is said to exhibit *absolute correlation* if their corresponding expression values are very close or very similar. Expression values of a gene can be obtained from corresponding expression values of the other gene by adding a constant if the pair of genes exhibits *shifting correlation*. If a pair of genes exhibits *scaling* correlation, expression values of one gene can be obtained by multiplying the corresponding expression values of the other gene by a constant. A new type of correlation, called *shifting-and-scaling* correlation was introduced in [15] because the mean squared residue proposed by Cheng and Church [82] was not precise enough to discover shifting and scaling patterns at the same time. A pair of genes is said to exhibit *shifting-and-scaling* correlation if expression values of a gene can be obtained by a series of multiplication and addition operations using constants. Absolute, shifting and scaling correlations are special cases of shifting-and-scaling correlation. These three basic types of correlations are described in detail in previous chapter. Figure 5.2 shows an example of a shifting pattern. Figure 5.3 shows an example of a scaling pattern.

5.2.2   An Example

We create a synthetic dataset consisting of expressions of genes $G1$ and $G2$ considering a group of samples, shown in Table 5.1. To show shift, scale and shift-and-scale patterns between a pair of genes, we use only 5 samples and for the remaining cases we use 10 samples. In Figures 5.5 and 5.6, we

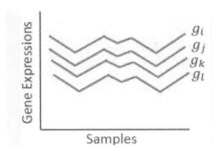

Figure 5.1: An example of co-expressed genes

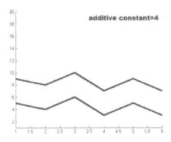

Figure 5.2: An example of shifting pattern

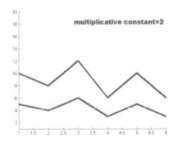

Figure 5.3: An example of scaling pattern

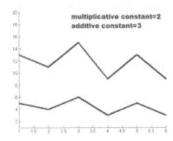

Figure 5.4: An example of shifting and scaling pattern

Table 5.1: Six cases of associations between a pair of genes.

Similarity types	Gene Names	S1	S2	S3	S4	S5	S6	S7	S8	S9	S10
Shift	G1	3	2	4	1	5					
	G2	7	6	8	5	9					
Scale	G1	3	2	4	1	5					
	G2	6	4	8	2	10					
Shift-and-scale	G1	3	2	4	1	5					
	G2	9	7	11	5	13					
Mixure of shift and scale	G1	3	2	4	1	5	3	2	4	1	5
	G2	7	6	8	5	9	6	4	8	2	10
Mixure of shift and Shift-and-scale	G1	3	2	4	1	5	3	2	4	1	5
	G2	7	6	8	5	9	9	7	11	5	13
Mixure of scale and Shift-and-scale	G1	3	2	4	1	5	3	2	4	1	5
	G2	6	4	8	2	10	9	7	11	5	13

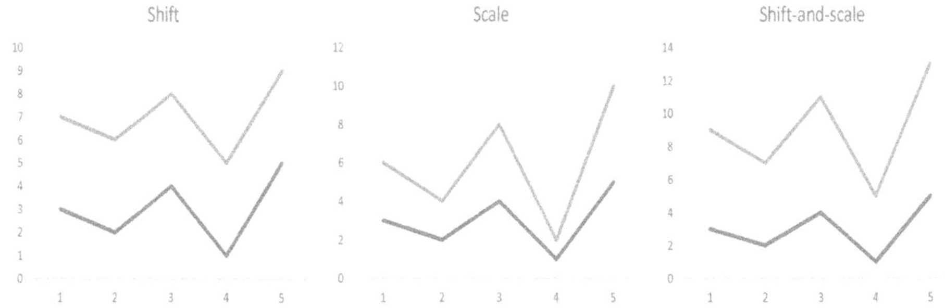

Figure 5.5: A pair of genes showing association in terms of shift, scale and shift-and-scale pattern.

use additive constant 4 for shift, multiplicative constant 2 for scale, and additive constant 3 and multiplicative constant 2 to realize different types of associations in a gene pair. The expression of a gene can be perfectly computed from another gene by modeling a linear regression $G2 = mG1 + c$ between a given gene pair. For the shift case, the expression of gene $G1$ can be computed using the expression $G2 = G1 + 4$. Similarly, for the scale case, the expression of gene $G2$ can be computed using the expression $G2 = 2 \times G1$. For the shift-and-scale case, the expression of gene $G2$ can be computed using the expression $G2 = 2 \times G1 + 3$. However, in the case of the mixture of shift, scale and shift-and-scale, it is not possible to perfectly compute gene expression of $G2$ from $G1$ with the linear regression. To do this, appropriate sample clustering followed by fitting the linear regression in the identified sample group can help compute intended gene expressions. For example, for

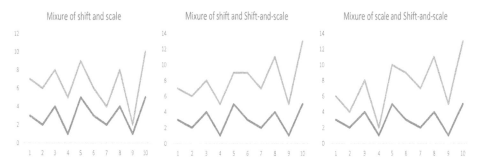

Figure 5.6: A pair of genes showing association in terms of a mixture of shift, scale and shift-and-scale patterns. First 5 samples for 1 case and remaining 5 samples for another case.

a mixture of shift and scale, for the initial 5 samples, the expression of $G2$ can be directly computed from $G1$ using the equation $G2 = G1 + 4$ and for the remaining 5 samples, it can be done using the equation $G2 = 2 \times G1$. Appropriate sample grouping is the main challenge here.

5.3 Measures to Identify Co-Expressed Patterns

A good proximity measure is one which can handle all the above types of correlations between a pair of genes and can help extract groups of genes of high biological significance. A measure which can handle not only shifting and scaling patterns in isolation, but also shifting-and-scaling correlation patterns is more appropriate in handling co-expression finding. An appropriate shifting-and-scaling correlation measure can help identify quality clusters based on functional similarity and co-expression analysis. Most correlation measures which are used in gene expression data analysis can perform well in detecting shifting and scaling correlations in isolation only. These measures fail to detect mixed correlation correctly, and as a consequence, some interesting co-expressed patterns with high biological significance are missed. Although there are some techniques which claim to be able to extract shifting-and-scaling cluster patterns, most do not consider gene expressions pair-wise. These have already been discussed at length in the previous section. Only a summary is repeated here. One such example measure is a *residue score-based* [16] measure, which attempts to handle mixed correlation patterns by operating both row-wise and column-wise, similar to mean square residue [82]. This measure estimates the amount of shifting-and-scaling correlation among the genes over the samples as an aggregate score (not normalized). There is another measure called *transposed virtual error* [327], which can also handle shifting-and-scaling correlation, but one cannot use it for estimation of pair-wise gene-gene correlation. Although several measures have been introduced to handle mixed correlation type, these are inadequate because (i) the measures are column-biased, and (ii) like Pearson Correlation, sensitive to noise due to use of the global mean. To address these issues and to handle shifting-and-scaling correlation, a mutual similarity measure, called SSSim (shifting and Scaling Similarity) was introduced. Unlike its close competitors, including Pearson Correlation, SSSim is least affected [16] by the presence of noise. This is due to the involvement of the local mean during computation of gene-gene correlation instead of the global mean. The score for any pair of gene expression ranges is always from 0 to 1, where 0 indicates that the genes are totally different, whereas 1 indicates that they exhibit perfectly similar shifting and scaling correlation.

5.4   Co-Expression Analysis Using Clustering

Gene expression analysis using clustering finds groups of co-expressed genes with high coherence or tight correlation that correspond to biological processes and/or functions. The groups of genes identified as co-expressed may be over the full space or sample spaces. The genes identified as belonging to a group may be correlated with shifted, scaled or shift-and-scaled associations. We define the problem of co-expression analysis using cluster analysis as follows.

Definition 5.4.1 *CEA: Given a gene expression dataset D consisting of n genes and m samples, co-expression analysis using cluster analysis is an approach to generate groups or modules of genes $M_1, M_2, \cdots, M_k$, where a module M_i is a group of tightly correlated genes $G_1, G_2, \cdots, G_p$ which are highly associated with biological processes $P_1, P_2, \cdots, P_i$ and/or functions $F_1, F_2, \cdots, F_j$.*

co-expression analysis is a multi-step process. It can be applied to both types of gene expression data, microarray and RNAseq. It is often necessary to apply appropriate pre-processing technique(s) specific to each type of data. A good number of techniques have been developed to support cluster analysis of gene expression data. Once an appropriate set of clusters are identified, depending on the biological question in mind, one can find a ranked list of clusters based on relevance for the subsequent downstream analysis. Cluster-specific co-expression analysis can be performed from three perspectives; topological, statistical and biological. We present a generic architecture for co-expression analysis for both microarray and RNAseq data.

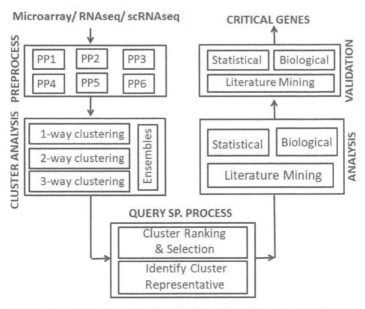

PP1: Discretize; PP2: Missing Value Estimate; PP3: Normalize; PP4: Remove Batch Effect; PP5: Remove Low Read count; PP6: Remove Outlier Gene;

Figure 5.7: Clustering in co-expression analysis: A generic architecture

5.4.1   CEA Using Clustering: A Generic Architecture

Figure   5.7 shows a generic architecture for clustering-based gene co-expression analysis. It includes five major steps, (i) preprocessing, (ii) clustering, (iii) query-specific cluster ranking and selection, (iv) cluster-based analysis, and (v) validation.

Preprocessing

Preprocessing activities transform the input raw gene expression data matrix into a content-rich, unbiased, and quality data matrix for subsequent downstream processing. It often includes multiple tasks depending on the condition, quality, and type of the input data.

A. Discretization

Discretization is a preprocessing task commonly used in microarray data analysis. Microarray data often include significant numbers of errors due to mostly inaccurate image acquisition, leading to inexact measurement of expression levels. Appropriate discretization can help improve the performance of a machine learning algorithm significantly. Discretization converts real values of gene expression data into a smaller subset of finite values.

When transforming, it maintains the variation (or trends) in the original data in the discretized dataset. Discretization reduces and simplifies the data, and absorbs a significant amount of the noise. Further, in data-driven analysis, learning from discrete data is simpler, cost-effective and efficient [73] [271].

Discretization methods used in co-expression analysis can be categorized into two groups: supervised and unsupervised. A supervised discretization method uses some form of prior knowledge while transforming the raw data into a discretized domain. It exploits domain-specific knowledge (class labels) and assigns a unique value to all gene expression objects that belong to the same class. On the other hand, an unsupervised method generally divides the range of values in a gene expression dataset into sub-ranges and assigns a unique value (discretized value) to all members in the same sub-range. Within these two categories, several discretization techniques have been developed for gene expression analysis. Figure 5.8 presents a broad categorization of discretization techniques. Unsupervised discretization techniques are further classified in two ways [271]: (i) using expression absolute values, and (ii) using expression variations between time points.

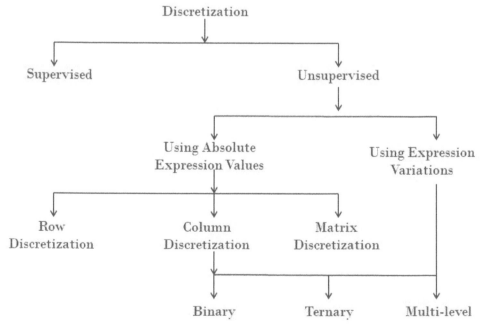

Figure 5.8: Discretization approaches

The definitions [271] given below help understand the categorization better.

Definition 5.4.2 *Binary Discretization: Binary discretization assigns only two discretized values, 0 and 1, based on a user-defined threshold. If the real expression value is less than the threshold value, it assigns 0, else 1.*

Definition 5.4.3 *Ternary Discretization: Ternary discretization uses three discretized values, say $0, 1, 2$ or $-1, 0, 1$. For example, if between a consecutive pair of samples (conditions or time points) an increasing trend is found, it may assign 1. If the trend is decreasing, it may assign -1, and in case of no change, it will assign 0.*

Definition 5.4.4 *Multilevel Discretization: Multilevel discretization uses multiple (≥ 4) discretized values when transforming real values into a discrete domain.*

Absolute discretization techniques discretize the continuous values directly using the absolute gene expression values. These techniques follow different approaches such as *row discretization, column discretization* and *matrix discretization*. A row discretization technique considers only the row values of the original data matrix during discretization, whereas column discretization uses the column values. In matrix discretization, the technique considers both row and column values simultaneously during discretization. Some popular discretization techniques [271], [73] of this category are: (a) Discretization using average value, (b) Mid-Ranged, (c) Max - X%Max, (d) Top X%, (e) Equal Width Discretization, (f) Equal Frequency Discretization, (g) k-means Discretization, (h) Column k-means Discretization, and (i) Bi-k-means.

An expression-variation-dependent discretization approach computes variations between two consecutive time points and assigns the discretized values based on the variation amounts. There are three categories of techniques: binary, ternary and multilevel. Binary discretization assigns two discrete values to the expression data, whereas ternary discretization generates three discretized values. Multilevel discretization uses more than three discrete values for closer prepresentation of the gene expression data. The higher the level of discretization, the lower is the chance of information loss, at the cost of computational complexity. This is because the fidelity of representation of the real-valued expression data increases with the increase in discretization levels. Some popular techniques of this category are: (a) Transitional State Discrimination (TSD) [267], (b) Erdal et al.'s method [117], (c) Kwon et al.'s method [267], and (d) the Ji and Tans method [195]. A detailed discussion of these techniques can be found in [73], [271].

The selection of an appropriate discretization technique for gene co-expression analysis is not straightforward. Since discretization transforms continuous real expression values into the discrete domain, there is information loss. If the transformation is not carried out properly, it will affect downstream processing [73]. To provide better representation, the discretization technique should understand and capture intrinsic characteristics of the biological data. The developer of the discretization technique should also consider the characteristics of the likely machine learning algorithms that will be applied on the discrete data during downstream analysis.

B. Normalization

co-expression analysis of microarray data helps understand the regulatory behavior of a gene as well as gene-gene interactions in a systematic manner [392]. Although continuous improvement in microarray technology has made it possible to study the behavior of thousands of genes simultaneously, due to inaccurate image acquisition and quantification, undesirable variations are often observed in the data. To eliminate such unwelcome variations and to facilitate unbiased co-expression analysis, normalization is essential. This preprocessing task removes possible sources of variations that influence gene expression quantification levels. A large number of normalization methods have been developed [392], [107], [193].

Yang et al. [392], [193] analyze a number of normalization methods and summarize them as (i) intra-slide, (ii) paired-slide for dye-swap experiments, and (iii) multiple slide normalization. Initially, one identifies the subset of genes to be normalized. There may be three possible approaches [193]: (i) the complete set of genes on the array, (ii) genes expressed constantly, and (iii) controls. In addition, one can also use the rank invariant genes, as suggested by Tseng et al. [411]. If $logG$ and $logR$ represent the *green* and *red* background-corrected intensities, respectively, M and A are defined by Yang et al. [193] as $M = log(R/G)$ and $A=1/2log(RG)$. There are three normalization approaches for the selected genes based on the (M, A) values: (i) global normalization (G) using the global median of log intensity ratios, (ii) intensity-dependent linear normalization (L), and (iii) intensity-dependent nonlinear normalization (N), using a LOWESS curve. In an ideal experimental scenario, although one can expect that for the selected genes $M = 0$, sometimes in practice, a different pattern may also be observed.

These methods of normalization consider the location shifts in logarithmically transformed intensities and attempt to control them. Since each grid is robotically printed by a different print-tip, one can use these methods individually for each microarray grid. Yang et al. [193] introduce another normalization method called *scale normalization*. This method controls the variability between two slides. It can also be applied separately to each print-tip. In addition, several extended versions of global and intensity-dependent normalizations have been proposed. One such method, introduced by Kepler et al. [215], estimates normalized intensities as well as intensity-dependent error variance based on local regression. In another effort, Wang et al. [429] present an iterative normalization approach for cDNA microarray data to estimate normalized coefficients and to identify control genes. A robust non-linear method for normalization using array signal distribution analysis and cubic splines has been proposed by Workman et al. [437]. Chen et al. [80] present a normalization method to adjust for location biases combined with global normalization for intensity biases. A non-linear LOWESS normalization in one-channel cDNA microarrays has been introduced by Edwards [114] to correct spatial heterogeneity. Several similar normalization methods have been proposed using statistical models [392]. Table 5.9 presents some of these methods in three categories namely, global, linear and non-linear.

Although it was initially thought that RNseq read-count data do not require normalization, later normalization has been found essential for read-count data also. To support effective downstream analysis of RNAseq data, factors such as transcript size, GC-content, sequencing depth, sequencing error rate, and insert size [388] are considered during normalization. It is beneficial to use multiple normalization methods and compare their performance using error models [388]. An important observation is that an approach like quantile normalization helps improve the quality of RNAseq data [165]. The NVT package [113] is able to identify the best normalization approach for an RNAseq dataset based on analysis and evaluation of multiple methods. Zyprych-Walczak et al. [464] introduce an approach to identify the optimal normalization for a specific dataset.

C. Missing Value Estimation

Missing values may occur in gene expression data due to various reasons. Two common reasons for occurrence of missing values in microarray data are the presence of (i) scratches, or (ii) dust in the microarray slides [1]. The presence of missing values often causes failure in generating inferences by co-expression analysis. Based on possible impacts on inference generation or prediction performance of a machine learning algorithm, researchers

Table 2: List of Normalization Methods

List of normalization methods including print-tip scale normalization

Method	Notation	Description
	O	Original data
Global	G	Global median normalization
	GP	Global median normalization on each print-tip
	GPS	Global median normalization on each print-tip with scale normalization
	G.s	Global median normalization and between-slide scale normalization
	GP.s	Global median normalization on each print-tip and between-slide scale normalization
	GPS.s	Global median normalization on print-tip with scale normalization and between-slide scale normalization
Linear	L	Intensity dependent linear regression normalization
	LP	Intensity dependent linear regression normalization on each print-tip
	LPS	Intensity dependent linear regression normalization on each print-tip with scale normalization
	L.s	Intensity dependent linear regression normalization and between-slide scale normalization
	LP.s	Intensity dependent linear regression normalization on each print-tip and between-slide scale normalization
	LPS.s	Intensity dependent linear regression normalization on each print-tip with scale normalization and between-slide scale normalization
Nonlinear	N	Intensity dependent nonlinear regression normalization (LOWESS)
	NP	Intensity dependent nonlinear regression normalization (LOWESS) on each print-tip
	NPS	Intensity dependent nonlinear regression normalization (LOWESS) on each print-tip with scale normalization
	N.s	Intensity dependent nonlinear regression normalization (LOWESS) and between-slide scale normalization
	NP.s	Intensity dependent nonlinear regression normalization (LOWESS) on each print-tip and between-slide scale normalization
	NPS.s	Intensity dependent nonlinear regression normalization (LOWESS) on each print-tip with scale normalization and between-slide scale normalization

Figure 5.9: Normalization methods

have introduced a theory to explain missing values. The theory considers incomplete data generation as a random process that can be understood in terms of two distinct processes [357]: (i) a complete data process, which generates a complete dataset and (ii) an incomplete or missing data process, which helps locate missing data elements. This theory assumes three categories of data [391]: (a) missing data values at random (MAR), (b) missing data values not at random (NMAR), and (c) missing data values completely at random (MCAR). If case (a), i.e., the MAR condition is true, one can eliminate instances with missing values, and inference generation should focus only on complete example data. On the other hand, if missing value occurrences are not random, i.e., condition (b) holds, we cannot ignore the missing values because it may lead to inappropriate or biased inferences. Treatment of missing values is a must to use any machine learning algorithm. Several imputation techniques have been developed to handle missing values. Most techniques estimate the missing values for a given object based on highly correlated neighboring objects. A popular imputing technique is k nearest-neighbor imputation (KNNImpute). Several other effective missing value estimation techniques [1] have been introduced for microarray and RNAseq data. Techniques useful for microarray data may not be effective for

RNAseq data. Another observation is that most nearest-neighbor techniques identify neighboring objects for a given object considering all samples (i.e., full feature space). However, it is understood that all samples or features do not contribute equally to the process of correlation computation. So, when identifying the closely correlated neighboring objects, one needs to consider only the relevant samples or features. Recently some effective techniques have been developed taking this into account. One such technique [1] uses biclustering to identify closely related neighboring objects over a relevant subset of samples or features. It takes a tightly correlated gene pair as the initial seed to obtain a quality bicluster with reference to a given object. The process continues until no more distant missing values exist. The performance of the method has been satisfactory for gene expression datasets of high dimensionality. However, its effectiveness for scRNAseq data remains to be explored.

RNAseq and scRNAseq data suffer from a different missing value problem. Most such datasets include a significantly high number of zeroes, an anomaly referred to as *zero inflation*. For effective handling of this problem, it is necessary to use an appropriate strategy. A common issue with scRNAseq data is dropout events. It requires a special imputation technique to avoid misleading downstream analysis. A good number of imputation techniques have been introduced to address the issue. scUnif [459], MAGIC [417], scImpute [244], DrImpute [422], SAVER [185], and BISCUIT [35] are some well-known imputation techniques for scRNaseq data. scUnif and MAGIC are effective for both scRNaseq and RNAseq data.

D. Outlier Gene Removal

A gene can be considered an outlier if it exhibits abnormal or peculiar expression values over a subset of samples or features of a given class. Such abnormality or non-conformity may be shown in the form of significant over-expression or under-expression over a subset of samples. Outliers may also be considered as observations that are distorted either during sampling or in the laboratory. Such abnormality or non-conformity in expression values of a sample often give an indication that it is a special case. If we do not remove such observations, it may lead to introduction of biases in subsequent analysis. Such non-conformity generally arises mostly due to three reasons: (a) *inherent variability* such as feature values within a population, which cannot be controlled, (b) *measurement error* due to rounding of obtained values or mistakes in recording compound measurement error, and (c) *execution error*, which is due to imperfect collection of data. It may happen when we inadvertently choose a biased sample, which is not a true representative of the desired population. Generally, outlying genes are eliminated using three approaches: (a) statistical, (b) clustering based, and (c) information theoretic.

Gene co-expression analysis often uses Pearson's correlation coefficient (PCC) as a proximity measure. Most co-expression analysis methods have been developed based on this measure. However, a major demerit of this statistical measure is its sensitivity to outliers. In the presence of outlying genes, PCC often fails to perform satisfactorily. So, outliers are eliminated in PCC-based co-expression analysis. Some researchers attempt to detect outliers using bi-weight mid-correlation [298]. These authors construct a graph based on the notion of density. The density for a given gene is defined as the ratio of the number of edges associated with it to the maximum possible number of edges in the graph. The approach defines a gene as an outlier, if it is not connected to any other gene; in other words, it considers a gene as an outlier if the sum of the row elements corresponding to the gene in the adjacency matrix is zero. One can modify the approach using a user-defined cutoff instead of using 0 to define a gene as outlier, if the density proportion corresponding to the gene is lower than the user cut-off value.

Wang and Rekaya [447], in their approach called Least Sum of Ordered Subset Variance t-statistic (LSOSS), emphasize using the mean expression value instead of the median value for normal and cancer samples. According to them, if outliers are present among the disease (say, cancer) samples, two peaks are exhibited in the distribution of their expression values. The higher peak corresponds to activated samples, whereas the lower peak indicates inactivated samples. One can take advantage of this phenomenon that uses the concept of *change point modeling*, while also addressing the outlier issue at the same time. The approach computes an optimal *change point* in its expression for each gene, which is useful in detecting potential outlier genes. It computes the sum of squares of two ordered subsets of cancer samples, uses it to estimate the squared sum of the t-statistic, and uses the mean value of the probable subset of cancer samples to estimate the mean value of cancer samples for the *t*-statistic.

Brechtmann et al. [123] present an outlier detection method for RNAseq read-count data, called OUTRIDER (Outlier in RNASeq Finder). The method attempts to compute the read-count expectations based on the gene co-variation that results from (i) technical, (ii) environmental, or (iii) common genetic variations, using an auto-encoder model. According to the authors, the RNAseq read counts follow a negative binomial distribution with a gene-specific dispersion. Based on this assumption, outliers are those read counts which deviate significantly from this distribution. The model has been validated using both synthetic data as well as benchmark data. OUTRIDER is an open source tool. It is equipped with essential functions that can be used to filter out genes aberrantly expressed in a dataset. OUTRIDER is not only suitable for identifying non-conformal or aberrantly expressed genes, but it has also been effectively utilized in rare-disease diagnostic platforms.

Barghash et al. [2] present an approach to identify outliers (samples and genes) in gene expression or methylation datasets. In this approach, a sample is considered an outlier for a given class of samples, if it deviates (in terms of Euclidean distance) significantly, with reference to a dataset-specific user-defined threshold, from the remaining samples belonging to the class (tumor or normal). To measure the deviation, the approach uses Average Hierarchical Clustering based on Euclidean Distance (AHC-ED). In addition, the approach also considers the functional similarity of a candidate outlier gene with other detected outlier genes. Based on data distribution, it uses different algorithms. For example, if the expression of a gene follows a normal distribution, it uses the Generalized Extreme Studentized Deviate algorithm (GESD) [331]. However, for non-normal distribution, it applies two distribution-free algorithms, namely, Boxplot and Median Absolute Deviations about the median (MAD) [183]. The approach uses GOSemSim [453] to test functional similarity among outlier genes. Functionally dissimilar candidate outlier genes are finally labeled for removal. To validate the clustering results, the method uses the Silhouette measurement. The approach performs satisfactorily for most test cases generated synthetically using hierarchical clustering. The solution is publicly available at GitHub.

Several other effective variants of the above-mentioned methods have also been developed. A clustering-based outlier detection approach attempts to recognize a gene as an outlier if when it is assigned to any cluster, the overall quality (in terms of say, homogeneity, coherence, or other validity indices) of the clustering results are degraded. One can recognize a gene as an outlier based on information theoretic measures also. If a gene is an outlier, the entropy of the associations of that gene with its k-nearest neighbors is significantly high.

E. Removal of Low Read Count

The existence of a high number of zeroes in the count data matrix often creates a major bottleneck in downstream analysis [87]. It can mislead the co-expression analysis and, consequently, inference generation. Empirical studies show that in the presence of such abundance of zeroes, an efficient correlation measure may fail to detect even a perfect correlation between a pair of genes. In addition, due to poor experimental conditions, some genes may be expressed with a very low number of mapped reads. Such low read mapping may occur at random during alignment and it does not provide any valuable information for subsequent co-expression analysis, rather it misleads the results. So, such low read-count entries need to be discarded during preprocessing [264].

F. Batch Effect Removal

Batch effect is a commonly occurring, yet powerful source of variation and systematic error in RNAseq data. It arises due to improper measurements, influenced by laboratory conditions. Batch effects can mislead RNAseq or scRNAseq data analysis, resulting in biased inferences [87]. Recent advancements in high-throughput sequencing technologies help detect and minimize this effect significantly by generating a large volume of data.

Nine factors are responsible for the generation of batch effects [192]: (a) array quality, chip type or platform may be different from one lot to the other, (b) operating procedures may vary in different laboratories, (c) methods used or protocols adopted to draw biological samples may not be the same in different setups, (d) procedures or kits or reagents used for RNA isolation may vary, (e) the synthesis of cRNA or cDNA is also another factor, (f) reagents used in the process of amplification or labeling or hybridization may be different, (g) the schedules for cleaning or conditions for washing (such as temperature, ionic strength, etc.) may be different from lot to lot, (h) there may be variations in the conditions (ozone levels or room temperature) during sample preparation or handling, and (i) types, settings, and calibration drift used in the scanner also may cause a batch effect.

One can avoid or minimize most of the above effects by taking appropriate precautions during experiment design. However, some effects cannot be avoided [192], especially, when sample sizes are large and experiments are carried out over a longer duration of time, e.g., several months. In experimental studies driven by clinical samples, where the samples are created from different clinics, such batch effects are inevitable. During integration of such specimen data, if the batch-specific biases are not eliminated, it may lead to misleading results or inferences.

Although sophisticated normalization methods (such as microarray signal intensity normalization) are in place to control the experimental artifacts among samples, the goal of such normalization is to improve the precision of multi-array measurements using calibration and/or homogenization of the distribution of signal intensity. Such normalization methods cannot handle systematic variations between multiple sample groups. Even after normalization, significantly high batch effects may still remain [192]. Several effective methods for batch effect removal have been introduced.

Principal Component Analysis (PCA) and Singular Value Decomposition (SVD), are two techniques that are useful in batch effect removal. Alter et al. [192] introduce a method based on these two techniques. It subtracts the principal component that represents the batch effect from the expression data and reconstructs the expression matrix using the remaining principal components. In another effort, Benito et al. [287] enhance the Support Vector Machine (SVM) approach to determine a separating hyperplane between two batches and then projects these batches onto the distance-weighted discrimination (DWD) plane, and compute the mean of the batch and finally subtract the DWD plane multiplied by this mean. Orthogonal Projections to Latent Structures are used by Bylesjo¨ et al. [281] to eliminate latent components that represent the batch effect.

In a study of scRNAseq data, Hicks et al. [176] observe that batch effect may cause a substantially high level of variability in cell-to-cell expression. To eliminate such variations, normalization is not adequate since normalization attempts to remove biases from individual samples by considering global properties of data. Batch effects affect specific subsets of genes differently, leading to significantly high variability and reduced ability to detect real biological signals [144]. To eliminate such effects for microarray data, several tools have been developed. Most such tools are not suitable for RNAseq or scRNAseq data. However, COMBAT [199] and ARSyN [303] are two batch effect removal solutions developed for microarray data, which are suitable for RNAseq data as well [91]. Reese et al. [343] introduce an extended version of PCA, called gPCA to assess the quantity of batch effect that occurs in RNAseq data. SVAseq [240] is another effective solution to remove such effects in RNAseq data. Based on an empirical study, Liu et al. [253] recommend SVAseq as the best solution to eliminate batch effect. Nygaard et al. report a detailed analysis of batch effect removal methods in a recent study [419].

5.4.2 Co-Expressed Pattern Finding Using 1-Way Clustering

Clustering is the process of grouping data objects into a set of disjoint classes, called clusters, so that objects within a class have high similarity to each other, while objects in separate classes are highly dissimilar. Cluster analysis of gene expression data can be performed in three ways.

(a) *Gene-based clustering:* In gene-based clustering, the genes are treated as the objects, while the samples are the features.

(b) *Sample-based clustering:* In sample-based clustering, the samples are treated as the objects, while the genes are the features. The samples are partitioned into homogeneous groups. Each group may correspond to some particular macroscopic phenotype, such as clinical syndromes or cancer types. Both gene-based and sample-based clustering approaches search for exclusive and exhaustive partitions of objects that share the same feature space.

(c) *Subspace clustering:* The current thinking in molecular biology holds that only a small subset of genes participate in any cellular process of interest and that a cellular process takes place only in a subset of the samples. This belief calls for subspace clustering to capture clusters formed by a subset of genes across a subset of samples. For subspace clustering algorithms, genes and samples are treated symmetrically, so that either genes or samples can be regarded as objects or features. Furthermore, clusters generated through such algorithms may have different feature spaces. One faces several challenges in mining microarray data using a subspace clustering technique.

- Subspace clustering is known to be an NP-hard problem and therefore many proposed algorithms for mining subspace clusters use heuristic methods or probabilistic approximations. This decreases the accuracy of the clustering results.

- Due to varying experimental conditions, microarray data is inherently susceptible to noise. Thus, it is essential that subspace clustering methods be robust to noise.

- Subspace clustering methods allow overlapping clusters that share subsets of genes, samples or time courses or spatial regions.

- Subspace clustering methods should be flexible enough to mine several (interesting) types of clusters.

- The methods should not be very sensitive to input parameters.

Microarray Data Analysis

A large number of 1-way clustering techniques, some of which have been especially developed for gene co-expression microarray data analysis, have been reported in the literature [15]. Clustering techniques group objects (or genes) into clusters or groups based on their proximity (similarity or dissimilarity) with the objective that similarity between a pair of objects from a cluster (intra-cluster similarity) is maximized, whereas similarity between objects from different clusters (inter-cluster similarity) is minimized. Based on the approach followed during formation of clusters, 1-way clustering techniques can be categorized as: *partitional, hierarchical, density-based*, and *model-based* [191], [58], [295]. This section presents some popular 1-way clustering techniques which perform well on microarray data. We present the algorithms according to the approach followed by them.

A. Partitioning Approach

Partitional 1-way clustering techniques create partitions (clusters) of objects in an iterative manner with an objective to maximize the intra-similarity among the objects belonging to a cluster and to minimize the similarity between objects from different clusters. k-means [191] is the pioneering algorithm from the partitional category that assigns objects to a cluster iteratively based on their distances from the respective cluster means. Over the years, k-means has been modified and a large number of variants have been developed. Although k-means, as a gene clustering algorithm, has advantages due to its simplicity, cost-effectiveness and scalability, it is sensitive to the initial selection of cluster means. Its random selection of initial cluster centers may easily lead it to be trapped at a local optimum. To address this issue, Jothi et al. [200] introduce a deterministic k-means algorithm, called DK-means. It explores a set of probable cluster centers based on a constrained bi-partitioning approach. DK-means shows stable and faster convergence on gene expression datasets compared to other close competing methods [200] such as k-means++ and MinMax algorithm. To address the local minima problem of the k-means algorithm, many other efforts have been made. One such effort is [210]. The authors use hierarchical sub-clustering to handle the local minima problem. The method is scalable and applicable to RNAseq and scRNAseq data. Even though k-means claims to handle voluminous gene expression data, to handle large microarray data, it often requires several hours to days based on the size of the dataset. This poses a serious challenge to computational biologists because it may lead to an incomplete solution when resources are restricted. To address this issue,

Hussain et al. [186] present an efficient solution which is implemented on Field Programmable Gate Arrays (FPGA) to provide an accelerated solution for microarray data. The proposed method works well as a server solution, and it can handle multiple datasets simultaneously.

Dembele et al. [101] have successfully applied fuzzy c-means (FCM) in microarray data analysis. A major issue of FCM is how to set an appropriate value for the fuzziness parameter, m. The authors in [101] show that the optimal value for m may vary from one gene expression dataset to another. They have demonstrated the effectiveness of their proposal to obtain the value of m in a yeast cell cycle dataset. In another effort, [131], the authors introduce an effective fuzzy clustering approach for gene expression data analysis, called FLAME (Fuzzy clustering by Local Approximation of MEmbership). Two important tasks performed by FLAME are (a) introduction of Cluster Supporting Objects (CSOs), objects with archetypal features, for formation of clusters, and (b) assignment of objects of a fuzzy membership vector, following an iterative converging process, in which membership spreads from the CSOs through their neighbors. The performance of FLAME is superior to traditional k-means, hierarchical clustering, FCM, and fuzzy Self-Organizing Maps (SOM) for different types of datasets.

B. Density-based Approach

Density-based 1-way clustering follows the notion of density to define a cluster. It considers highly dense areas, typically of any shape, as the source of clusters. A popular density-based clustering technique is DBSCAN [118]. In [182], a robust and scalable density-based clustering method, called FWCMR for microarray data analysis, was presented. The method takes advantage of the distributed processing schema inspired by MapReduce, which enables the parallel execution of gene expression clustering for any arbitrary size data. The method can handle high-dimensional data in reasonably low execution time. FWCMR performs well in the presence of noise. The scalability of FWCMR is almost similar to the standard Spark machine learning library.

C. Model-based Approach

Model-based 1-way clustering has been found to be effective in gene expression clustering. It uses a cost-effective computational model to address clustering problems. In addition to popular model-based clustering algorithms, some specialized clustering techniques have been proposed to analyze high-dimensional gene expression data. CLICK [367] is a gene expression clustering technique that iteratively splits proximity graphs using minimum cuts. Another notable solution, reported in [336], uses the weighted Chinese restaurant process to cluster microarray gene expression data. One of the major distinguishing features of this method is its ability to infer the number of clusters with high accuracy. It is also able to recognize genes with complex correlation associations such as time-shifted and/or inverted with others and assign them into the appropriate cluster. Two additional features of this algorithm are (i) its ability to handle missing data along with clustering and (ii) its ability to provide multiple cluster strength measurements such as *tightness* and *stability*. Such information can help biologists rank clusters and also perform subsequent downstream studies.

In addition to the above, many other clustering approaches have been developed for microarray data analysis, including graph-based, heuristic-based, and other hybrid approaches. Xu et al. [445] report a graph-based high- dimensional clustering technique. They introduce a minimal spanning tree (MST)-based representation for high-dimensional gene expression data, and define the clustering problem as a tree partitioning problem. This representation defines clusters as subtrees in the MST. Two distinguishing features of this method are (a) cost-effective implementation of clustering and (b) the ability to recognize clusters of any shape. The effectiveness of the approach was evaluated on three data sets, (i) yeast Saccharomyces cerevisiae, (ii) human fibroblasts serum, and (iii) Arabidopsis expression data. The test results were highly encouraging. In [51], Ben et al. report two algorithms for gene expression data clustering. One ensures an asymptotic guarantee of correctness under a reasonable error model and the other is heuristic, i.e., it does not give a formal proof of convergence. The algorithms have been evaluated on both (i) simulated noisy data and (ii) real biological data. Elyasigomari et al. [116] introduce a hybridized gene selection and cancer identification approach. The authors combine the Cuckoo Optimization Algorithm and Genetic Algorithm (COA–GA) for effective gene selection to identify interesting genes from cancer data. The proposed hybrid method has been able to extract several biologically relevant significant genes from cancer expression data.

RNAseq Data Analysis

The evolution of high-throughput sequencing (HTS) technologies has revolutionized the scope and depth of understanding of the genome, epigenome, and transcriptome of a large number of organisms [340]. RNA sequencing has been more popular due to its cost-effectiveness and acceptability as shared-resource HTS cores. The recent advancement of HTS technologies in sequencing ribonucleic acid content (RNAseq), has almost replaced the application of microarrays for transcriptomic studies, because RNAseq does not require prior knowledge of the genome sequence to quantify gene expression in terms of counts of transcripts. It has become essential for every computational biologist to understand the advantages and limitations of the intensive process of RNAseq analysis. Compared to microarray gene expression analysis, RNAseq analysis can extract many additional intertesting and exciting facts. One can acquire more detailed information about the transcriptome such as allele-specific expression and transcript discovery, which cannot be expected from microarrays. The pre-processing required for RNAseq data is mostly the same as with microarray expression data.

Clustering of RNAseq data identifies patterns of gene expression by grouping genes based on their distance in an unsupervised manner. It can be used as an exploratory tool that allows the user to organize and visualize relationships among groups of genes, and to select certain genes for further consideration. This section focuses on 1-way cluster analysis of high-throughput sequence data (counts). Gene co-expression analysis of RNAseq data enables prediction of gene function at a genome level. It is also useful to discover critical genes for a given disease and understand the functions of these genes and their associations with diseases. Co-expression analysis for microarray and RNAseq data have several similarities and differences. Unlike microarray gene expression data, these count-based measures are positive, discrete, and substantially skewed, with a significantly large dynamic range. Further, for genes expressed from weak to moderate levels, the precision is often low. It is due to the sampling nature of sequencing [340]. We focus on extraction of interesting biological entities that share common profiles across different conditions (control to disease), such as co-expressed genes that are associated with the same biological processes. Like microarray data analysis, recent works on cluster analysis of RNAseq data [308], [372], [421] can be categorized into partitioning-based, hierarchical, density-based, model-based, graph-based, and hybrid.

A. Hierarchical Clustering

A hierarchical clustering algorithm [339] builds a nested sequence of clusters based on either a divisive (or top-down) or an agglomerative (bottom-up) approach. Such a nested sequence of cluster patterns can be visualized using a dendrogram, where the height of a node represents the dissimilarity between the two clusters merged together at the node. This dendrogram representation reveals associations among clusters and such association information is often useful in analyzing functions of genes. Hierarchical clustering has been successfully applied to both microarray and RNAseq data analysis. Between the two basic hierarchical clustering approaches, the divisive approach is less cost-effective than the agglomerative approach, and hence, is not so popular in analyzing gene expression data.

Anders and Huber [23] use hierarchical clustering with Euclidean distance, following a variance-stabilizing transformation in gene expression data analysis. Severin et al. [364] cluster fourteen diverse tissues of soyabeans. They normalize the data using a variation of the Reads Per Kilobase per Million (RPKM) mapped reads measure and apply hierarchical clustering with the Pearson correlation measure. Witten [433] reports analysis of hierarchical clustering of samples with a modified log likelihood ratio statistic as the distance measure based on a Poisson log linear model.

Agglomerative nesting (AGNES) [213] is a bottom-up hierarchical clustering method that accepts a gene expression dataset as input and uses a fusion approach to generate a set of quality clusters. It executes a series of successive fusions of the individual gene expressions in the dataset. It uses the Euclidean distance for dissimilarity computation and the Manhattan distance for fusion. AGNES constructs a dissimilarity matrix using the Unweighted Pair Group Method with Arithmetic Mean (UPGMA) [396] to generate refined clusters following a cost-effective merging process with an updated dissimilarity matrix. Initially, it takes two most similar objects and joins them to form a cluster of two objects. A constraint followed by AGNES is that once it forms a cluster with two candidate objects, they are not separated later. Although it looks like a rigid decision, this helps the method obtain good results in low computation time. Another advantage of AGNES is its ability to handle missing data. To estimate missing values, it takes the present values and uses them in computing the average and mean absolute deviation. The rigid decision-making process that it cannot correct later, sometimes leads it to inappropriate cluster results. In addition, an application of a different proximity measure when merging the candidate clusters, may lead to different results.

Divisive Analysis (DIANA) [318] is another cost-effective hierarchical clustering algorithm for gene expression data analysis. Unlike AGNES, DIANA adopts a top-down or divisive approach. It initiates cluster generation with the whole population (of say, n instances), and divides the data into two components, and then goes further to split them into groups of smaller numbers of instances until at step $(n-1)$, all instances are individually apart in n clusters, each with a single instance. Like AGNES, a split-up cluster cannot be joined later. Like AGNES, DIANA handles missing data following a similar approach. A common drawback of this algorithm is that it cannot repair what was done earlier. In other words, it cannot reunite split clusters. In addition, it requires the computation of a diameter while splitting clusters, limiting its effective application in gene expression analysis.

Another effective hierarchical clustering method is Clustering Using Representatives (CURE) [150]. CURE is a compromise between centroid-based and all-point extreme clustering approaches. CURE begins with a fixed number of scatter points that represent the extent and shape of the clusters. These scatter points are shrunk toward the representatives of clusters, i.e., the centroids. This scattering approach helps it address the limitations of all-point and centroid-based clustering methods, and to generate clusters of any shape. A major limitation of CURE is that it is insensitive to outliers. This is because the shrinking of the scattered points toward the cluster representatives makes it less effective on the adverse effect of outliers. CURE has been applied by Guha et al. [150] in gene expression data analysis.

Chameleon [211] is another effective hierarchical clustering algorithm that has been applied in RNAseq data analysis. It overcomes the inappropriate merging problem of other hierarchical clustering such as AGNES, DIANA, and ROCK [191], especially in the presence of noise, by exploiting a dynamic modeling technique. It includes two major steps. (i) It uses a graph-based approach to partition the data into a set of smaller groups or sub-clusters with stronger links (highly related) among the elements within a cluster. During partitioning, it ensures that the presence of noise cannot influence the process. (ii) It extracts an appropriate set of clusters by applying an iterative merging process. To identify appropriate pairs of sub-clusters for merging, it considers both relative closeness and interconnectivity, enabling it to address the issue of inappropriate merging. The performance of Chameleon is usually better than most of its competitors.

Among popular hierarchical clustering algorithms, Balanced Iterative Reducing and Clustering using Hierarchies (BIRCH) [460] is an algorithm that uses a tree-based approach. It includes four major steps. (i) It scans the dataset and constructs a height-balanced cluster feature (CF) tree. A CF represents a triple $< n$ (number of data points), ls (linear sum) and ss (square sum)$>$ that summarizes the elements of a cluster. (ii) It scans all leaf entries to reconstruct a smaller CF tree that eliminates the outliers and groups crowded subclusters into larger ones. This step is optional. (iii) It applies agglomerative clustering to the subclusters to obtain a set of clusters that almost represent the distribution of data. (iv) Finally, it eliminates the presence of any minor or localized errors present in the output of the previous step. This step is also optional. Due to the provision of incremental updates, BIRCH can handle large datasets efficiently. The output of BIRCH is independent of the order of the input data. Its ability to handle (i) outliers, (ii) large datasets, and (iii) arbitrary input order, makes BIRCH suitable for gene expression data analysis.

B. Partitional Clustering
The 1-way clustering algorithms discussed above for microarray data analysis have also been applied to RNAseq data. However, most classical partitional algorithms are inadequate in analyzing RNAseq data, especially in the extraction of biologically enriched cluster extraction.

An extended version of the k-means algorithm is the Intelligent Kernel K-means (IKKM) [164] algorithm. To handle non-linearly separable human gene expression data, IKKM constructs a kernel matrix using a linear kernel function. An important advantage of IKKM is that it does not require the number of clusters as input. It determines it by estimating the center of mass of the dataset and then locating the farthest objects from the center of mass as well as from each other. It forms the clusters centered around these farthest objects. The process is repeated until it converges to stable cluster results. When no object needs to change its cluster membership, a stable number of clusters is obtained. IKKM is able to handle a gene expression dataset, although in general, it often fails to perform well with high-dimensional data. The performance of IKKM has been found superior to intelligent k-means when evaluated on real-life datasets such as tumor data, lymph node metastasis data, and other gene expression datasets.

C. Density-based Clustering
The density-based 1-way clustering algorithms discussed above may also be applied to RNAseq data analysis. However, because RNAseq data are discrete and sampling-biased, there is need for improved versions for generation of biologically enriched clusters. Several improved versions of classical density-based clustering algorithms such as DBSCAN have been developed.

DENsity-based CLUstEring (DENCLUE) [178] is effective and efficient in extracting clusters of any shape. DENCLUE has been successfully applied in gene expression data analysis and the performance has been found better than many other contemporary algorithms.

D. Model-based Clustering:

Self-Organizing Maps (SOMs) [191] are effective in both microarray and RNAseq data analysis. These neural network-based methods are widely used in gene clustering. SOMs have already been discussed in the previous chapter. Gene expression data are generally highly connected and it is possible that a single gene has a high correlation with multiple clusters. In such cases, model-based methods are more suitable due to their probabilistic nature. Other important features of SOMs that make them suitable for gene expression data analysis are as follows: (i) The single-layered neural network structure of SOMs helps generate an intuitively appealing map of a dataset with large dimensionality in 2-D or 3-D space. Such a map places similar clusters as neighbors to each other. (ii) Visualization and interpretation are easier due to the association of each neuron of the network with a reference vector. Further, each data point is mapped to the neuron with the closest reference vector, making inference generation easy. (iii) SOMs can handle large gene expression datasets compared to competing methods such as k-means or hierarchical clustering.

Chinese Restaurant Clustering (CRC) [336], discussed in the previous chapter, is a model-based Bayesian clustering method found effective in gene expression analysis. It can recognize and extract groups' genes with strong, yet complicated correlations (shifted and/or scaled and/or inverted). CRC achieves it by computing the probability of each gene joining each of the existing clusters as well as being alone in its own cluster. CRC takes advantage of the Chinese Restaurant Process (CRP) and uses a predictive updating technique to minimize the influence of nuisance and as well as missing data.

Si et al. [372] introduce a promising model-based clustering approach for RNAseq data, using an effective hybridization of Poisson or NB models. This method significantly improves the performance of the EM algorithm. The four distinguishing features of the method are: (i) NB and Poisson models are well suited to RNAseq data due to the discrete and skewed nature of count data, (ii) the performance, based on both simulation and benchmark data analysis, is superior to heuristic methods such as k-means and SOM, (iii) it is highly flexible, and (iv) the number of clusters is selected in a unified way.

E. Evolutionary Algorithm-based Clustering

Most unsupervised clustering approaches discussed above face difficulty in the estimation of the accurate number of clusters in a given dataset. So, determining an optimal number of clusters is a challenging task. Any slight deviation from the accurate or optimal number of clusters may lead to significant errors. One possible way to address this problem is to use a combination of multiple clustering methods to estimate the Pareto-optimal cluster size and several cluster validity indices as objectives. The goodness of a solution, in case of a multi-objective optimization problem [278], [40], is typically evaluated through dominance. In such problems, the goal is to identify the non-dominating solutions, which is the set of solutions that are not dominated by any other member of the solution set. Such non-dominated solutions form the Pareto-optimal set [141] of the whole feasible decision space. The boundary defined by the set of all the points that are mapped from the Pareto-optimal set, is referred to as the Pareto-optimal front [141]. A recent robust estimator of Pareto-optimal cluster size is MOCCA (Multi-Objective optimization for Collecting Cluster Alternatives) [227]. It estimates the cluster size based on aggregation of the best cluster numbers of various clustering algorithms and by exploiting multiple cluster validation indices as the objectives. MOCCA ranks the Pareto-optimal cluster sizes by considering the domination factor.

In addition to the above, many other methods have been developed for gene signature identification. Mitra et al. [284] use machine learning techniques such as the Random Forest and Random Survival forest algorithms to identify gene signatures for multiple myeloma. Aziz et al. [34] apply GeneSpring and other R plug-ins to identify gene signatures in colorectal cancer. Chen et al. [79] identify gene signatures on nonsmall-cell lung cancer using the decision tree algorithm and survival analysis. However, there have been only a few attempts to develop Pareto-optimal techniques for gene signature identification. For gene selection, Basu et al. [47] introduce the Strength Pareto Evolutionary Algorithm. In another effort, Awad and Jong [30] introduce an optimization method called Spectral Signatures Selection through MOGA Multi-Objective Genetic Algorithm.

Clustering in scRNAseq Data

The revolutionary development of single-cell RNA sequencing (scRNAseq) technology has enabled the generation of sequence data with millions of cells. It is now possible to accurately profile the transcriptomes of a large number of cells. The pioneering single-cell RNA sequencing (scRNAseq) experiment was performed in 2009. The experiment profiled only eight cells [160]. However, seven years later, an organization called 10X Genomics was successful in creating a dataset of more than 1.3 million cells [289].

With the release of such rich, voluminous data, the need for efficient computational methods has increased. The computational analysis of scRNAseq data includes multiple steps such as (i) preprocessing, which includes quality control, mapping, quantification, and normalization, (v) grouping using a clustering method, (vi) determining trajectories, and (vi) identifying a subset of differentially expressed genes. The analysis of scRNAseq data is helpful in the following ways.

- It is able to divulge complex and rare cell populations,

- it can extract heterogenous gene regulatory relationships among cells, and

- the trajectories of distinct cell lineages can be tracked during development, in terms of both temporal and spatial information.

Downstream analysis of scRNAseq data is challenging due to the presence of a large number of zero-inflated counts because of dropout or transient gene expression. It often misleads downstream analysis. Preprocessing techniques such as normalization applied to handle noise, often lead to elimination of interesting genes, and consequently lead to poor performance. In addition, confounding factors such as the influence of a large number of biological variables and the presence of technical noise also contribute to inappropriate read counts [335].

To help analyze single-cell data, clustering techniques have been used in both unsupervised as well as unsupervised frameworks. A number of clustering methods have been introduced for downstream analysis of such data. This section presents some prominent clustering methods developed for scRNAseq data analysis [335].

SC3 is an effective single-cell clustering method which uses k-means clustering in addition to consensus clustering based on multiple clustering solutions [219]. It identifies a small subset of principal components (using PCA) to support cost-effective clustering based on the random matrix theory with high precision. SC3, implemented in R, is popular due to its user-friendly interface and non-dependency on any prior knowledge. This fast clustering method avoids the need for user-specified parameters. It was validated using 12 scRNAseq datasets and was superior to several other competing methods in terms of accuracy and stability. In an empirical study [335] conducted using five toolboxes, namely, SC3, SNN-Cliq, SINCERA, SEURAT and pcaReduce, the performance of all the clustering methods was found to be similar. However, SC3's powerful integration feature makes it a stable performer for a wide range of datasets.

DropClust is another effective algorithm for cluster analysis and visualization of ultra-large single cell RNAseq (scRNAseq) data [374]. The two main distinguishing features, (i) speed and (ii) ability to delineate both major and minor cell types, make DropClust appropriate for analysis of large-scale scRNAseq data generated using droplet-based transcriptomics platforms. An important advantage of this method over other methods is its ability to detect minor cell sub-populations. DropClust has made this possible through (i) structure preservation of data sampling and (ii) subjective identification of informative principal components. This clustering method is useful for downstream analysis of droplet-seq data.

Clustering through Imputation and Dimensionality Reduction, or CIDR, is an ultrafast cost-effective clustering method which exploits a simplified, yet robust implicit imputation approach that can ease the impact of dropouts in scRNAseq data in a righteous manner. Experimental study on synthetic and benchmark data show that CIDR can improve the performance of standard principal component analysis and is superior in terms of clustering accuracy than competing methods like t-SNE, ZIFA, and RaceID. CIDR is also much faster than most competing methods [354].

SCENIC [17] identifies stable cell states in tumor and brain scRNAseq data based on the activity of the GRNs in each cell to handle large dimensionality. SCENIC uses two complementary methods to address high dimensionality in single-cell data: (i) small sample extraction to infer the gene regulatory network and (ii) gradient enhancement, instead of Random Forest, to achieve better performance. The method shows that for effective GRN analysis, single-cell data are more suitable, and one can use the genomic regulatory codes to guide the identification of transcription factors and cell states [335].

SEURAT, introduced by Satija et al. [363], is an effective toolbox that helps localize spatial cells. It attempts to predict spatial cell localization by combining scRNAseq data with in situ RNA patterns. Based on experimentation on RNAseq data of 851 single cells from Danio rerio embryos, SEURAT was able to predict the spatial location of a complete transcriptome and correctly locate unusual sub-populations. In the above dataset, four cell types are included: 'NPC', 'GW16', 'GW21' and 'GW21+3'. SEURAT facilitates observation of genes that fluctuate significantly. The user can consider such a subset of genes rather than the complete set for subsequent analysis to minimize the cost of computation [335].

SIMLR is a kernel-based similarity learning method introduced by Wang et al. [423] for dimensionality reduction of scRNAseq data. This single-cell interpretation via a multi-kernel learning (SIMLR) approach can be successfully applied to large-scale datasets. The authors considered four datasets for single-kernel comparison without weight terms, and their observation was that the addition of weight terms significantly enhances SIMLR's performance. The robustness of SIMLR has been further experimentally investigated by Zhang et al. [459] for drop-out events in single-cell data. Further, they explored the application of SIMLR to low-rank imputed data and the performance was superior to the other general clustering algorithms [335].

SNN-Cliq is a Shared nearest-neighbor-based approach, introduced by Xu et al. [441], for grouping cells of the same type. scRNAseq data are, generally, of very high dimensions, and out of thousands of genes, only a few are significantly expressed in different types of cells, making the task of clustering challenging. SNN-Cliq determines the number of clusters automatically in association with an SNN similarity metric [335].

NMF, introduced by Shao et al. [366], identifies subgroups in scRNAseq data using non-negative matrix factorization (NMF). Generally, single-cell data are too noisy, and hence even some popular techniques like PCA, cannot be directly used. Experimental study showed that the first few principal components given by PCA hardly help in the discrimination process, explaining the differences only partially. However, NMF is effective in this case. It is helpful, especially in the identification of single parts, and hence can detect the natural subtypings of individual cells and functional cell subsets [335].

Monocle [409] is an unsupervised algorithm to analyze single-cell gene expression data to discover the expression sequences of key regulatory factors. Trapnell et al. introduced this method to investigate mouse myoblasts. They identified eight transcription factors that were not detected earlier. This tool expects users to prepare phenotype data and feature data. It includes the general functions of a single-cell toolkit as well as the functions for quality control and difference analysis. This tool requires dimensionality reduction before clustering in order to be visualized. It also infers the development trajectory [335].

TF-IDF [289] is another effective statistical method that can be used for scRNAseq data clustering. This *term frequency-inverse document frequency* approach for data transformation and scoring is effective in text analysis. Recently, researchers have exploited *TF-IDF* using two different approaches for scRNAseq data clustering. In one approach, it was used to identify a subset of the most informative genes for effective clustering. In another approach, it uses all the genes for clustering, but initially binarizes the gene expression data based on a *TF-IDF cutoff*. A major bottleneck with this approach is the quadratic time requirement for distance calculations for datasets with millions of single cells.

A. Hierarchical Clustering in scRNAseq Data Analysis

Hierarchical clustering (both *agglomerative* and *divisive*) is used for a wide range of clustering applications, including scRNAseq data clustering. Agglomerative clustering sequentially merges individual cells towards formation of bigger clusters, while a divisive approach divides a bigger cluster into smaller ones. A major disadvantage with this clustering approach is the high memory and time requirement, which grows quadratically with the number of instances.

CIDR [249] or Clustering through Imputation and Dimensionality Reduction, reduces dimensionality by performing PCoA on the dissimilarity matrix to alleviate the impact of dropouts in scRNAseq data. It clusters the first few principal coordinates hierarchically, and finds an appropriate number of clusters based on the *Calinski Harabasz index* [280]. Recently, a number of clustering tools for scRNAseq data have been developed using hierarchical clustering over reduced dimensional space.

SINCERA, proposed by Guo et al. [152], is a scRNAseq profiling analysis pipeline. This pipeline is able to identify cell types, gene signatures and crucial nodes. Working with mouse lung cells, *SINCERA* was able to distinguish the main cell types of the fetal lung. SINCERA also introduces an effective gene sequence prediction model based on logistic regression. This hierarchical clustering tool converts data to z-score before clustering. It finds the number of clusters, k, by finding the first singleton in the hierarchy [335].

An appropriately designed hierarchical clustering algorithm can handle large gene-gene networks and can effectively extract modules of high biological significance, irrespective of number of nodes in the network.

B. Partitioning-based Clustering in scRNAseq Data Analysis

Among the various clustering approaches, partitioning-based clustering is simple and popular. The k-means clustering algorithm is an excellent example of this approach. The k-means algorithm operates on scRNAseq data iteratively to identify k cluster representatives (centroids), where each cell is assigned to the closest cluster representative. Lloyd's algorithm [161] is a variation of k-means, with the ability to scale linearly with any number of instances or nodes.

A major disadvantage of k-means is that it does not guarantee obtaining the global minimum. To address this issue, one can apply k-means repeatedly with distinct initial conditions and finally find a consensus as in the case of SC3 [219]. k-means can find clusters of spherical or globular shapes. So it often misses the rare cell types in a large dataset. RaceID [149] attempts to overcome the issue by augmenting k-means with outlier detection to identify rare cell types. Similarly, SIMLR [423] adapts k-means by simultaneously training a custom distance measure [218].

C. Graph-based Clustering in scRNAseq Data Analysis

Graph-based clustering is a well-established alternative to unsupervised analysis of gene expression data. Because it overcomes the scalability issue of k-means and hierarchical clustering, this approach has become increasingly popular in handling large-scale scRNAseq data. *Community detection* is a variant on the idea of clustering that is well-suited to graph-based clustering. This approach attempts to identify communities or groups of nodes that are densely connected. It starts by constructing a k-nearest-neighbor graph to represent the scRNAseq data. However, in such a graph-based approach, the selection of an appropriate number of nearest neighbors, i.e., k, is essential. Inappropriate selection of k may lead to inaccurate number and size of the detected final communities. Such an approach can handle outliers by reweighing the shared nearest neighbors of a pair of cells [218].

An important advantage of graph-based clustering is that unlike hierarchical clustering, this approach generates only a single solution at low cost. This approach does not require the number of clusters as user input. The Louvain algorithm [289], [60] is an effective effort in this regard and has been successfully applied to scRNAseq data. In another effort, shared-nearest-neighbor graphs have been combined with the Louvain algorithm to handle scRNAseq data efficiently. PhenoGraph [243] is an example of such effort. SEURAT [363] and Scanpy [434] incorporate this approach.

Another effective graph-based algorithm is Shared Nearest Neighbor (SNN)-Cliq that attempts to cluster high-dimensional single-cell transcriptomes using the shared nearest-neighbor concept. SNN-Cliq performs well with both synthetic and benchmark datasets.

5.4.3 Subspace or 2-way Clustering in Co-Expression Mining

Gene-based clustering for functional analysis groups gene expressions indicating co-expressions or co-functions in terms of mutual correlation. Traditional 1-way clustering approaches cluster data by operating on the entire set of samples. An important observation is that biologically significant genes exhibit high correlation over subsets of samples, rather than the full sample space. Based on this observation, a new category of clustering algorithms has evolved, known as subspace or 2-way clustering (or biclustering), due to the dimensionality of the considered expression data.

Microarray Data Analysis

Correlation between the expressions of a pair of genes can be absolute, shifting, scaling, and shifting-and-scaling. A shifting-and-scaling correlation has closer correspondence to functional similarity and co-regulation in gene expression data analysis than pure shifting and scaling correlations. A residue-score-based measure handles shifting-and-scaling correlation by operating on both rows and columns in a bicluster just like the mean squared residue. Such a measure produces an aggregate score that reflects the amount of shifting-and-scaling correlation among both genes and samples. The produced residue score is not normalized. Transposed virtual error [327] is another bicluster quality assessment measure that can detect shifting-and-scaling correlations. However, the measures are not effective in computing pairwise correlation computation due to the involvement of columns (with just two elements in each column) and produce column-biased-result. Like Pearson correlation, the measures are sensitive to noise due to direct use of means. To address these issues, two other measures, BioSim [41] and SSSim [15], have been proposed to handle shifting-and-scaling correlation effectively even in the presence of noise. BioSim assumes values between -1 and $+1$ corresponding to negative and positive dependence and 0 for independence. BioSim aggregates step-wise relative angular deviation of the expression vectors. Unlike, BioSim, the SSSim score [16] ranges from 0 to 1 for a given pair of gene expressions. A value of 1 indicates that the expressions perfectly exhibit shifting-and-scaling correlation. Empirical analysis has shown that both measures are analytically suitable for modeling co-expression because they account for expression similarity, expression deviation and relative dependency.

Next, we discuss subspace clustering approaches [297], initially proposed by Agrawal et al. [10]. Subspace clustering algorithms are further divided into two categories: (a) bottom-up search and (ii) top-down search.

A. Bottom-Up Subspace Search Methods

Methods in this category leverage the benefits of the *downward closure property* to reduce the search space. According to this property, if a subspace S contains a cluster, for any subset of S will also a cluster exist. An algorithm creates bins (for each dimension) to determine locality finally forming a multi-dimensional grid. To create such a grid, one can use any of two approaches: (i) a static grid-sized approach and (ii) data-driven strategies to determine the cut-points. In the static grid-sized approach, there are two well-known clustering algorithms, CLIQUE [10] and ENCLUS [81]. CLIQUE uses a grid-density-based approach to find clusters within the subspaces, whereas ENCLUS attempts to define clusters based on entropy using a frequent itemset mining approach like Apriori [13]. The algorithms belonging to the second approach handle large datasets using a grid-based, adaptive, and parallel approach. MAFIA [145] and CBF [76] belong to this group. CLTree [251] is another significant development that separates high- and low-density areas using a modified decision tree algorithm. Another useful algorithm of this category is DOC [332]. It combines both bottom-up and top-down approaches to discover optimal projective clusters with strong clustering tendency using an iterative approach over a subset of dimensions. This category of algorithms is suitable for clustering gene expression data because of the (i) ability to handle high-dimensional data, (ii) ability to detect arbitrary sized clusters, (iii) ability to detect clusters which are embedded, intersected or disjoint, and (iv) ability to handle voluminous data. Although these algorithms have several advantages, they (i) do not scale well with the increase in dimensionality, (ii) often eliminate small clusters, and (iii) have high running time complexity.

B. Top-Down Subspace Search Methods

A top-down or divisive approach initiates cluster formation with an approximation of clusters considering an equally weighted full feature space. It uses an iterative process to modify the weights to form refined clusters. With full feature space, this approach cannot be considered cost-effective. One can use sampling techniques for improvement. Two factors influence the performance of this approach: (i) subspace size and (ii) number of clusters. A number of clustering methods have been developed under this category, including PROCLUS [8], ORCLUS [9], FINDIT [436], δ-Clusters [462], and COSA [130]. These have already been discussed in detail.

In addition to the above two categories of subspace clustering algorithms, to extract biologically enriched clusters with subset of instances over subset of relevant dimensions, biclustering algorithms are used. Biclustering algorithms can be divided into four categories: (a) greedy iterative search, (b) divide-and-conquer, (c) exhaustive bicluster enumeration, and (d) iterative row and column clustering combination. Biclustering methods have already been discussed in detail in the previous chapter.

A biclustering algorithm attempts to group genes that exhibit correlation, possibly over the subspace of samples [297]. Cheng and Church [82] define a bicluster as a sub-matrix of size $I \times J$ that exhibits high coherence, where I and J represent the genes (rows) and conditions (columns), respectively. Here, $|I| \leq |N|$ and $|J| \leq |M|$, and M and N specify the size of the original matrix. A biclustering algorithm aims to find the largest biclusters with the minimum the mean squared residue, which is an NP-hard problem [82]. Based on the approach used to extract biclusters, the biclustering algorithms are classified [266] into the following categories.

C.1. Greedy Iterative Search

A greedy iterative search biclustering algorithm forms clusters by iteratively adding or deleting rows/columns, with an objective to maximize local gain. Cheng and Church, in their pioneering work [82], demonstrate the application of this greedy approach in gene expression analysis. This approach cannot detect overlapping/embedded clusters because the elements of the already identified biclusters are masked by random noise. Yang et al. [448] developed a probabilistic algorithm called FLOC to discover a set of k-possible overlapping biclusters simultaneously. FLOC can easily be extended to support additional features at low cost. FLOC performs satisfactorily on yeast gene expression data. Ben-Dor et al. [50] introduce another probabilistic model that handles large Order-Preserving Sub-Matrices (OPSMs) with high statistical significance. Its purpose is to compute a set of conserved rows and a set of columns that give the largest *xMotif*, i.e., the one that includes the largest number of rows. OPSM was successfully applied on breast cancer data to extract significantly interesting patterns.

C.2 Divide-and-Conquer

Divide-and-conquer algorithms have the advantage of being fast. However, they also have the drawback of missing good biclusters that may be split before they can be identified. The Block Clustering [169] algorithm begins with the entire data in one block (bicluster) and iteratively finds the best split. A major advantage of this type of divide-and-conquer approach is that it can help interpret the clusters. However, a major issue has still remained: how to identify an optimal number of clusters. To address this issue, Duffy and Quiroz [111] used permutation tests to determine when a given block split is not significant. Later, Tibshirani et al. [406] added a backward pruning method to the block splitting algorithm and designed a permutation-based method called Gap Statistics, to induce the optimal number of biclusters.

C.3 Exhaustive Bicluster Enumeration

This approach is based on the assumption that the best biclusters can only be extracted using an exhaustive enumeration of all possible biclusters in the data matrix. Although this approach performs well on synthetic as well as benchmark datasets, it has a serious limitation. Due to high complexity, it can only be executed by imposing restrictions on the size of the biclusters. Tanay et al. [397] introduce a biclustering algorithm called SAMBA. This graph-based method coupled with statistical modeling is polynomial and finds the most significant biclusters. SAMBA has been tested on a collection of yeast expression profiles and on a human cancer dataset, and the cross-validation results show high specificity in assigning function to genes based on their biclusters. The extracted biclusters were helpful in detecting new concrete biological associations. SAMBA was able to identify and associate finer tissue types than was previously possible.

C.4 Iterative Row and Column Clustering Combination

This method operates iteratively on the rows and columns of the gene data matrix and combines the results to generate an optimal set of biclusters. CTWC [137], which is a successful effort belonging to this category, generates a dendrogram. An advantage of CTWC is that it can handle noise effectively and can recognize clusters of any shape successfully. Additionally, it gives a stability index to each cluster, based on which one can decide the goodness (how real) of the clusters. CTWC has performed satisfactorily on a number of benchmark datasets including colon cancer, leukemia, breast cancer, glioblastoma, skin cancer, and antigen chips.

Tang and Zhang [399] introduce an unsupervised framework to analyze gene expression data, based on the Integrated Two-Way Clustering (ITWC) method, discussed in the previous section. An advantage of ITWC is that while clustering iteratively through genes and sample groups, one can dynamically manipulate the associations between them. ITWC has been used on three gene expression datasets, where two are from multiple-sclerosis patients, and the third data set is a collection of leukemia patient samples.

Busygin et al. [72] develop a two-way, node-driven double-conjugate clustering algorithm called DCC, using self-organizing maps (SOM) and angle-metric as the similarity measure. This algorithm has been applied to whole genome microarray data. DCC has also been applied to leukemia data. DCC could separate sample classes with 98% accuracy with the corresponding gene clusters containing only a fraction of the genes.

In a recent exhaustive empirical study by Mandal, Bhattacharyya, et al. [275], biclustering approaches were divided into two categories: *frequency-based* and *network-based*. The authors compared 17 popular biclustering algorithms using 6 microarray gene expression datasets for cancer. For systematic evaluation of the algorithms, the authors performed enrichment analysis, subtype identification, and biomarker identification. As an outcome of the study, 103 gene biomarkers were identified using a frequency-based method. Out of these, 19 were for blood cancer, 36 for lung cancer, 25 for colon cancer, 13 for multi-tissue cancer, and 10 for prostate cancer. The network-based approach detected 41 gene biomarkers, of which 15 are from blood cancer, 12 from lung cancer, 6 from colon cancer, 7 from multi-tissue cancer, and 1 from a prostate cancer dataset.

RNAseq Data Analysis

This section presents a set of effective and popular biclustering algorithms that have been successfully applied in the analysis of RNAseq data for extraction of biologically enriched biclusters to support downstream analysis.

Ge, Son, and Yao [135] developed an integrated web application for differential expression and pathway analysis of RNAseq data, called iDEP. It seamlessly connects (i) 63 R/Bioconductor packages, (ii) 2 web services, and (iii) comprehensive annotation and pathway databases for 220 plant and animal species. One can download the R code and relevant pathway files and customize the code to reproduce the workflow. The authors apply iDEP on an RNAseq dataset of lung fibroblasts with Hoxa1 knockdown. An outcome of the study is the discovery of the possible roles of SP1 and E2F1 and their target genes (including microRNAs) in blocking G1/S transition. In another experiment, the authors discover an interesting fact related to mouse B cell. They demonstrate that without functional p53, ionizing radiation activates the MYC pathway and its downstream genes involved in (i) cell proliferation, (ii) ribosome biogenesis, and (iii) non-coding RNA metabolism in that cell. iDEP was able to unveil the multifaceted functions of p53 and the possible involvement of several microRNAs such as miR-92a, miR-504, and miR-30a.

Kaiser and Friedrich [204] present a user-friendly R-package, called Biclust that includes (i) a collection of bicluster algorithms, (ii) preprocessing methods for two-way data, (iii) validation of bicluster results, and (iv) facility to visualize the results. This package is popular among computational biologists and data analysts to conduct empirical studies on RNAseq data as well as to compare their own method(s) with popular algorithms.

Murali and Kasif [291] introduce an effective representation for gene expression data called conserved gene expression motifs or xMOTIFs. It defines the expression level of a gene as conserved along a subset of samples, if the gene is expressed with the same abundance over all the samples. In other words, a conserved gene expression motif can be defined as a subset of genes that are simultaneously conserved over a subset of samples. The authors develop a method to extract significantly conserved gene motifs that represent all samples and classes in the data. In an experimental study using benchmark RNAseq datasets, the method was effective in constructing xMOTIFs that help differentiate the various classes without ambiguity.

We know that genes participate simultaneously in multiple biological processes and hence their expressions are a composition of the contributions from active processes. Henriques and Madeira [174] present a biclustering approach called BiP (Biclustering using Plaid models) that allows expression levels to vary in overlapping areas based on biologically meaningful assumptions such as weighted and noise-tolerant composition of contributions. BiP provides the benefits of existing biclustering techniques to enable detection and recovery of noisy as well as excluded areas. It generates an explanatory model to support analysis of overlapping transcriptional activity. The authors establish the effectiveness of BiP based on experiments on synthetic data.

Xie at al. [439] present a biclustering algorithm called QUalitative BI-Clustering algorithm Version 2 (QUBIC2) to handle both microarray and RNAseq data. The major strengths of this algorithm are (i) the use of an improved Gaussian model for accurate estimation of multimodality in zero-enriched expression data, (ii) the use of information divergence for functional gene module optimization based on an effective dropout-saving expansion strategy, and (iii) exhaustive statistical validation of the extracted biclusters. Experiments on both benchmark and synthetic data showed that QUBIC2 identified more biologically enriched biclusters than the other competing algorithms from both microarray as well as RNAseq (both bulk and single cell) data.

Kluger et al. [220] introduce an efficient spectral biclustering method to identify checkerboard structures in gene expression data. Such checkerboard structures present in a gene expression data matrix can be recovered with the help of eigenvectors that correspond to characteristic expression patterns across genes or conditions. Additionally, one can detect such eigenvectors easily using linear algebraic approaches, based on singular value decomposition (SVD) in association with an integrated normalization process. The authors validate the method using publicly available cancer expression datasets. BCSpectral compared well in performance against a number of competing methods.

Prelic et al. [329] present a systematic approach to evaluate biclustering techniques using a simple binary reference model. It is their observation that the model is able to capture the essential features of most biclustering approaches. However, it may not be able to determine all the optimal groupings exactly. To address this issue, the authors introduce a fast biclustering method called Bimax based on divide-and-conquer. They use several synthetic and benchmark datasets to evaluate the performance of a hierarchical clustering method against five significant biclustering algorithms together with the reference model. They observed that biclustering is generally better than hierarchical clustering in gene expression analysis.

ScRNA Data Analysis

Biclustering for scRNAseq data analysis is still in its infant stage. There are only a few efforts in this evolving area. We present three significant biclustering approaches developed to identify interesting genetic drivers based on scRNAseq data analysis. Zeisel et al. [456] present a divisive biclustering method that Sorts Points into Neighborhoods (SPIN). The method discovered molecularly distinct classes of cells over large-scale-single-cell RNA sequencing of mouse somatosensory cortex and hippocampal CAI region. It was successful in discovering 47 molecularly distinct subclasses that include complete sets of known major cell types in the cortex. They identified a number of class- specific marker genes for nine major classes of cells that help alignment with known cell-type location and morphology. Shi and Huang [371] present a simplified, yet faster shared nearest-neighbor biclustering approach, referred to as BiSNN-Walk, to identify cell sub-populations and associated genetic drivers from scRNAseq data. BiSNN-Walk builds upon the popular shared nearest-neighbor algorithm called SNN-Cliq [442]. It uses an iterative approach using an inner loop and an outer loop. The inner loop includes three major steps: (a) cell clustering, (b) gene finding, and (c) expression matrix update. The outer loop removes cell clusters extracted in the inner loop from the input matrix. It updates the input matrix and feeds it to the inner loop, so that more stable cluster results are obtained. This process repeats until further clustering is not fruitful. This is ascertained if all walk-trap clusters obtained are of size 2 or less or all candidates have zero transitivity. BiSNN-Walk is known for its following distinguishing features.

- It generates an ordered (ranked) set of reliable clusters.

- It gives a ranking of the genes in a cluster based on their level of affiliation to the associated cell clusters.

- It exhibits improved performance over SNN-Cliq in terms of extraction of biologically significant clusters in terms of GO enrichment.

Xie et al. [201], in their comprehensive review, emphasize the applicability of biclustering in scRNAseq data analysis. There may be abundant zeros in the scRNAseq data, and hence, the developers of 2-way clustering algorithms for analysis of such data should be careful. The algorithms successfully applied in microarray data may not be applicable directly to such data. Both bulkRNA and scRNAseq data require a unique design of 2-way clustering algorithms. Generally, a 2-way clustering algorithm in isolation is not enough to analyze RNAseq data for identification of interesting genetic drivers. It needs to use dimensionality reduction, result annotation processes, as well as visualization, for subsequent hubgene-centric analysis. To develop an effective unified pipeline by integrating 2-way biclustering with other appropriate result annotation tools is not straightforward. It requires deeper understanding of the expressiveness of and constraints in sequence data, expected biological inferences and the motivation as well as the abilities of the algorithm(s) or tool(s) used.

5.4.4   Co-Expressed Pattern-Finding Using 3-Way Clustering

As discussed in the preceding sections, 1-way and 2-way classical clustering approaches are focused on analysis of gene-sample (GS) or gene-time (GT) microarray gene expression data. Biclustering approaches are effective in handling such GS or GT microarray data in extracting interesting patterns. But three-dimensional microarray expression data, i.e., gene-sample-time (GST) data are relatively more complex and hence need more sophisticated clustering techniques for effective extraction of three-dimensional clusters that represent a subset of co-expressed genes over a subset of samples across a subset of time points. We refer to such methods as triclustering methods and the generated three-dimensional coherent patterns are called *triclusters* [15]. A tricluster represents information that is often found useful in identifying useful phenotypes, potential genes related to these phenotypes and their expression rules. There are a few triclustering techniques that extract genes which have similar expression patterns across a set of samples over a set of time points. For a triclustering algorithm developer, one of the major challenges is to take into account both inter-temporal and intra-temporal gene coherence. Other important challenges in tricluster analysis are to avoid time-dominated and sample- dominated results effectively and to detect time-latent triclusters of high biological significance. Triclustering has already been discussed in detail in a previous chapter. An example of GST data is shown in Figure 5.10. An example of triclusters is shown in Figure 5.11.

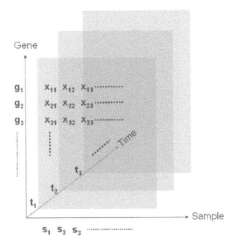

Figure 5.10: GST data

Triclustering in Microarray Data Analysis

A triclustering method extracts subsets of co-expressed genes by grouping genes over subsets of samples across subsets of time points. Such a method considers both (i) *inter-temporal coherence*, which refers to similarity of expression patterns of the same gene across different time points, and (ii) *inter-gene coherence*, which refers to the similarity among different genes at the same time point. A triclustering method should not suffer from the curse of time dominance or sample dominance [15]. Time dominance refers to a situation where more time points are included in a partial tricluster, which later hinders the inclusion of relevant samples in the tricluster. Similarly,

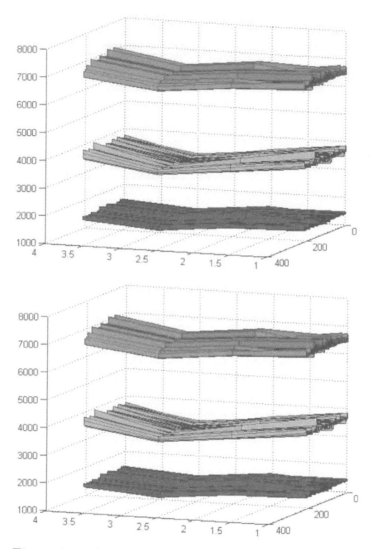

Figure 5.11: Visualization of some example triclusters

sample dominance refers to a situation where samples are given preference over time while adding samples and times into a partial tricluster. A triclustering method should be able to extract all possible time latent triclusters.

This section presents four popular triclustering techniques that have performed satisfactorily on gene-sample-time or GST data. The techniques are: (i) Mining Coherent gene clusters from GST microarray data [94], (ii) TRI-CLUSTER [230], (iii) gTRICLUSTER [158], and (iv) ICSM [15]. We discuss their application to microarray data analysis here.

Mining Coherent gene clusters from GST microarray data [94]:
This triclustering method attempts to extract subsets of co-expressed genes
over subsets of samples across the entire time range of the GST dataset. The
distinguishing features of this method are as follows.

- Use of set enumeration trees and pruning operations for efficient extraction of all possible subsets of samples or genes.

- Consideration of inter-temporal coherence during extraction of triclusters.

- Ability to detect inter-temporal coherence of both shifted and scaled forms.

- Ability to extract all possible triclusters in two ways.

- Ability to extract overlapped triclusters in a deterministic manner.

This method's drawbacks include the following. First, it extracts triclusters
over the entire time span. Second, although the second approach extracts
the complete set of coherent gene clusters, it does so at a very high computational cost. Third, although the method considers inter-temporal coherence,
it avoids consideration of coherence among the genes participating in a tricluster. Finally, this method is not able to extract all possible triclusters,
since the simultaneous implementation of both approaches is not practically
feasible.

TRICLUSTER [230]: Zhao et al. introduce an efficient and deterministic graph-based triclustering method to extract coherent triclusters over
the $G \times S \times T$ space. It is dependent on four user input parameters. Some
distinguishing features of TRICLUSTER are as follows.

- It extracts only maximal triclusters that satisfy certain pre-specified homogeneity criteria.

- The extracted triclusters may be located anywhere in the input data matrix. The triclusters may have overlap of various proportions.

- TRICLUSTER extracts triclusters of various types.

TRICLUSTER's limitations include the following. First, it has too many,
i.e., four input parameters. Second, it is not cost effective. Third, with a single threshold value, it is often inadequate in extracting all relevant biclusters,
rather it requires different thresholding. This is due to the aggregate nature

of comparison. Finally, the method extracts clusters with scaling patterns successfully. However, it often fails to detect clusters with shifting patterns.

gTRICLUSTER   [158]: This enhanced triclustering method generates a set of highly coherent triclusters over the GST space. The important features of this method are summarized below.

- Like TRICLUSTER, it is dependent on four input parameters.

- It can extract triclusters of various types.

- It considers inter-temporal coherence when generating triclusters.

In addition to dependence on too many input parameters, gTRICLUSTER suffers from several other limitations. Although it addresses one major issue with TRICLUSTER by considering inter-temporal coherence, it fails to handle inter-gene coherence, which may be more significant. Further, the use of the Spearman correlation coefficient does not guarantee the ability to capture coherence effectively along the time dimension.

ICSM   [15]: ICSM addresses a major issue with gTRICLUSTER by considering both inter-temporal as well as inter-gene coherence by exploiting order-preserving sub-matrices. It finds a set of triclusters of high biological significance from GST data. ICSM operates sequentially on all unordered pairs of GS planes corresponding to two time points, and produces a set of triclusters. When processing an unordered pair of GS planes, ICSM takes two steps. The first step generates a set of initial modules which are considered partial triclusters. These initial modules are enhanced to form complete triclusters in the second step. During step 1, ICSM generates initial modules for an unordered pair of GS planes using dynamic ranges over expression levels of genes for each pair of samples. It identifies the dynamic ranges corresponding to each pair of samples and then traverses these ranges to generate the initial modules. The method is cost-effective in terms of time requirement because it uses the concept of order-preserving sub-matrices when traversing through the ranges. An expression submatrix is order preserved if each row has an increasing sequence of values based on permutation of its columns [15]. It extends the initial modules to include time points from the remaining time points using the PMRS similarity measure. PMRS is a planar similarity measure between a pair of planes. It is measured by arranging the rows of the matrices representing the planes into two vectors. The correlation between these two vectors gives similarity between the planes.

PMRS is inspired by NMRS [15], which also detects similarity between a pair of planes. Two important features of ICSM are the following.

- It considers both inter-temporal as well as inter-gene coherence.

- It introduces two statistical indices for evaluation of inter-gene coherence and inter-temporal coherence: (i) intra-temporal homogeneity and (ii) inter-temporal homogeneity.

In terms of limitations, ICSM requires significantly high resources during extraction of triclusters from GST microarray data.

RNAseq Data Analysis

Due to limited work on triclusters, there are only a few works on RNAseq data analysis using the triclustering approach.

Tchagang et al. [400] develop a 3-way clustering algorithm, known as Order-Preserving Triclustering (*OPTricluster*), for 3D time-series gene expression data analysis. OPTricluster fuses two methodologies: (i) a combinatorial approach on the sample dimension and (ii) the order-preserving (OP) concept on the time dimension. The algorithm is able to extract three-dimensional clusters with high coherence. It investigates proximities (both closeness and differences) between samples over the time domain. Experiments on real datasets show that OPTricluster is robust in the presence of noise and can compute proximities between biological samples accurately.

Jung et al. [202] present a triclustering algorithm called TimesVector. The algorithm extracts clusters from 3D gene expression data by capturing co-expressed patterns over two or more sample conditions. The three steps of TimesVector are: (i) reducing dimensionality and clustering time-condition concatenated vectors, (ii) extracting co-expressed and distinct expression patterns by post processing, and (iii) exploring unclassified genes for rescuing genes. TimesVector satisfactorily extracts enriched triclusters from both microarray and HTS data.

David and Cristina [156] present an evolutionary algorithm, called TriGen to extract enriched triclusters from gene expression data. TriGen introduces three distinct fitness functions, MSR3D, LSL and MSL . However, an effective measure to evaluate the triclustering algorithms was still missing. TriGen introduces a measure called TRIQ for evaluation of triclustering algorithms. TRIQ uses three types of information: (i) how the genes, conditions and timepoints are correlated, (ii) visualization of extracted gene patterns and their graphical validation, and (iii) functional annotations for

the extracted genes. TRIQ has been used to extract triclusters in real benchmark datasets including (i) yeast cell cycle (*Saccharomyces cerevisiae*) and
(ii) humans (*Homo sapiens*).

5.5   Network Analysis for Co-Expressed Pattern-Finding

The development of high-throughput microarray and sequencing technologies have provided a range of opportunities to systematically characterize
diverse types of biological networks [272]. Biological networks can be broadly
classified as protein interaction networks [15], [452], [37], [49], metabolic
networks [341], [124], gene co-expression networks [229], [190], and gene
regulatory networks [373], [348]. These networks provide an effective way
to summarize gene and protein correlations. In this section, we focus on gene
co-expression networks, which are undirected graphs where nodes represent
genes and pairs of nodes are connected by an edge if the pair is significantly
co-expressed. Gene co-expression networks provide association between individual genes in terms of expression similarity and a network-level view of
the similarity within a set of genes. In co-expression networks, two genes
are connected by an undirected edge if their activities have significant association, as computed using gene expression measurements such as Pearson
correlation, Spearman correlation, or mutual information. Compared to gene
regulatory networks, a gene co-expression network is built upon gene neighborhood relations, which give interesting geometric interpretations of the
network. One of the most important applications of gene co-expression networks is to identify functional gene modules or network modules, which are
represented by the strongly connected regions of the co-expression network.

Due to the non-transitive nature of connections among genes, genes form
a very complicated connectivity network with respect to a particular similarity measure in a gene expression dataset. Such a connectivity network
is often referred to as a co-expression network. A major use of this co-
expression network is extraction of network modules that represent strongly
connected regions in the co-expression network. These modules may present
highly co-expressed genes, which are functionally similar.

In unsupervised learning, gene expression data are usually analyzed to
find groups of genes associated with the same biological functions. As discussed in the preceding sections, clustering helps group the genes based
on their similarity [15], [58]. Gene co-expression network analysis is a
special form of clustering, where grouping of genes always starts with the

preparation of a network called a co-expression network [62]. This network is further analyzed to generate regions which are highly connected groups of nodes in the network. A gene co-expression network provides the association among individual genes in terms of expression similarity and a network-level view of similarity within a set of genes. A gene co-expression network provides a good approximation of the complicated web of gene functional associations [457]. In co-expression networks, two genes are connected by an undirected edge if their activities have significant association computed by using gene expression measurements such as Pearson correlation.

5.5.1   Definition of CEN

A gene co-expression network can be defined as a graph, $G = \{V, E\}$, with the following properties.

- Each vertex $v \in V$ represents a gene.

- Each edge $e \in E$ represents a connection between a pair of vertices v_1, v_2 where $v_1, v_2 \in V$.

- There is an edge between two vertices $v_1, v_2 \in V$ if the similarity between the genes corresponding to the vertices is more than a user-defined threshold.

Definition 1: A CEN co-expression network is defined by an undirected graph $G=\{V,E\}$ where each $v \in V$ corresponds to a gene, and each edge $e \in E$ corresponds to a pair of genes $d_i, d_j \in D$ such that d_i is connected to d_j by an amount $G(d_i, d_j)$.

Definition 2: A gene d_i is connected to a network module C_i, if $d_i \in C_i$ and $f_m(C_i, d_i) > \tau$, where f_m is a similarity measure.

Definition 3: A network module C_i is a set of genes forming a dense region in the co-expression network and TOM $(C_i) \in$ any top TOM values corresponding to the extracted network modules, where TOM is a topological overlap matrix.

5.5.2   Analyzing CENs: A Generic Architecture

co-expression network analysis can be performed on gene expression (microarray or RNAseq) data. It extracts biologically enriched co-expressed modules or subsets of genes to facilitate subsequent downstream analysis for identification of interesting (i) disease biomarkers or (ii) gene-gene associations. Figure 5.12 shows a generic architecture for co-expression network analysis. It includes five major components: (C_1) preprocessing, (C_2) co-expression network construction, (C_3) module extraction, (C_4) module-based analysis, and (C_5) validation of outcomes. Each component has a definite role to play and the output of one component, say C_1, becomes the input of component C_2, and so on.

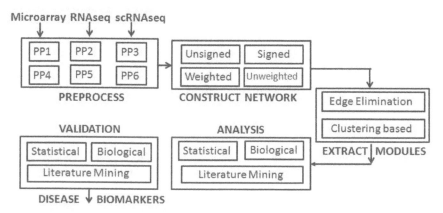

PP1: Discretize; PP2: Missing Value Estimate; PP3: Normalize; PP4: Remove Batch Effect;
PP5: Remove Low Read count; PP6: Remove Outlier Gene;

Figure 5.12: A generic CEN analysis architecture

C_1: **Preprocessing**

The preprocessing steps for gene expression data prior to CEN analysis are similar to those steps discussed in the preceding sections.

C_2: **Network Construction**

co-expression networks are useful in investigating the functionality of genes at the system level [457]. A simplified and straightforward approach for CEN construction is to create an adjacency matrix based on proximity values between each pair of nodes. The nodes of the network represent genes and the connectivity between a pair of nodes represents an association when both are significantly co-expressed across appropriately chosen tissue samples. To construct the co-expression network, computational biologists threshold the value of the co-expression similarity measure. One can pick a threshold in two distinct ways: (i) using a *hard* thresholding approach, where the gene co-expression similarity is encoded using binary information (1 for connect-edness and 0 for unconnectedness), and (ii) using a *soft* thresholding approach, which assigns the weight for a connection to be a number between 0 and 1. Generally, hard thresholding results in unweighted networks while soft thresholding results in weighted networks.

When following hard thresholding to construct the adjacency matrix, one has to select a gene expression similarity measure to compute a proximity matrix, where each entry in the matrix indicates the proximity between the corresponding genes' expressions. Based on the proximity matrix, to construct the adjacency matrix, one may use a thresholding approach. Two genes d_i and d_j are connected if $Similarity(d_i, d_j) > \tau$, a userdefined threshold. Thus, for each connected pair of genes, the corresponding adjacency matrix can be generated using Equation 5.1.

$$A(i;j) = \begin{cases} 1, & \text{if } d_i \text{ and } d_j \text{ are connected} \\ 0, & \text{otherwise} \end{cases} \tag{5.1}$$

One can definitely ask whether it is biologically meaningful to encode gene co-expression using only binary values. A limitation of hard thresholding is that it leads to significant loss of information due to the quantification of gene-gene connectivity strength and sensitivity to the selection of the threshold. As response, Zhang et al. [457] introduce a more meaningful way for CEN construction. To construct a biologically meaningful gene co-expression network, they introduce the notion of a weighted gene co-expression network [457]. They define a general framework for soft thresholding that assigns a weight to each gene pair connectivity. To generate a connection weight based on the co-expression similarity measure, the authors propose several adjacency functions. They introduce a biologically motivated scale-free topology criterion to estimate the parameters of the adjacency function. Initially, the authors conduct and provide important empirical evidence for prediction of the biological significance of a gene. They also provide theoretical and experimental evidence to establish that the gene modules defined using the weighted topological overlap measure are more cohesive than the modules obtained by the unweighted approach. Finally, the clustering coefficients for the weighted networks are generalized. The authors apply their methods to both synthetic and benchmark data (cancer and yeast microarray data) to demonstrate effectiveness.

In a similar effort, Mahanta et al. [270] improve the performance of weighted gene co-expression network construction by introducing a spanning tree-based soft-thresholding approach in association with a topological overlap similarity measure (TOM) [341]. The authors introduce a similarity measure called *NMRS* (Normalized Mean Residue Score) for construction of the initial proximity matrix. Several proximity measures have been used for gene expression data analysis, including Euclidean distance, Pearson correlation coefficient, Spearman correlation coefficient and mean squared residue. A common limitation of these measures is that they cannot detect the three correlation patterns, i.e., shifting, scaling, and shifting-and-scaling patterns, especially in the presence of noise. However, the effectiveness of the NMRS measure has been established in [270]. The authors use the proximity matrix obtained based on the NMRS measure for generation of the adjacency matrix following a soft thresholding approach, which is subsequently used to construct the CEN. The method performs well for several benchmark datasets.

Types of CEN

CENs are classified based on the way correlation is computed and recorded for a given pair of genes. The four distinct ways a CEN can be constructed [87] are given below.

(a) **Unsigned Network Construction:** Many correlation measures return correlation values between a pair of genes in the range from -1 to +1. An unsigned co-expression network uses absolute values of correlation measures. In other words, a negatively correlated pair of genes is also considered co-expressed because these genes may positively co-regulate other genes. However, such pairs may disrupt the structure of the co-expression network. This problem is handled in the following types of networks.

(b) **Signed Network Construction:** In signed networks, the correlation values between pairs of genes are transformed to a range from 0 (perfect negative correlation) to 1 (perfect positive correlation). A correlation value < 0.5 is considered negatively correlated and a correlation value > 0.5 is considered positively correlated.

(c) **Weighted and Unweighted Network Construction:** In weighted co-expression networks, nodes in the network are connected to one another like complete networks, and continuous weights ranging from 0 to 1 are assigned to each edge based on the tendency of a pair of genes to be co-expressed. Un-weighted co-expression networks are similar to weighted co-expression networks. The only difference is that the weights assigned to the edges are either 1 or 0, obtained by transforming (also by comparing with the threshold) the edge weights. 1 and 0 are used to represent whether a pair of genes is connected or disconnected.

Examples of CEN Types

As examples of CENs, consider a synthetic gene expression dataset with 4 genes, $G1$, $G2$, $G3$ and $G4$, and let the correlation values between the genes using the Pearson correlation measure be as shown in the Table 5.13. Only the gene pairs $(G2, G3)$ and $(G2, G4)$ have negative correlations.

In an unsigned CEN, absolute values of Pearson correlation ($A_{ij} = |Corr(G_i, G_j)|$) represent the amount of association between pairs of genes. We see in Figure 5.14 that negative correlations between the gene pairs $(G2, G3)$ and $(G2, G4)$ are now positive. In a signed co-expression network, the association between a pair of genes is calculated using the equation $A_{ij} = (0.5 + 0.5 \times Corr(G_i, G_j))$. We observe in the figure that all the correlation values are changed.

In a weighted CEN, a continuous weight ranging from 0 to 1 is assigned to each edge based on the quantification of the tendency of a pair of genes to be co-expressed. The weight of an edge is estimated based on the recommendations of shared neighbors. In an unweighted CEN, the weights are either 1 or 0, and are obtained using a threshold. We observe in Figure 5.14 that edge weights below 0.5 are removed in an unweighted network.

Gene names	G1	G2	G3	G4
G1	1	.77	.37	.61
G2	.77	1	-.80	-.45
G3	.37	-.80	1	.68
G4	.61	-.45	.68	1

Figure 5.13: A synthetic data matrix to demonstrate CEN types

C_3: Module Extraction

A module of a co-expression network is a subset of nodes with tight connectivity or high cohesiveness. There is high functional similarity among the genes belonging to a module. Accurate detection of highly cohesive modules

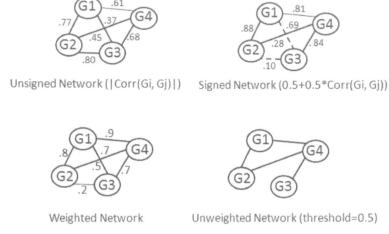

Figure 5.14: Types of co-expression networks

is an important objective of co-expression network analysis. It helps subsequent downstream analysis to uncover interesting gene(s) or biomarker(s) that may have close association with a given disease. Modules may be extracted from microarray or RNAseq data using two approaches: (i) edge elimination and (ii) clustering. We discuss each of these approaches.

(a) **Edge Elimination Approach**

Extraction of a set of modules or subsets of tightly connected nodes from a CEN is the objective of co-expression network analysis. A module can also be defined as a tightly connected component of a graph or tree. So, module extraction can be thought of as a graph component identification problem, where the members of a component show high similarity among themselves. Edge elimination extracts tightly connected components using a top-down approach. It recognizes a module by disconnecting it from the rest of the network by eliminating a weak edge.

To extract functionally similar modules, one can use a spanning tree-based [330] [270] approach in association with a topological overlap similarity measure (TOM) [341]. A TOM is used to effectively identify biologically enriched network modules. A spanning tree T_G is a minimally connected graph G where each pair of nodes is connected minimally, i.e., with a single edge. In other words, a spanning tree has (i) no cycles and (ii) no nodes that are disconnected from the rest. For each connected and undirected graph, there exists at least one spanning tree. From a graph G, one can extract spanning trees using various approaches. Prim's algorithm [330] extracts spanning trees from any undirected connected graph. Several variants of Prim's algorithm have also been developed [48].

A weighted spanning tree (where each edge is assigned a weight based on the computation of proximity between the expressions of two genes) is an effective means for CEN representation as well as for extraction of modules. Identifying spanning tree(s) with maximum weights helps extraction of biologically enriched modules. In addition, one can also use the TOM similarity measure for better module extraction because it exploits node connectivity information in a graph. To implement this approach, one initially builds an adjacency matrix from the undirected graph, and subsequently constructs the TOM by defining topological weights, using Equation 5.2,

$$T_{ij} = \frac{l_{ij} + w_{ij}}{min(k_i, k_j) + 1 - w_{ij}} \qquad (5.2)$$

where $l_{ij} = \sum_u w_{iu} w_{ij}$ and $k_i = \sum_u w_{iu}$ is the connectivity of a node. To construct the adjacency matrix, we choose a similarity measure that effectively detects shifting, scaling and shifting-and-scaling patterns in gene expressions. Using the adjacency matrix and a variant of Prim's algorithm [330], a maximum spanning tree is constructed from the connected components using the weight information in the topological overlap matrix. In the next step, one extracts the network modules by eliminating the weaker edges (with reference to a user-defined threshold) from the spanning tree. In other words, a network module can also be defined as a component of the maximally weighted spanning tree. Therefore, the module extraction process using edge elimination is a three-step process: (i) adjacency matrix construction with reference to a similarity threshold, (ii) maximally weighted spanning tree finding, and (iii) module extraction by eliminating weaker edges. The efficiency of this approach in terms of scalability and accuracy depends largely upon (i) the selection of an appropriate similarity measure, (ii) implementation of a maximally weighted spanning tree finding algorithm, and (iii) heuristics used to eliminate weak edges.

Mahanta et al. [270] apply this approach to five benchmark gene expression datasets, Yeast Sporulation, Yeast Diauxic Shift, Subset of Yeast Cell Cycle, Arabidopsis Thaliana, and Rat CNS. The authors claim that

their approach, called Module Miner, is able to extract highly enriched
network modules in terms of both p-values and q-values.

(b) **Clustering Approach**

Clustering-based module extraction from large high-dimensional gene
expression data has many applications in "omics" biology. Unsupervised extraction of enriched modules from such data is challenging due
to the complexity of the structures that underlie these datasets. In
module-based analysis, we consider a group of tightly connected co-
expressed genes as a single functional unit. To map associations of such
gene modules to phenotypes, we need an appropriate representation
and analysis pipeline. One such effective pipeline for module-based co-
expression network analysis is Weighted Gene Co-expression Network
Analysis (WGCNA) [235], [62], [132], [236].

Langfelder et al. [235] introduce weighted gene co-expression network analysis as an effective systems biology approach to describe co-
expression patterns across gene samples. This approach is useful because
it (i) extracts tightly connected clusters (modules) from gene expression
data, (ii) summarizes modules with reference to module eigengene or
intra-module hubgene to find associations among the modules or between a module and an external sample trait, and (iii) identifies interesting biomarker(s) for a given disease. It characterizes the modules
as highly cohesive, dense and interconnected clusters of genes. To extract such clusters, WGCNA uses an unsupervised approach based on
the TOM measure [341]. It facilitates the use of multiple clustering
approaches for network module detection. The hierarchical approach is
most commonly used. It uses dendrograms for cluster representation,
where the branches correspond to the modules. For identification of
modules, it facilitates branch cutting techniques such as (i) constant-
height cut and (ii) dynamic branch cut. Although hierarchical clustering is the default with WGCNA, facilitating module detection, it cannot
extract the exact number of clusters from the dataset.

WGCNA has been used in many real-life biological situations such as (i)
causal gene identification for deadly diseases like cancer, (ii) yeast data,
(iii) mouse data, and (iv) brain digital data. WGCNA is available as an
R package that includes a large family of R functions for (i) network
construction, (ii) cluster (module) extraction, (iii) gene selection, (iv)
computation of topological properties, (v) simulation and visualization
of data, and (vi) inter-operation with other packages.

Botia et al. [62] improve the standard WGCNA pipeline by incorporating an additional k-means clustering step. The approach works in two steps, where initially it constructs the network A as an $n \times n$ matrix, in which each entry $A_{i,j}$ indicates the interaction strength between the corresponding pair of genes. It generates another square proximity matrix A' based on A, to subsequently use for the extraction of clusters. The difference of this version from the standard WGCNA pipeline is that it includes an additional k-means clustering step which takes the output of standard WGCNA clustering as input and generates a stable set of clusters. This hybridization helps improve the overall performance of the method by generating enriched clusters. It alleviates the three major drawbacks of hierarchical clustering, (i) failure to generate an appropriate number of clusters, (ii) dependency of final results on how proximities between the clusters are compared, and (iii) inability to alter the structure once a gene has been assigned to a branch in the dendrogram. Although the improved version generates better clusters, it is unable to estimate the accurate value of k without prior knowledge. Improved variants of the k-means algorithm (for example, k-means++ [28]) attempt to determine the k value automatically, following different strategies. These approaches have been used with several real-life gene expression datasets for validation.

In addition to the above, several other clustering-based tools for gene expression analysis using the module-based approach have been proposed. Watson et al. [430] introduce an R package called *coXpress* to identify groups of genes that are differentially co-expressed. The approach was effective in two publicly available microarray datasets. *coXpress* extracts gene clusters that are tightly connected across a set of biologically related experiments. *coXpress* also facilitates resampling-biased performance evaluation for the extracted groups of genes based on p-value. BFDCA handles all types of correlation patterns discussed in Section 5.2.1. Results obtained based on both simulation and real biological data analysis show that BFDCA detects disease-related gene clusters and assesses the regulatory impact of disease-related genes with high accuracy. In a similar recent effort, Russo et al. [356] develop a user-friendly tool called CEMiTool as a Bioconductor package for discovery and analysis of co-expressed modules. This tool generates user-friendly reports with high-quality graphs. CEMiTool has been successfully applied to over 1000 transcriptome datasets, and to an RNAseq patient dataset infected with Leishmania, producing promising results.

C_4: Cluster- or Module-Based Analysis

The results generated by clustering or network-based approaches, as discussed above, can be analyzed in two ways, (a) hubgene-centric topological analysis and (b) biological analysis. We present these two approaches, highlighting their features.

(a) **Hubgene-centric Topological Analysis**

The structural properties or functional behaviors of a co-expression or other type of biological network is usually dependent on just a few genes. The removal of such gene(s) may cause significant functional changes in addition to structural changes [369]. Such genes are often referred to as *hubgenes*, because they play key roles in gene expression analysis. Hubgenes are generally highly connected or high-degree genes in a co-expression network. A module in a CEN often includes a large number of genes. To investigate the behavior of such modules topologically, hubgenes are essential. These genes help investigate the functional as well as topological behavior of a set of genes at a module level. Depending on the inter-gene connectivities among the modules, the hubgenes are categorized into two groups: intra-hubgene and inter-hubgene. An *intra-hubgene* is defined as a gene belonging to a module which is (i) highly connected and is (ii) central to the module. On the other hand, an *inter-hubgene* is a gene that is (i) highly connected and (ii) connected with genes from more than one module. Figure 5.15 demonstrates examples of intra- and inter-hubgenes. It shows two gene modules represented with dotted ellipses. Gene G_{16} interfaces with both modules in addition to being highly connected. On the other hand, G_{14} and G_{11} represent the modules, being intra-hubgenes. Several approaches can be used to identify both types of hubgenes from CEN modules. These approaches identify the hubgenes by analyzing the gene-gene connectivity information as well as various centrality measures such as (i) betweenness centrality, (ii) eigenvector centrality, (iii) pagerank centrality, (iv) closeness centrality, and (v) radiality [369]. Some methods also use pathway information for identification of representative hubgenes to support downstream analysis. The performance of these approaches in identifying hubgenes depends on size, shape, density, and topology of the network modules.

The evolution of advanced, high-throughput gene expression data gathering technologies facilitate the investigation of the status of a cell's components and simultaneous identification of the causes of formation

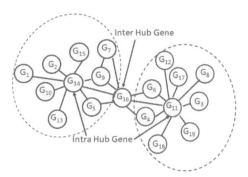

Figure 5.15: Inter- and intra-hubgene in a CEN. Dotted ellipses represent two modules.

of gene groups and their interactions with each other [43]. Network analysis is an effective means for comprehension of such high-throughput genomic data [237] [43] [20], [19]. A feature of a genomic data network is its scale-free topology or connectivity distribution [20], [19]. In network-based analysis, hubgenes are central to the network's architecture, and it has been demonstrated that such crucial genes are essential for survival in lower organisms such as the fly, worm, and yeast [43]. Substantial research has also revealed the significance of hubgenes in the genomic analysis of higher organisms such as humans and mice.

In network-based analysis of gene expression data, it is helpful to focus on intra-modular hubs instead of whole network hubs, especially for co-expression network applications [43]. Several researchers have come to the conclusion that properly selected intra-modular hubgenes from clusters of interconnected genes or proteins have significant associations with traits such as disease status or survivability. In other words, such crucial genes of diseased modules may have significant clinical importance.

To identify intra-modular hubgenes from a control or disease module and to perform module-centric analysis focused on such hubgenes, one can use weighted correlation network analysis tools such as WGCNA. Such a tool extracts modules of tightly interconnected nodes (highly enriched in a specific functional category) from a gene expression dataset, represented as a weighted correlation network. Next, it correlates the genes belonging to a module with specific clinical traits such as the status of a disease and survival time. Finally, it uses an effective intra-modular connectivity measure or an appropriate centrality measure (as discussed above) to extract intra-hubgene(s) from the modules. This approach successfully addresses the scalability issue that typically arises in most other approaches when attempting to handle large CENs. It is cost-effective and the analysis is goal-oriented because it avoids irrelevant analysis tasks.

A number of methods have been developed [87] to support hubgene-finding and hubgene-centric analysis. DHGA [97] identifies hubgenes effectively and performs differential hubgene centric analysis in a CEN. WiPer [36] is a user-friendly package that helps hubgene-finding based on connectivity scores. Langfelder et al. [237] empirically justify the superiority of hubgene identification compared to meta-analysis because (i) hubgenes help gain more biological insight, and (ii) are more relevant to basic research. According to them, intra-hubgene identification is more useful in understanding biological insights than meta-analysis with respect to consensus modules. Ballouz et al. [39] compare CENs constructed from microarray and RNAseq data. Two important observations of the study are that (i) the correlation between the degrees of nodes from the two CENs is negative and (ii) the networks have different hubgenes. MRHCA [87] is a tool that uses a non-parametric approach for both microarray and RNAseq data. This tool (a) identifies hubgenes in a CEN and (b) extracts modules associated with each hubgene.

The following are the steps in intra-modular hubgene centric analysis.

(a) Modules of co-expressed genes are extracted from a co-expression network. These modules are strongly enriched in biological functions.

(b) Biologically relevant modules are filtered by correlating the genes with progression of a disease or a trait.

(c) An intra-modular connectivity measure, such as module membership, is used to select genes with the highest degree from each biologically relevant module.

Empirical assessment of these hubgenes can provide significant biological insights with respect to the progression of disease or a trait.

Although we have highlighted advantages of hubgenes in CEN-based gene expression analysis, one may ask the question whether (or when) hubgene identification always leads to extraction of additional meaningful genes compared to statistical analysis based on significance testing. To respond to this crucial query in the context of multiple genomic data analysis with the availability of ground truth, Langfelder et al. [237] conducted an empirical study using hubgene-centric CEN analysis and a standard statistical approach. They were interested in comparing (i) the significant of the biological insights gained by the approach in terms of its relevance and (ii) the validation successes (reproducibility) for each independent gene expression dataset and in clinical diagnostic or prognostic applications. The study focused on three criteria: (i) the ability to find predictive genes for lung cancer survival, (ii) the ability to find methylation markers related to age, and (iii) the ability to find mouse genes related to total cholesterol. Their observation with reference to the first criterion was that intra-hubgene-based gene expression analysis was more effective than standard statistical analysis based on p-values or q-values, especially in the detection of biologically significant genes. However, for the other two criteria, both approaches performed almost equally well.

A major issue with hubgene-centric analysis is to associate the modules between the control (or normal) state and disease state for effective, yet efficient analysis. As the numbers of modules extracted by computational methods from both states often become high, to map a module from the control state to one or more modules from the disease state for subsequent investigation to find interesting genes, requires special effort. We discuss such an effort called X-module that associates condition-specific CEN modules effectively.

Association Finding among Condition-Specific CEN Modules

Extraction of modules from gene co-expression networks is of biological interest, since genes in a module share common functions and biological processes or pathways. Finding accurate associations or relationships among such modules across conditions (control and disease) is essential for subsequent downstream analysis. Accurate identification of association between modules across stages helps understand inter-modular preservation and biological behaviors among modules from the same or two distinct datasets. Such association(s) among the modules extracted from the control and disease samples in microarray or RNAseq data may be one-to-one or one-to-many.

X-module [207] is a method equipped with an effective fusion measure to identify possible mapping(s) among two sets of co-expressed modules extracted from control and disease stages. X-module takes two sets of condition-specific co-expressed modules as input, extracted from corresponding CENs, and generates corresponding module pairs. After necessary preprocessing, it considers a pair of co-expressed modules in a given dataset, and estimates topological and biological similarities. For such similarity assessment between two network components (modules), it uses four parameters: (i) semantic similarity, (ii) eigengene correlation, (iii) degree difference, and (iv) number of common genes. The method identifies consensus modules based on module membership and eigengene correlation assessment of the paired modules. For confirmation, it validates their significance both statistically as well as biologically. Some important features of X-module are given below.

- It performs well on both microarray and RNAseq data.
- It does not depend on any specific module extraction approach used in a CEN method.
- Identification of mapped modules from both biological and topological associations across different stages, provides insights into the common behavioral changes in the modules during progression of a disease.

However, X-module performs well only for single species. There is need to validate its performance in finding module associations for one species against multiple species across stages.

Cluster-based or module-centric co-expression analysis can extract interesting patterns or behaviors of genes that would help identify crucial genes, participating in a common biological process. However, such network analysis methods in sequential processing mode are time-consuming, even for a moderately sized dataset. To address this issue, special efforts are necessary to construct and analyze CENs and extract significant modules using parallel processing.

Parallel Network Module Extraction

A generalized version of the TOM measure, called *GTOM* [450] has been useful in developing parallel CEN analysis. A major limitation of TOM is that it discovers smaller CEN modules, whereas the generalized TOM overcomes this issue. Jaiswal et al. [194] present a parallel network module extraction method called *PNME*, implemented utilizing general-purpose computing on graphics processing units (GPGPU), to facilitate efficient and parallel (i) construction of CENs, (ii) analysis of CENs, and (iii) extraction of highly correlated modules. PNME uses GTOM to extract significant modules corresponding to metastasis and non-metastasis stages of breast cancer in significantly less time. It works on both microarray as well as RNAseq data.

In co-expression analysis using either the clustering or CEN approach, it is essential to consider the occurrence of heterogeneity among samples. Condition or tissue-specific co-expressed network module extraction may not be possible in a CEN generated from multiple tissues or conditions, if the correlation signals among the tissue or condition-specific modules are not properly captured or are diluted due to lack of correlation with other tissues or conditions. On the other hand, restricting co-expression network analysis to a specific tissue or condition may also reduce sample size, and hence, may also decrease the ability to detect and analyze shared co-expression modules statistically. Therefore, we believe that CEN analysis methods that do not distinguish between tissues or conditions are applicable for identification of common co-expression modules, although a differential co-expression network analysis approach that compares condition- or tissue-specific behavior is preferable for extraction of modules specific to a condition or tissue.

Example 1: Consider an example dataset (shown in Table 5.16) that includes two genes g_1 and g_2 with expression values for seven samples. Let us use the Pearson correlation coefficient to measure the similarity between a pair of genes. The correlation value for the pair (g_1, g_2) is 0.422. If we consider a correlation threshold of 0.5 for a pair of genes to be connected, no edge exists between (g_1, g_2), since the correlation does not cross the threshold. In Figure 5.17, we show the connectivity with a dotted line, representing low correlations. However, for the same synthetic dataset, one can see that the PCC value for three subsets of samples, $\{s_1, s_2, s_4\}$, $\{s_3, s_7\}$ and $\{s_5, s_6\}$ are 0.67, 0.89, and 0.82, respectively. Each of these is higher than the threshold value. Hence, considering three subsets of sample genes, g_1 are g_2 should be connected. Figure 5.18 depicts this scenario.

It is our observation that most methods for CEN analysis consider the full space, i.e., all samples or features, when computing similarity for a given pair of genes whether heterogeneity exists among the samples or not. But, it is essential to consider subspace similarity in CEN construction and subsequent downstream analysis for more biologically relevant

Gene name	S1	S2	S3	S4	S5	S6	S7
g_1	0.33	0.29	0.129	0.72	0.11	0.03	0.86
g_2	0.45	0.14	0.34	0.56	0.56	0.5	0.59

Figure 5.16: A synthetic gene expression data

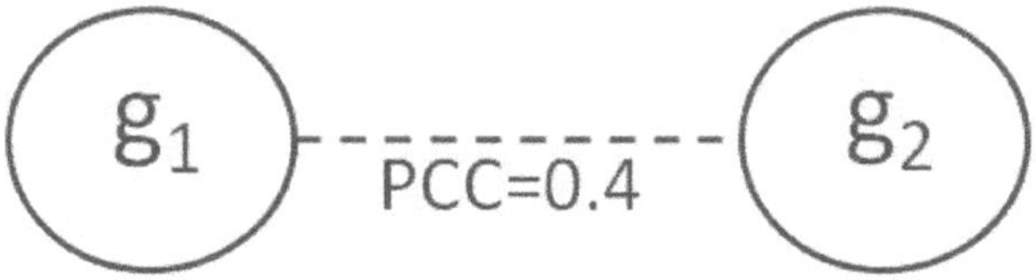

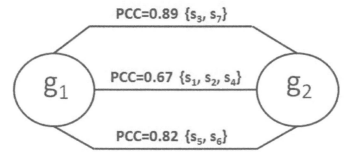

Figure 5.17: A pair of genes with low correlation value

Figure 5.18: A pair of genes connected via different subsets of samples

module extraction and comparison. It is another observation that most correlation measures are influenced by the order of samples in the gene expression matrix. However, for non-time series gene expression data analysis, such a constraint should not exist since there is no explicit ordering of samples. Therefore, we emphasize that one must be careful in choosing a similarity measure. The measure should not be order sensitive. One possible way is to operate on every possible pair of samples rather than only consecutive pairs of samples.

A technique called Subspace Module Extraction (SME) constructs a CEN that represents similarity among genes over a subset of samples. The CEN is a multigraph where multiple edges between a pair of nodes correspond to subspaces of samples or features over which genes are co-expressed. This special CEN can be thought of as a subspace co-expression network. Once the subspace CEN is constructed, one can extract highly coherent network modules from it. When extracting network modules, one needs to use a measure that can effectively exploit the topological properties of the graph. SME uses a measure called TSOM (Topological Sub-space Overlap Measure) [15] that considers both direct as well as neighborhood connectivity to compute similarity between a given pair of genes in a subspace CEN.

SME in Subspace CEN Construction

A subspace CEN can be thought of as an enhancement of the traditional CEN. An SME traces subsets of samples for which a pair of genes is similar. It treats the genes in the expression data as nodes of the graph, and it labels an edge between a pair of nodes or genes with a subset of samples. SME uses angular similarity to compute correlation between a given pair of genes, and hence, there can be multiple subsets for which a pair of genes are similar. A subspace CEN is defined as follows.

Definition 5.5.1 *Subspace CEN: A subspace CEN is a variant of a co-expression network that clearly identifies the subset(s) of samples over which a given pair of genes are similar. A subspace CEN is represented by a multigraph G=(V,E) where V and E represent the set of genes (unordered pair of vertices) and edges, respectively. Each edge in a subset CEN corresponds to a subset of samples.*

Figure 5.20 shows an example of a traditional co-expression network. Unlike such a network where there is only one edge between a pair of genes or nodes (as shown in Figure 4.1), a subspace co-expression network may contain more than one edge between a pair of genes or nodes.

SCEN construction is initiated with the extraction of possible subspaces of samples over which a pair of genes is similar. Consider two genes g_i and g_j to be similar over an ordered sample subset $\{s_1; s_2; \cdots ; s_n\}$, based on the following expression:

$abs(arctan(g_i(s_k) - g_i(s_{k-1})) - arctan(g_j(s_k) - g_j(s_{k-1}))) \geq \alpha$; for k $=2, \cdots , n$

Here, $g_m(s_n)$ is the expression value of gene g_m for sample s_n, $arctan(x)$ is the inverse tangent of x, $abs(x)$ is the absolute value of x, and α is a user-defined similarity threshold.

SCEN construction extracts relevant sample subset(s) for a pair of genes. SME starts with the first sample and traverses in an ordered way through the rest of the samples to identify other samples to form the sample subset. It includes a sample in the subset using an arctan threshold of α.

Once a subset is formed, it starts with another sample (non-selected) and iterates through a similar traversal process to include more samples. Construction continues until no sample is left for consideration. It does not consider subsets of samples with cardinality less than three for further analysis. Finally, it produces a multigraph where multiple edges between a pair of genes or nodes correspond to the subspaces over which the genes are similar.

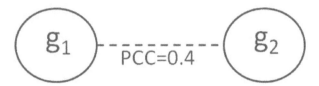

Figure 5.19: An example of a traditional co-expression network

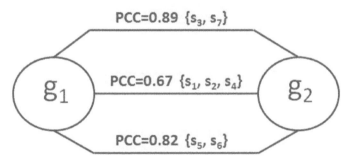

Figure 5.20: An example of a subspace co-expression network

(b) **Biological Analysis**
Three approaches have been discussed here for biological analysis of the outcomes of co-expression analysis.

Regulatory Network Analysis: Co-expression networks are useful in analyzing gene expression data for identification of interesting genes that play important roles in disease progression and biological functions. Using only gene-centric or gene-gene correlation-based association information, such networks are unable to infer causality relationships among genes. Causality information helps uncover many interesting facts about Transcription Factors (*TF*) and Target Genes (*TG*) involved in a given disease in progression. Tools have been developed to support causality relationship extraction. ARACNE [276] and GENIE3 [415][17] are two tools that construct regulatory networks from co-expression networks and support causality analysis. ARACNE [276] presents associations that are found to be regulatory by eliminating indirect associations among genes. It focuses on strong and direct associations among the genes. GENIE3 includes TF information that helps construct the regulatory network by identifying the expression patterns of TFs that best explain the expression of each TG. A major constraint with this tool is that TF information used in the network are not to help perform well by random chance. Several other tools have also been proposed in [362],[172],[83], each with its own advantages as well as disadvantages.

Pathway Analysis: This analysis approach obtains a set of significant genes or proteins based on the results generated by a clustering or module-centric approach. Such analysis identifies the association of a given gene or protein with a pathway and helps interpret omics data from the point of view of canonical prior knowledge structured in the form of pathway diagrams. It also helps find distinct cellular processes, diseases or signaling pathways that are associated with the identified gene(s) by a clustering or module-centric method. Genes associated with a particular disease have a high probability of being functionally connected within the processes or pathways associated with the corresponding disease. Pathway analysis can validate such genes or proteins.

(c) **Literature Mining**
Results obtained using downstream topological or biological analysis, following clustering-based module-centric approaches may not be authentic enough to be claimed as biologically significant. Literature mining and investigating well-established facts related to the concerned biological question can further strengthen the claim(s). Such mining activity may identify relevant established (c_1) wet-lab results, (c_2) ground truths, (c_3) knowledge bases, or (c_4) results obtained by other computational or statistical approaches to corroborate gene expression analysis. We can either (a) confirm our findings against established results or inferences, or (b) consider established finding(s) as reference(s) or support(s) to substantiate our findings as significant. For example, if the finding is a recommended set of hubgenes as causal for a given disease query based on module-centric analysis, case (c_1) may refer to similar results obtained by other researchers in wet-lab experiments. Similarly, c_2, c_3, and c_4 refer to established ground truths (used as benchmarks), relevant knowledge repositories and similar results obtained through other computational approaches, respectively.

C_5: **Validation**

This section presents three approaches to validate the results generated using clustering-based and network-based approaches from microarray or RNAseq gene expression data: (a) statistical, (b) biological and (c) literature mining.

1. **Statistical Validation**

    A number of statistical measures, tools, and indices have been introduced to analyze and validate results generated by clustering-based or network-based approaches from gene expression data.

    P-value: The P-value evaluates the quality of gene groups extracted by a clustering or CEN module extraction technique with reference to the groups of genes already defined by gene annotations in the Gene Ontology repository. One computes the p-value using the hypergeometric test or Fisher's Exact Test [54]. For a given group of genes, this test assumes a null hypothesis that the genes included in the group are random samples from the population. This test prepares a contingency table as shown in Figure 5.21 using annotations in the Gene Ontology. To compute the p–value using Fisher's Exact Test from the

	In gene ontology list	*Not in gene ontology list*
In generated gene set	$x1$	$x2$
Not in generated gene set	$x3$	$x4$

Figure 5.21: Contingency table for Fisher's Exact Test

contingency table, it uses the hyper-geometric distribution given in Equation 5.3.

$$p\ value = \frac{\binom{x1+x2}{x1}\binom{x2+x3}{x2}}{\binom{n}{x1+x3}}. \qquad (5.3)$$

With $n = x1+x2+x3+x4$. The p-value is useful in understanding how well a gene group matches pre-specified GO categories. A group of genes belonging to a (bi)cluster with low p-value is an indication that the genes belong to enriched functional categories identified in annotations of Gene Ontology. Such gene groups with significantly low p-values are considered biologically significant.

Q-value: In both cluster-based and gene-gene network analysis, Q-value is an effective statistical validity measure to assess performance in terms of biological significance of identified gene groups. The Q-value for a given gene is the proportion of false positives among the genes that are as or more highly differentially expressed. In other words, it is nothing but the minimum False Discovery Rate (FDR) at which this gene appears biologically significant. One can find the GO categories and Q-values from an FDR-corrected hypergeometric test for enrichment using benchmark web-based tools such as GeneMANIA [288]. To estimate the Q-values for a given gene, the Benjamini Hochberg procedure has been used. This procedure attempts to decrease the FDR. It avoids incorrect rejection of the null hypothesis by controlling the occurrence of smaller p-value by chance.

Precision, Recall and F-measure

Three statistical validity measures, Precision, Recall and F-measure [313] are typically used in the evaluation of the quality of gene groups obtained from a clustering or network module extraction or PPI complex- finding technique. *Precision* is defined as the percentage of match between the predicted groups of genes and the pre-specified or known gene groups. *Recall*, on the other hand, is the fraction of known gene groups that has been detected within the predicted clusters using the gene group finding method.

Assume that a co-expressed pattern-finding method predicts P_c clusters. Let the benchmark ground truth include B_c clusters. Using *overlapping score*, one can find the match between the predicted clusters and benchmark gene groups. Let N_{pc} be the number of predicted clusters that match at least one benchmark gene group and let N_{bc} be the number of benchmark gene groups that match at least one predicted cluster. One defines Precision and Recall for this method using Equations 5.4 and 5.5, respectively,

$$Precision = \frac{N_{pc}}{P_c}, \tag{5.4}$$

$$Recall = \frac{N_{bc}}{B_c}. \tag{5.5}$$

In addition, we can use another index, called the *F-measure* to compare a method against other methods. *F*-measure is the harmonic mean of

Precision and Recall. F-measure is given by Equation 5.6,

$$F--measure = \frac{(2 \times Precision \times Recall)}{(Precision + Recall)}. \tag{5.6}$$

A good co-expressed pattern-finding method is expected to exhibit high precision and recall values, and consequently, a high F-measure. A high precision value indicates better coverage of the cluster-finding method with reference to a set of benchmark clusters. Due to the increasing availability of the knowledge of genome, some researchers are of the opinion that the overlapping score threshold should be increased to 0.75 for some real biological datasets such as the yeast dataset. However, the existence of false positives in a dataset makes it difficult to use such threshold values for the dataset. Another observation is that inconsistencies observed during evaluation of a method can be handled using other means, which do not use thresholding. This may lead to the use of *Positive Predictive Value, Sensitivity* and *Accuracy* to determine the performance of these methods.

Positive Predictive Value (PPV): This measure assesses the amount of match of the predicted cluster set against the benchmark cluster set. One estimates PPV for the jth cluster using Equation 5.7,

$$PPV_{cluster j} = max_{i=1}^{P_C} PPV_{ij}. \tag{5.7}$$

The assessment of PPV for an individual cluster represents how closely the predicted cluster resembles the best matching benchmark cluster. For the estimation of PPV over all cluster sets w.r.t. all annotated complexes, it is desirable to obtain an overall PPV. The overall PPV of a method is computed by averaging PPV's of all clusters. A high PPV value specifies a higher fraction of correspondence between the predicted cluster and the ground truth clusters, implying better quality results.

Sensitivity: This measure indicates the fraction of the benchmark cluster set included in the predicted cluster set. Sensitivity S_j for a given j^{th} cluster is defined as the maximum value of sensitivity obtained for the jth cluster over all B_c real clusters. Mathematically, it is given as $S_j = max_{j=1}^{B_c} S_{i,j}$, where $S_{i,j} = X_{i,j}/N_i$, where N_i represents the cardinality of complex i. The overall sensitivity is the weighted average of the individual sensitivities, defined in Equation 5.8,

$$S = \frac{\sum_{i=1}^{B_C} N_i \times S_i}{\sum_{i=1}^{B_C} N_i}.$$ (5.8)

The higher the sensitivity value, the larger is the coverage of the predicted clusters by the benchmark clusters.

Accuracy: It is defined as the geometric mean of *PPV* and *S*. Accuracy is considered a compromised, yet effective evaluation measure for a gene group-finding method over the other two measures, i.e., *PPV* and *S*. Mathematically, it is shown in Equation 5.9,

$$Acc = \sqrt{PPV \times S}.$$ (5.9)

Accuracy is useful for unbiased evaluation of any co-expressed pattern-finding method with reference to the ground truths.

5.6 Chapter Summary and Recommendations

(a) *Clustering based on improper heuristics:* A major limitation of most traditional clustering approaches is that they attempt to partition the data without considering real biological associations. A clustering approach often creates multiple groups, even though such biologically meaningful groups do not exist. Most clustering methods form clusters using heuristic optimization, which may lead to wrong cluster outcomes.

(b) *Subspace module extraction:* Most CEN construction and analysis methods operate on gene expression data over the full space, i.e., all samples or features, whether heterogeneity exists among the samples or not. However, it is essential to consider the occurrence of heterogeneity among the samples. Condition- or tissue-specific co-expressed network module extraction may not be possible in a CEN generated from multiple tissues or conditions, if the correlation signals of the tissue or condition-specific modules are not properly captured. In such cases, it is necessary to consider subspace similarity computation in the process of CEN construction and subsequent downstream analysis for biological validity.

(c) *Heterogeneity in tissues:* A major issue with scRNA data clustering occurs due to the presence of heterogeneity in tissues. One such example is human blood, where cell-type frequencies span at least two orders of magnitude. Recent developments in sequencing technologies have enabled deeper sequencing, revealing additional rare cell types, pushing the range of frequencies to three or four orders of magnitude. Most clustering methods perform well when clusters are approximately equal in size, and thus perform poorly in the presence of rare cell types. Although some tools have been introduced to handle rare groups, unfortunately, they are unable to handle both frequent as well as rare cell types simultaneously.

(d) *Lack of customization:* Most clustering methods introduced to handle scRNA data are less flexible, and hence cannot be customized easily to use in a wide range of applications. So, developing a generic clustering method for handling a wide range of scRNA data is necessary.

(e) *Dependency on user input:* Most scRNA data clustering methods are dependent on one or more user input parameters. The performance of such clustering algorithms is largely dependent on appropriate selection of these parameters. For example, in the case of partitional methods or nearest-neighbor methods, the selection of k plays an important role. Most methods depend on some heuristics to decide an appropriate value for k. However, such heuristics often fail to support biologically enriched cluster identification.

(f) *High dimensionality reduction:* Generally, scRNAseq data include a large number of instances, samples or dimensions. Handling data with large dimensionality with the ability to detect clusters of both rare and frequent cell types within a reasonable time, is challenging. Visualizing such cluster results and interpreting meaningfully are difficult and time consuming as well. Traditional dimensionality reduction methods based on linear transformations (such as PCA) often fail to capture associations among cells accurately because of high levels of dropout and noise. Although nonlinear techniques are more flexible, a major difficulty with these methods is that they depend on one or more user input parameters which influence the results significantly. These parameters have to be tuned manually, which is difficult.

(g) *Lack of validation methods:* Another major issue with scRNA data clustering is the lack of validation techniques. A common approach used by most researchers is to acquire some known cell types (say, by selecting some well-studied and understood cells from distinct cell lines) and use them as ground truth to validate cluster results. However, these tissue samples (used as ground truth) are often simple, and hence, are unlikely to be helpful in validating cluster results for current large-scale and complex experimental studies.

Chapter 6

Differential Expression Analysis

6.1 Introduction

Microarrays have been widely used to analyze the transcriptome to understand the molecular basis of the variations of phenotypes in general, and for a given disease, in particular. Such analysis has been more effective with RNA sequencing. Both technologies have advantages and limitations. With the recent introduction of scRNA-seq, it is now possible to quantify the distribution of expression levels for each gene across a population of cells [304], [88]. A major advantage of RNA-seq data analysis is that it enables the study of the whole transcriptome in a single experiment.

An objective of transcriptome profiling is to identify genes that are differentially expressed (DEG), i.e., to find whether for a given gene, an observed difference in expression levels is significant between two conditions (i.e., normal and disease) [88]. A difference in expression levels between two conditions is considered significant, if it is greater than what is expected due to natural random variations. Such analysis that enables identification of differentially expressed genes is referred to as Differential Expression (DE) analysis, which is carried out on RNA-seq data based on fold changes of read counts or due to the presence of significant differences. The significance of the DEGs can be estimated using statistical analysis. Transcriptome profiling is helpful in characterizing molecular functions of the DEGs or in identifying the pathways in which DEGs are involved. For effective interpretation of the extracted knowledge, researchers use appropriate data visualization technique(s).

DOI: 10.1201/9780429322655-6

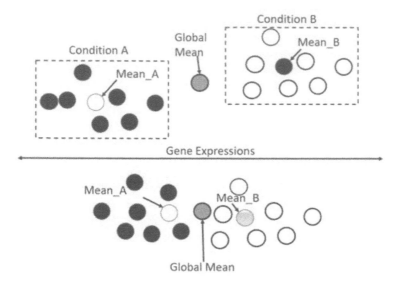

Figure 6.1: Differential expression of a gene: Local means vs Global mean

In top part of Figure 6.2, we consider information regarding conditions when estimating means. It can be clearly seen that the estimated local means have significant deviations from the global mean. In the lower part, we do not consider conditions when estimating the means. The local means are closer to the global mean and they do not have significant deviations from the global mean. So, based on deviations of means, we decide whether a given gene is differentially expressed or not.

6.1.1   Importance of DE Analysis

Differential expression analysis has numerous applications in transcriptomic data analysis. Generally, RNA-seq count data contains the expression of coding and non-coding genes expressed in non-targeted tissues as well as genes which cause increase in the dimensionality of the dataset. This analysis helps extract differentially expressed genes which are of interest for downstream analysis. It also eliminates the inherent curse-of-dimensionality problem in RNA-seq count data. Another important application of RNA-seq is the comparison of transcriptomes across developmental stages, and across disease states, compared to the normal cells [88]. The scRNA-seq data analysis helps identify the marker genes associated with a cell type by finding differentially expressed genes across the groups of cells identified in the previous analysis step using clustering [218].

6.2 Differential Expression (DE) of a Gene

During the transition of a gene from one state or condition to another, the amount of produced mRNA may change significantly. Such differentially expressed gene(s) play a significant role in the identification of interesting biomarkers for a given disease. There is always need for methods that use an effective metric to identify differentially expressed genes with high precision for transcriptome profiling.

Definition 1: *Differential Expression:* Differential expression refers to the amount of variation in expression across two states or conditions (e.g., normal and disease).

Definition 2: *Differential Expression of a Gene:* Differential expression of a gen is defined as the variation in the quantity of produced mRNA across states

6.2.1 Differential Expression of a Gene: An Example

Table 6.1 shows an example of synthetic RNA-seq count data for six genes over two conditions. For each condition (normal or disease), there are five samples. Figure 6.2 presents the patterns of the two genes, $G1$ and $G2$. We observe that the expression of $G1$ increases significantly in the samples in the disease condition, whereas, gene $G2$ shows almost steady expression in the disease condition. So, gene $G1$ may be considered differentially expressed gene.

Table 6.1: Synthetic RNA-seq Count Data: An Example

	Normal					Disease				
	S1	**S2**	**S3**	**S4**	**S5**	**S6**	**S7**	**S8**	**S9**	**S10**
G1	21	16	29	24	22	31	26	39	34	32
G2	13	8	21	16	14	10	5	18	13	11
G3	162	7	74	0	12	0	109	1	9	926
G4	57	26	77	91	51	34	50	31	49	9
G5	23	374	36	23	28	1494	10	42	4	0
G6	5	6	14	4	31	1	8	7	6	1

Definition 3: *DE Gene Patterns:* A differentially expressed gene pattern indicates the trend in the amounts of mRNA across conditions. Figure 6.2 shows an example of DE gene pair patterns.

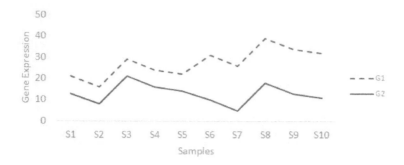

Figure 6.2: An example pair of DE gene patterns

6.3   Differential Expression Analysis (DEA)

Differential expression analysis is an established and effective approach for transcriptome profiling for identification of crucial gene(s) to support disease diagnosis. It identifies the crucial gene(s) by determining whether the differences between gene expression levels from two states or conditions are greater than the expected natural random variations. The basic tasks of DE analysis tools are the following [88].

a. Determine the magnitudes of differences in expressions between two or more conditions based on expression levels from replicated samples.

b. Apply statistical testing (such as t-test, F-test, Likelihood Ratio Test (LRT), Fisher's Exact Test, or Wilcoxon Signed Rank Sum Test) to estimate the significance of differences in expression levels.

On RNA-seq data, DE analysis recognizes crucial genes for a given disease such that differences in read counts between two conditions are greater than expected by chance. Such a method makes some assumptions regarding the distribution of read counts. In this analysis, the null hypothesis between two conditions states that the means computed for each feature (say, *gene, transcript*, etc), for the two conditions A and B are equal. Poisson distribution is commonly used to model read counts because it is a better fit for counts arising from technical replicates [277], especially, when a small number of biological replicates are available.

Note that this model predicts smaller variations than those seen among biological replicates [350]. Such data show a over-dispersion that is often missed (not captured) by the equal mean and variance assumptions of a Poisson distribution. Robinson et al. [350] introduce a modeling approach based on the Negative Binomial (NB) distribution for counts across samples to capture the over-dispersion, i.e. extra-Poisson variability. The NB distribution is more widely used than Poisson due to the introduction of a scaling factor for the variance. It has parameters, which are uniquely determined by mean and variance [306]. A bottleneck with the NB distribution is that the number of replicates in most RNA-seq dataset is often too small for reliable estimation of variance and mean for each individual gene. There are several empirical studies that evaluate the performance of DE analysis methods [92], [379], [381].

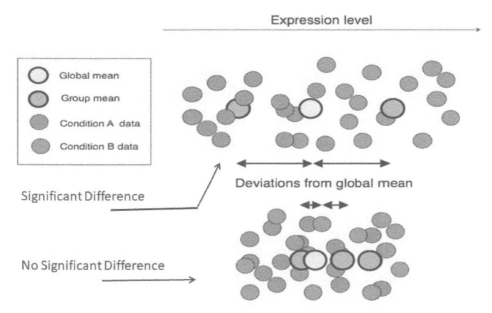

Figure 6.3: Expression deviation helps identification of DE genes: An illustration

6.3.1 A Generic Framework

In Figure 6.4, a generic framework of DE analysis hs been depicted. The four major tasks, i.e. (i) preprocessing, (ii) DE gene identification, (iii) DE gene analysis, and (iv) validation, performed during analysis are described next.

6.3.2 Preprocessing

Both microarray and RNA-seq technology generated gene expression data can be analyzed using similar statistical approaches. A major issue with both types of data is the presence of significant amounts of bias and noise.

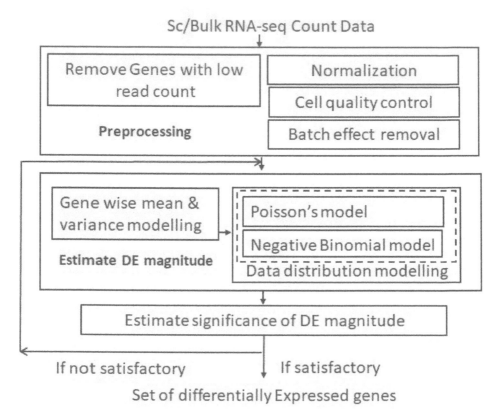

Figure 6.4: Differential expression analysis workflow

Guo and Sheng [153] observe that correlation between gene expression profiles generated by Affymetrix one-channel microarray and RNA-seq has a
Spearman correlation coefficient of 0.8. The authors are also of the opinion that the presence of background noise makes it challenging to measure
gene expressions of low abundance genes. Empirical studies establish that
in comparison to high abundance genes, low abundance genes have significantly lower correlations. Hussain, et al [88] state that on an average only
20% of all genes are covered by a microarray experiment, whereas RNA-seq
provides a comprehensive view of the transcriptome. In addition, the analysis of microarray data is simpler than that of RNA-seq data. Although the
cost-effectiveness of microarray analysis is higher than RNA-seq, one can
minimize the cost by lowering the desequencing depth in RNA-seq. An advantage of RNA-seq is that even without a reference genome, it is possible to
carry out de novo analysis. Another difference between these two data types
is that in RNA-seq, it is assumed that the read counts follow *Poisson* or *Negative Binomial* distribution, whereas the gene intensities across experimental

replicates in microarray data are assumed to follow a normal distribution or a heavy-tailed distribution [323]. Most researchers can achieve better precision in differentially expressed gene identification in RNA-seq data due to higher sensitivity. Without quality gene expression data, one cannot expect accurate or high-precision DEG identification.

Preprocessing transforms the input raw gene expression data matrix into a content-rich, unbiased and quality data matrix for subsequent downstream processing. It includes multiple tasks depending on the condition, quality, and type of input gene expression data. We discuss a few commonly used preprocessing tasks on microarray and RNAseq data.

Normalization

Normalization is necessary for both types of gene expression data for effective downstream biological analysis. To design a normalization method, we consider factors related to the expression data generation such as transcript size, GC-content, sequencing depth, sequencing error rate, and insert size [88]. Based on the approach and the number of factors considered, a number of normalization methods have been introduced. Each method has its drawbacks. To overcome this issue, several normalization methods can be used along with a consensus function to generate an unbiased normalized representation of the gene expression data by eliminating the bias in the individual methods. Such integrative approaches (like quantile normalization) perform satisfactorily for both microarray and RNA-seq data. There are several tools that help control the factors responsible for bias, either partially or totally. EDASeq [347] is an R package that controls GCcontent bias. One can also use the NVT package [113] to evaluate and identify the best possible normalization method for RNA-seq datasets using visualization. A method developed by Zyprych-Walczak et al. [464] finds the optimal normalization for a given dataset. However, normalization cannot eliminate batch effects, which affect specific subsets of genes differently and cause strong variability and decreased ability to identify interesting biological signals [241]. Normalization considers global properties of data to eliminate biases from individual samples.

Missing Value Estimation

Missing values occur in microarray data due to the presence of (i) scratches or (ii) dust in the microarray slides [1]. As discussed in the previous chapter, missing data values may occur (i) at random, (ii) not at random, or (iii) completely at random. If missing values are not handled properly prior to downstream processing, misleading inferences may be generated. If missing values occur randomly, one can remove the instances with missing values. In such a situation, for inference generation, one should consider only instances with complete example data. However, if missing values do not occur randomly, one should not ignore the missing valued instances because it may lead to inappropriate or biased inferences. Some imputation techniques to handle missing values in microarray data have already been discussed in previous chapter. Although the similar missing values problem does not occur in RNA-seq and scRNA-seq data, they suffer from the presence of a significantly high number of zeroes, a phenomenon referred to as *zero inflation*. Some imputation techniques for handling such zero-inflated instances are presented in previous chapter.

Pre-Test Filtering

Although RNA-seq technology quantifies gene expressions using a dynamic range, most count data matrices include large numbers of genes with low expression values [365]. Such low-expression genes are often inseparable from sampling noise. The presence of these genes in a count data matrix does not contribute to the downstream analysis, rather deteriorates the performance of DEG identification. Detection and filtering of low-expression genes greatly improve the cost-effectiveness as well as the performance of DEG identification.

Most DEG identification methods recommend elimination of low-expression genes that have average counts lower than an empirically determined threshold [350], [23], [259], [238]. There are other approaches to estimate the values of filtering thresholds. Two popular such estimation approaches are: (i) based on the distribution of intergenic reads per kilobase per million mapped reads (RPKM) values [167] and (ii) based on external RNA controls consortium (ERCC) spike-in controls [290].

Some authors claim that independent filtering improves the detection power of high-throughput experiments [365], [64]. Experiments with a transcriptome-wide qPCR dataset by Sha et al [365] have shown that there is significant improvement in sensitivity and precision of DEGs after filtering. However, there are no universal optimal filtering thresholds for all pipelines. If such an universal filtering threshold can be determined with an exhaustive empirical study, it may be possible to identify large sets of crucial DEGs.

Normality Test

Prior to performing statistical analysis of gene expression data, one important requirement is to understand the distribution pattern(s) of the expression values in different stages [311], [140]. Such understanding helps apply statistical techniques to generate meaningful inferences. Although the application of statistical techniques, assuming the normal distribution is common, it may not be appropriate in all situations. There may be significant violations of the normality assumption. In such cases, statistical analysis cannot be expected to generate results with high precision. So, normality test(s) should be performed at an initial stage to check if the gene expression data conform to the normal distribution. A normality test can provide answers to the question whether the samples are drawn from a normally distributed population.

One common approach to test conformity of a given gene expression dataset to the normal distribution is to evaluate graphs in association with a normality test. Researchers use various statistical tests [311] to assess the normality assumption. These include (i) Anderson-Darling test, (ii) chi-square goodness of fit test, (iii) Cramer-von Mises test, (iv) Kolmogorov-Smirnov (K-S) test, (v) Lilliefors corrected Kolmogorov-Smirnov test, (vi) Shapiro-Wilk test, (vii) D'Agostino skewness test, (viii) Anscombe-Glynn kurtosis test, (ix) D'Agostino Pearson omnibus test, and (x) Jarqua-Bera test. Each test has its advantages and limitations. Instead of one, use of an appropriate combination of tests selected based upon empirical studies can give more accurate results for any gene expression data.

Quality Checking

The occurrence of low-quality reads is a major issue with the RNAseq count data. Such reads typically occur due to lack of appropriate (i) library preparation and (ii) sequencing. There are three additional sources which contribute to low-quality reads, (i) sequence-specific bias, (ii) PCR artifacts, and (iii) untrimmed adaptor sequences. To rectify the situation prior to downstream processing, one applies quality checking to identify biases and minimize (it is not possible to eliminate totally) information loss so that an improved DE analysis can be achieved. To support such activities, we need technique(s) to handle parameters, viz. (i) GC content, (ii) presence of adaptors, (iii) read quality, (iv) over prepresentation of k-mers, and (v) duplicated reads. A number of tools have been introduced to handle this issue. Some example tools [88] are *FastQC. NGSQC, HTQC, PRINSEC, cutadapt*, and *SolexaQA*.

Quality Control for Mapping Reads

The quality of RNA-Seq count data is affected by the presence of biases caused by sources like *hexamer, GC, amplification, mapping, sequence-specific* and *fragment-size biases*. Application of quality control mechanisms reduces such biases. The quality of reference-based alignment is influenced by the quality of the reference genome. Such alignment processes are impacted by missing or erroneous information, because they depend on the reference genome and annotation information [338], [426]. Some unwanted biases may also be introduced in the alignment data by the mapping algorithm or the sequencing technology. Such biases need to be eliminated or minimized carefully and it requires quality control mechanisms. A number of tools have been proposed to handle this issue, and most facilitate viewing of the results graphically. Some popular quality control tools are *RSeQC, MultiQC, QuaCRS, Picard, RNA-SeQC*, and *SOAPdenovo-Trans*.

Quality Control for de Novo Assemblies

An appropriate quality control mechanism for de novo transcriptome assemblies improves precision of downstream analysis. Some methods use only the input reads to evaluate the quality of de novo assemblies based on accuracy and completeness. One such tool is TransRate [377]. It introduces two statistical scoring metrics, (i) *TransRate contig score* and (ii) *TransRate assembly score* to assess quality. The contig score gives an accuracy assessment for each contig assembly, whereas assembly score assesses the assembly in terms of completeness and accuracy. Another effective tool is DETONATE [139]. Empirical studies [88] show that around 70% of the genes identified as differentially expressed are common between de novo assemblies and reference-based sequences. The remaining 30% is different due to incomplete gene annotations, exon-level differences, and transcript fragmentation. Even when a reference genome is available, de novo analysis can still be performed to eliminate biases.

Batch Effect Removal

In RNA-seq data, batch effects are powerful sources of variation and systematic error. These variations occur because measurements are influenced by laboratory conditions. Such unavoidable batch effects affect DE analysis of RNA-seq or scRNA-seq data and mislead the findings [380]. However, the effects can be reduced to a great extent in high- throughput sequence data because the large volume of data assists in detection and minimization of batch effect. Hicks et al. [176] observed that batch effect causes significant cell-wise variations in scRNA-seq data. Normalization methods remove biases from individual samples only, not from a batch because they consider global properties of the genes. On the other hand, batch effect typically causes variations or errors in specific subsets of genes [160]. Two commonly used batch effect handling tools are modified ComBat [316] and ARSyN [303]. Although these tools were initially developed for microarray data, they can handle such variations in RNA-seq data also. Other effective tools [88], [343], [240] include gPCA, Swamp, and SVAseq.

Dimensionality Reduction

Among the three types of gene expression data, microarrays, RNA-seq and scRNA-seq data, the number of dimensions in scRNA-seq data is typically very high. Most downstream analysis methods fail to perform satisfactorily over such high-dimensional scRNA-seq data due to (i) the curse of dimensionality and (ii) the inclusion of many irrelevant dimensions when computing proximities. The use of an appropriate algorithm to select only those dimensions that are useful in capturing the underlying structure of single-cell RNa-seq data and to embed the expression matrix over a lower-dimensional space helps achieve high-precision downstream analysis [261]. Single-cell RNA-seq data are inherently low dimensional [173], and as stated by Luecken et al. [261], the biological manifold where cellular expression profiles lie can be adequately described with only a few dimensions. A dimensionality reduction technique facilitates a better visual representation of high-dimensional scRNA-seq data over an optimal reduced space for easy understanding. Scatter plot is one such means to represent the data visually over reduced space. Another objective of dimensionality reduction is to reduce data to its essential components by summarization to support effective downstream analysis. The accomplishment of both objectives helps analyze the variability of high-dimensional scRNA-sq data across the states towards generation of meaningful inferences.

Cell Quality Control

In addition to the aforementioned preprocessing tasks, this task is specially relevant for scRNA-seq data. Typically, scRNA-seq data are high dimensional, with a significant number of low-quality cells. Detection and removal of such poor-quality cells is essential to achieve high-precision downstream processing. Some researchers treat such low-quality cells as dead cells. Ilicic et al. [189] present a generic approach to detect dead cells using more than 20 curated biological and technical features. The authors claim that preprocessing (filtering dead cells) improves classification accuracy by over 30%. Several recent tools treat such poor-quality cells. SCell [104] is an integrated tool for cell quality filtering. This tool is open source, with an intuitive graphical interface. It detects low-quality cells for a given sample based on the difference between the number of genes expressed at the background level and the average number of expressed genes. Cells in which the number of expressed genes is significantly lower than the average number of genes expressed, are treated as dead cells and are filtered. Another flexible data preprocessing pipeline called scPipe [405] enables multiple preprocessing steps including removal of low-quality cells. This pipeline is developed using R and C++. It outputs the count data matrix as well as QC plots and other data summaries in HTML format. SinQC [197] is another cell quality control tool that detects low-quality samples by integrating gene expression patterns and data quality information.

Discussion

Although preprocessing generally improves the quality of input gene expression data, not all preprocessing steps may be relevant for a given microarray or RNAseq dataset for DE analysis. Some datasets are already preprocessed, and hence may not need most preprocessing steps. Some preprocessing techniques such as discretization (relevant for microarray data), normalization, and dimensionality reduction may lead to information loss. Depending on the nature of downstream processing, some preprocessing techniques may not be useful. However, the use of appropriately designed preprocessing technique(s) can significantly help in achieving the desired goals of differential expression analysis.

6.3.3   DE Genes Identification

Differential expression analysis (DEA) identifies significantly changing expression patterns of transcripts across experimental conditions using statistical methods (e.g., mean or standard deviation based) [42]. A gene expression pattern's behavior are caused by genetic and epigenetic factors during diseased condition. Differently expressed genes play crucial roles in disease biomarker identification. Genes with non-conforming expressions in diseased conditions appear in both microarray and RNA-seq data, This chapter focuses on the issue of DEG identification for RNA-seq data. Analysis using statistical method(s) helps identify crucial genes. However, the use of inappropriate statistical methods may lead to deteriorated performance of the DE gene finding methods. Further, an inadequate number of samples may also be a bottleneck in differentiating such genes using statistical methods. Furthermore, the success of most statistical methods largely depends on data distribution patterns. Although most statistical methods assume the distribution of data to be normal, not all gene expression datasets follow the normal distribution.

Fast and accurate normality tests are available (discussed in the previous section) to check whether the input data follow the normal distribution. If the data follow normal distribution, we can apply parametric tests for analysis. In case of non-normal distributions, one can apply non-parametric tests. However, with a very small number of samples, none of the parametric or non-parametric tests may be effective. In such a case, we can apply other methods such as *Fold Change* [42]. Tusher et al. [412] define Fold Change for a gene as the ratio of the means of two groups of samples. If the means of two groups of samples are represented as μ_{g1} and μ_{g2}, it is expressed as $\frac{\mu_{g1}}{\mu_{g2}}$. Choe et al. [85] define *Fold Change* in a different way. They define it for a dataset with all *log*2-transformed values as the difference between the means of two groups of samples. Although this measure is simple, it may not give good results for all datasets. Some parametric and non-parametric tests which are commonly used during statistical identification of DE genes are discussed next.

DE Gene Identification Using Parametric Methods

A parametric test assumes that (i) the input data follow a normal distribution and (ii) the same variance property is exhibited acreoss groups. In other words, parametric tests can be applied when the gene expression data are expressed in (i) absolute numbers or (ii) values, rather than ranks. A number of parametric tests have been developed for gene expression data that follow the normal distribution to identify crucial genes for a given disease.

(a) *t*-Test: This parametric hypothesis test considers a test statistic that follows a Student's *t*-distribution under the null hypothesis. Student's *t*-distribution (or simply the *t-distribution*) is a continuous probability distribution. It arises in estimating the mean of a normally distributed population where the sample size is small and the population standard deviation is unknown. Thus, it is a continuous probability distribution function, strongly related to the normal distribution.

If we select a sample of n independent observations from a normally distributed population with mean μ and variance σ^2, the standard normal distribution, Z, is given as

$$Z = \frac{\bar{X} - \mu}{\sigma/\sqrt{n}}$$

where, $\bar{X}$ is the mean of the selected sample. In practice, we may not know the value of σ (population standard deviation). So, instead, we use the standard deviation of our selected sample, which is denoted as S. The value of S changes with the sample selected. This distribution is termed t-distribution, with $(n-1)$ degrees of freedom.

$$t = \frac{\bar{X} - \mu}{S/\sqrt{n}}$$

Example 1: Consider a sample of 8 values from a normally distributed population with $\mu = 0$. The values are
$-4, -2, -2, 0, 2, 2, 3, 3$.
To find the t-distribution for the sample, we need mean and standard deviation of the sample, which are $\bar{X} = 0.25$ and $S = 2.65$, respectively.
Therefore,
$t = \frac{0.25-0}{2.65/\sqrt{8}} = 0.266$

A t-test compares the average values of the two data sets and determines whether they are from the same population or not.
Types of t-tests include the following.

- *Paired t-test:* It is performed when the samples typically consist of matched pairs of similar units, or when there are cases of repeated measures. Paired t-tests are of a dependent type, as these involve cases where the two sets of samples are related, like a comparative analysis involving children, parents or siblings.

- *Equal variance t-test:* It is used when the number of samples in each group is the same, or the variance of the two data sets is similar.

- *Unequal variance t-test or Welch's test:* It is used when the number of samples in each group is different, and the variance of the two data sets is also different.

(b) *ANOVA:* This parametric statistical tool is used to compare two or more datasets. It calculates the variance among the variables in the dataset and determines whether a relationship exists between the datasets. ANOVA is based on the following assumptions,

(a) *Independence of cases*: It assumes the samples are randomly selected.

(b) *Normality*: It assumes the samples follow a normal distribution.

(c) *Homogeneity*: It assumes the variance between the groups or datasets are the same.

Three types of ANOVA tests are commonly used.

(a) *One-way ANOVA:* This test compares multiple groups (> 2) based on one factor or parameter. For example, it can be used to compare the output of three groups of workers based on *working hours* of each group.

(b) *Two-way ANOVA:* It can compare more than two groups based on two factors. For example, comparing the output of three groups of workers based on two factors, such as *working hours* and *working conditions* of each group.

(c) *k-way ANOVA:* This test is useful in comparing more than two groups based on more than two factors. For example, comparing the output of three groups of workers based on several factors such as *working hours, working conditions, skills,* and *distance between home and workplace.*

The following example explains the working of this tool.
Example: Consider a dataset as given in Table 6.2 that includes three groups of students in a class based on the food intake. The dataset represents the number of questions on an examination solved by each student. We want to find whether there is any relationship between "how much food a student takes" and "how many questions the student solves."

The general observation is that as food intake increases, the number of questions solved also decreases. Using ANOVA, we find whether there is any correlation between food and number of questions solved (X). Table 6.3 shows the intermediate results to illustrate the working of ANOVA.

Table 6.2: Example Data to Illustrate Working of ANOVA

Low Food Intake		Medium Food Intake		High Food Intake	
Student	Questions (X)	Student	Questions (X)	Student	Questions (X)
1	10	5	8	9	4
2	9	6	4	10	3
3	6	7	6	11	6
4	7	8	7	12	4

Table 6.3: Intermediate Results to Illustrate Working of ANOVA

Low Food Intake			Medium Food Intake			High Food Intake		
St	Qns (X)	X^2	St	Qns (X)	X^2	St	Qns (X)	X^2
1	10	100	5	8	64	9	4	16
2	9	81	6	4	16	10	3	9
3	6	36	7	6	36	11	6	36
4	7	49	8	7	49	12	4	16
	$\sum X_1 = 32$	$\sum X_1^2 = 266$		$\sum X_2 = 25$	$\sum X_2^2 = 165$		$\sum X_3 = 17$	$\sum X_3^2 = 77$

- Total number of samples (students), $N = 12$.
- Number of samples in each group, $n = 4$.
- Number of groups, $k = 3$.
- Correction term, $C_x = \frac{(\sum X)^2}{N}$. Here,
  $C_x = \frac{(32+25+17)^2}{12} = \frac{5476}{12} = 456.33$.
- Sum of squares of total, $SS_T = \sum X^2 - C_x$. Here,
  $SS_T = (266 + 165 + 77) - 456.33 = 51.67$.
- Sum of squares among groups, $SS_A = \frac{(\sum X)^2}{n} - C_x$. Here,
  $SS_A = (\frac{32^2}{4} + \frac{25^2}{4} + \frac{17^2}{4}) - 456.33 = 28.17$.
- Sum of squares within groups, $SS_W = SS_T - SS_A$. Here,
  $SS_W = 51.67 - 28.17 = 23.5$.
- Mean of sum of squares among groups, $MSS_A = \frac{SS_A}{k-1}$. Here,
  $MSS_A = \frac{28.17}{3-1} = 14.085$.
- Mean of sum of squares within groups, $MSS_W = \frac{SS_W}{N-k}$. Here,
  $MSS_W = \frac{23.5}{12-3} = 2.611$.
- F ratio: This value signifies the confidence level of the correlation in the datasets based on the given factor, i.e., the relation among

the groups of students based on the food intake. Mathematically,

$$F_{ratio} = \frac{MSS_A}{MSS_W}$$

$F_{ratio} = \frac{14.085}{2.611} = 5.394$

Table 6.4 presents the sources of variations based on the example data.

Table 6.4: Sources of Variations in the Example Data

Source of variance	Degrees of freedom (df)	SS	MSS	F Ratio
Among Groups	$k - 1 = 3 - 1 = 2$	28.17	14.085	5.394
Within Groups	$N - k = 12 - 3 = 9$	23.5	2.611	
Total	11			

Null Hypothesis: H_0: There is no significant effect of food intake on the number of questions solved.

Alternate Hypothesis: H_a: There is significant effect of food intake on the number of questions solved.

From the F-ratio table for $\alpha = 0.05$, $F = 4.2565$, we can find that-calculated F (5.394) > F($\alpha = 0.05$). Therefore, null hypothesis is rejected. There is 95% confidence in the correlation between food intake and questions solved.

From the F-ratio table for $\alpha = 0.01$, $F = 8.022$, we can find that-calculated F (5.394) < F ($\alpha = 0.01$). Therefore, null hypothesis is accepted. There is 99% confidence that there is no correlation between food intake and questions solved.

(c) *Pearson correlation test:* Correlation is a statistical measure that estimates the degree of relationship between two variables. If two or more quantities vary so that movements in one tend to be accompanied by movements in the other, they are said to be correlated. Correlation analysis obtains the degree and direction of the relationship between the two variables under study. The measure of correlation is called the *correlation coefficient*. The value of the correlation coefficient varies between +1 and −1, where +1 indicates positive correlation and −1 indicates negative correlation. As the correlation coefficient value goes towards 0, the relationship between the two variables becomes weaker.

Pearson's correlation coefficient is a parametric test that measures the statistical relationship between two continuous variables. It is based on the method of covariance. It computes the magnitude of the association or correlation, as well as the direction of the relationship.

Pearson correlation between variables in datasets X and Y is calculated by

$$r_{XY} = \frac{\sum_{i=1}^{n}(X_i - \overline{X})(Y_i - \overline{Y})}{\sqrt{\sum_{i=1}^{n}(X_i - \overline{X})^2}\sqrt{\sum_{i=1}^{n}(Y_i - \overline{Y})^2}}$$

where X_i and Y_i are the mean values of datasets X and Y, respectively. This is obtained by dividing the covariance by the product of the standard deviations. Pearson's correlation coefficient makes the following assumptions.

- Both variables are normally distributed.

- There are no significant outliers. Pearson's correlation coefficient is very sensitive to outliers and can lead to misleading results.

- Each variable is continuous.

- The two variables have a linear relationship.

- The observations are paired observations. For every observation of the independent variable, there is a corresponding observation of the dependent variable. For example, for correlation between age and weight, there must be a weight value corresponding to each age value.

- Homoscedascity holds because the variables have equal variances,

DE Gene Identification Using Non-Parametric Methods

In this section, we introduce some well known non-parametric methods that are commonly used to assess the significance of DE genes. Test statistics obtained from a single method are often inadequate for unbiased assessment of the significance. So, the knowledge of multiple testing techniques and their applicability is essential.

(a) *Wilcoxon Rank Sum Test*: This non-parametric test is also known as the *Mann Whitney-U Test*. It tests the equality of the means of two independent samples. One can consider it as the non-parametric version of the two-sample t-test, where unlike the t-test, the sample size is small and not normally distributed. This test compares the outcomes of two independent samples and check whether the samples are likely to belong to the same population or not, based on comparison of the medians between the two samples. If the test statistic for *Wilcoxon Rank Sum Test* is U, we can calculate U_1 for sample 1 and U_2 for sample 2 as follows.

$$U_1 = n_1 n_2 + \frac{n_1(n_1 + 1)}{2} - R_1$$

$$U_2 = n_1 n_2 + \frac{n_2(n_2 + 1)}{2} - R_2$$

where, n_1 and n_2 are the numbers of data examples in samples 1 and 2 respectively and R_1 and R_2 are sums of the ranks for samples 1 and 2 respectively.

If $U_1 < U_2$, then $U = U_1$, else $U = U_2$.

We can form the hypotheses for Wilcoxon Rank Sum Test as –

Case 1- H_0 : The two samples are equal.

Case 2- H_1 : The two samples are not equal.

As an example, let us consider a situation where 10 people are to be tested for a medication for asthma. Five people are selected in random and are given the actual drug. The other 5 are given a placebo. The people take the medication for 5 days and record the number of times shortness of breath is experienced per day. The records of the volunteers are reported in Table 6.5.

Table 6.5: Experiment with 10 Volunteers.

Placebo	7	5	6	4	12
Drug	3	6	4	2	1

The next step is to give ranks to the data examples. For a total of 10 examples, rank is given from 1 to 10. We first order all samples and compute the ranks, as shown in Table 6.6.

Here, $n_1 = 5, n_2 = 5, R_1 = 4.5 + 6 + 7.5 + 9 + 10 = 37, R_2 = 1 + 2 + 3 + 4.5 + 7.5 = 18$.

$U_1 = 5(5) + \frac{5(6)}{2} - 37 = 3$; $U_2 = 5(5) + \frac{5(6)}{2} - 18 = 22$.

Therefore, $U = 3$.

Using the critical value table for U is seen in Figure 6.5 with significance value 0.05, $U = 2$. Since $3 > 2$, the null hypothesis is not rejected.

Table 6.6: Ordering and Ranking of the Total Samples

Placebo	Drug	Total sample (ordered)		Ranks	
		Placebo	Drug	Placebo	Drug
7	3		1		1
5	6		2		2
6	4		3		3
4	2	4	4	4.5	4.5
12	1	5		6	
		6	6	7.5	7.5
		7		9	
		12		10	

n_2	α	3	4	5	6	7	8	9	10	11	12	13	14	15	16	17	18	19	20
3	.05	--	0	0	1	1	2	2	3	3	4	4	5	5	6	6	7	7	8
	.01	--		0	0	0	0	0	0	0	1	1	1	2	2	2	2	3	3
4	.05	--	0	1	2	3	4	4	5	6	7	8	9	10	11	11	12	13	14
	.01	--	--	0	0	0	1	1	2	2	3	3	4	5	5	6	6	7	8
5	.05	0	1	2	3	5	6	7	8	9	11	12	13	14	15	17	18	19	20
	.01	--	--	0	1	1	2	3	4	5	6	7	7	8	9	10	11	12	13
6	.05	1	2	3	5	6	8	10	11	13	14	16	17	19	21	22	24	25	27
	.01	--	0	1	2	3	4	5	6	7	9	10	11	12	13	15	16	17	18
7	.05	1	3	5	6	8	10	12	14	16	18	20	22	24	26	28	30	32	34
	.01	--	0	1	3	4	6	7	9	10	12	13	15	16	18	19	21	22	24
8	.05	2	4	6	8	10	13	15	17	19	22	24	26	29	31	34	36	38	41
	.01	--	1	2	4	6	7	9	11	13	15	17	18	20	22	24	26	28	30
9	.05	2	4	7	10	12	15	17	20	23	26	28	31	34	37	39	42	45	48
	.01	0	1	3	5	7	9	11	13	16	18	20	22	24	27	29	31	33	36
10	.05	3	5	8	11	14	17	20	23	26	29	33	36	39	42	45	48	52	55
	.01	0	2	4	6	9	11	13	16	18	21	24	26	29	31	34	37	39	42

Figure 6.5: Critical value table for U using n_1 and n_2

(b) *Permutation Test*: The Permutation test constructs a sampling distribution, for any test statistic, by resampling the observation data. It is used for experimental data where the distribution is built under a strong null hypothesis that the set of observed data has no effect on the outcome, i.e., after random shuffling of the data as many times as we like, the response value of the shuffled data looks exactly like the response value for the real data. The hypothesis is formed as follows:

Case 1- H_0 : There is no difference between the two groups of data.

Case 2- H_1 : There is a difference between the two groups of data.

The statistic for the Permutation Test is the mean-difference of the response.

Test statistic $= \mu_{treatment} - \mu_{control}$.

Mean-difference for all shuffles is calculated to form a distribution to find the p-value. Numerous shuffles are conducted and a histogram of all the test statistic values is formed. To calculate the p-value for a Permutation Test, we count the number of test-statistics that are equal to or more than the initial test statistic, and divide that number by the total number of test-statistics we calculate.

As an example, assume we test a new medication for asthma on a group of 10 volunteers. A random selection is done such that 5 volunteers are given the drug and 5 volunteers are given a placebo. According to the null hypothesis, there is no difference when the data are shuffled, which means the drug does not work. In the first step, two groups are formed, one with the drug, called the *treatment* group, and the other with the placebo is the *control* group. The response value is the number of asthma attacks faced by the volunteers. The recorded asthma attacks and the corresponding shuffled data are as shown in Table 6.7.

Table 6.7: Recorded and Shuffled Data of 10 Volunteers. μ Represents *Mean of Columns*

Placebo (Control)	Drug (Treatment)	Real data (ordered)		Shuffle 1		ordered Shuffle 1	
		Control	Treatment	Group1	Group2	Group1	Group2
7	3		1	6	3	1	
5	6		2	4	6	2	
6	4		3	2	4		3
4	2	4	4	7	5	4	4
12	1	5		1	12		5
		6	6			6	6
		7				7	
		12					12
		$\mu_c = 6.8$	$\mu_t = 3.2$			$\mu_c = 4$	$\mu_t = 6$

$T_{real} = 3.2 - 6.8 = -3.6$; $T_{shuffle1} = 6 - 4 = 2$, so on.
If 50 shuffles are conducted for this experiment, then to find the p-value, we count the number of test statistics that are equal to or more than -3.6 and divide the value by 50.

(c) *Modified Wilcoxon Rank Sum Test*: As in the Wilcoxon Rank Sum test, outcomes of two independent samples are compared to decide whether the samples are likely to belong to the same population. Consider a random sample bivariate population, $(X_1, Y_1), \ldots, (X_n, Y_n)$, having a continuous distribution function $F(x, y)$. The marginal distribution of X_i and Y_i are $F_X(x_i)$ and $F_Y(y_i)$ respectively. The hypotheses are stated as –
H_0 : The two samples are equal, i.e., $F_X = F_Y$.
H_1 : The two samples are not equal, i.e., $F_X \neq F_Y$.
For a sample X, let the data be ordered as $X_1 < \cdots < X_n$. The ranks of the data are $R_i(where, i = 1, \ldots, n)$, for each X_i in the combined sample.

A value Z_n is defined as asymptotically normal with asymptotic mean equal to 0 and asymptotic variance given by $\sigma^2 = (1 - \rho_G)/12$. ρ_G is a correlation value such that $\rho_G = corr\{F_X(X_1), F_Y(Y_1)\}$.

$$Z_n = \sqrt{(2n+1)}[\frac{1}{n}\sum_{i=1}^{n} a_n(\frac{R_i}{2n+1}) - \int_0^1 a(u)dF_x\{H^{-1}(u)\}]$$

. where $a_n(u)$ (for $0 < u < 1$) is a score function such that $a_n(u)$ converges to $a(u)$ and $n \longrightarrow \infty$ and

$$H(y) = \frac{1}{2}F_X(y) + \frac{1}{2}F_Y(y), \{H^{-1}(u) = inf\{y : H(y) \geq u\}$$

. where $a_n(u) = u$ is the Wilcoxon score and $F_X = F_Y$. Let $\hat{\rho}_G$ be the Spearman coefficient of rank correlation, such that,

$$\hat{\rho}_G = \frac{12}{n(n^2 - 1)}\sum_{i=1}^{n} S_i T_i - \frac{3(n+1)}{n-1}$$

. where S_i is the rank of X_i in the X sample and T_i is the rank of Y_i in the Y sample. $\hat{\rho}_G$ is a consistent estimator of ρ_G, and therefore $\sigma_n^2 = (1 - \hat{\rho}_G)/12$ is a consistent estimator of σ^2. Thus, the Modified Wilcoxon Rank Sum statistic (W_p) under pairing is defined as –

$$W_p = \sqrt{(2n+1)}\{\frac{1}{n(2n+1)}\sum_{i=1}^{n} R_i - \frac{1}{2}\}/\sigma_n$$

. The consistency of σ_n leads to asymptotic standard normality of W_p. At significance level α, the null hypothesis $H_0 : F_X = F_Y$ is rejected [233].

(d) *Significance Analysis of Microarrays*: It is a non-parametric, permutation-based method used for large-scale microarray data analysis, such as microarray experiments where 10,000 proteins identify 100 proteins using a $-value cutoff of 0.01. SAM uses t-test at the individual gene level to determine the significance of the expression pattern of the gene. It is used when the samples may not be independent of each other and are not normally distributed. SAM can identify expression patterns that have little difference between control and test groups,but are still significant. The input to SAM is gene expression measurements along with a response variable for each gene experiment. The response variable is a grouping like, treatment and control, or a multiclass grouping like types of cancer, or a quantitative variable like blood pressure.

SAM computes a statistic d_i, for each gene i, to measure the strength of the relationship between the gene expression and response variable. Repeated permutations of the data are conducted to determine whether the expression of a gene is significantly related to the response. The cut-off value for significance is determined by a parameter δ, chosen by the user based on the false positive rate [89].

(e) *Shrinkage-t*: It is a non-parametric test that does not require prior information regarding the population covariance matrix. It is an estimator for population mean μ under the "*large p small n*" setting, where p is the number of genes or dimensions and n is the sample size. The shrinkage estimator works on population mean under the quadratic loss function with an unknown covariance matrix, $\sum_p$ [424].

In addition to the parametric and non-parametric tests discussed above, there are a few others which also help the identification of DE genes statistically. The readers are referred to [42] for the tests. Before applying any statistical technique or measure on gene expression data, one has to understand the distribution patterns of the data.

Data Distribution Modeling

Some measures are applicable only when the data follow relevant distribution pattern(s). There are well defined method(s) to check whether a gene exptession dataset follows a given data distribution model. This section briefly introduces data distribution models which most gene expression data follow. A detailed discussion of these data distribution models is available in the *Chapter* 4.

(a) *Poisson Model*: This model refers to a discrete probability distribution for a given number of events occurring in a given interval of time. The average number of times an event occurs in the same interval of time is also given. A gene expression dataset is said to follow Poisson distribution, if the four conditions are satisfied: (i) during the interval, an event may occur any number of times, (ii) the occurrences of events are independent and not influenced by the probability of occurrence of other events during the same time interval, (iii) the rate of occurrence of an event is constant, and (iv) the probability of occurrence of an event is proportional to the length of the time period.

For any (i) discrete random variable X representing the number of events in gene expression data occurring over a given time period and (ii) λ, the expected value or the average of X, if X follows a Poisson distribution, the probability of observing k events over the time period is–

$$P(X = k) = \frac{\lambda^k e^{-\lambda}}{k!} \qquad (6.1)$$

where e is Euler's constant [232].

(b) *Negative Binomial Model*: Statistical approaches to support identification and assessment of DE genes are complicated because the number of genes in a gene expression dataset is higher than the number of samples, and count-variability of RNA-seq data cannot be properly modeled with common probability distributions, such as binomial and Poisson distributions. Negative Binomial distribution parameterized towards a constant value or mean value is a powerful model to calculate the dispersion estimates across two conditions. The Negative Binomial (NB) distribution offers a more practical model for handling RNA-Seq count variability than the Poisson distribution model [353], [352]. One major reason is that the NB model allows an exact (non-asymptotic) test for comparing two groups, saving statistical power significantly by estimating the negative binomial dispersion parameter for all genes. An experimental study reported in [103] claims that the assumption of constant dispersion parameter is not realized for the RNA-Seq data sets. The experiments suggest that there are several intermediate possibilities – between the assumption of a constant dispersion parameter and separate dispersion parameters against each gene. These possibilities are often more realistic and at the same time still retain the power-saving feature and permit the exact test. In another approach, Robinson et al. [350], in developing their tool *edgeR*, adopt an empirical Bayes approach to shrink the gene-wise dispersion estimates towards (i) a constant value or (ii) a trend estimate using nonparametric regression. Another study [23] introduces a technique to model the dispersion parameter as a smooth function of the mean and use nonparametric regression to better fit the negative binomial variance as a function of the mean, and uses predicted dispersion parameters from this fit in place of the unknown values for comparing means. Based on this technique, the authors introduce a tool called *DESeq*.

Yanming et al. [103] use the negative binomial distribution to incorporate a non-constant dispersion parameter for all genes to measure the variability across two conditions and identify differentially expressed genes because Binomial or Poisson distribution models cannot appropriately model the variability of count data due to overdispersion. The small sample size (commonly found) in this type of data often prevents the uncritical use of tools, mostly derived from large-sample asymptotic theory. The proposed test is based on parameterization of the negative binomial distribution. The test is an extension of an exact test introduced by Robinson and Smyth [352], [353]. In the test of Robinson and Smyth, a constant dispersion parameter is used to model the count variability between biological replicates. Di et al [103] introduce an additional parameter to allow the dispersion parameter to depend on the mean. This parametric method complements nonparametric regression approaches for modeling the dispersion parameter. The test was applied to an Arabidopsis dataset and the results substantiated that the test is simple, powerful and reasonably robust against deviations from model assumptions.

6.3.4 DE Gene Analysis

Once the DE genes have been found using statistical tests, one can further analyze these genes in two ways.

(i) They can be considered as potential crucial genes for a disease and further biological tests can be carried out to test their acceptability as biomarkers. Further analysis can be (a) *P-value* (discussed in the subsequent section) analysis to evaluate the quality of genes with reference to the genes already annotated as associated with the disease in the Gene Ontology repository and (b) *pathway* analysis to identify the association of a given gene or protein with a pathway and to interpret omics data in the context of canonical prior knowledge structured in the form of pathway diagram. Pathway analysis also helps find distinct cellular processes, diseases or signaling pathways that are statistically associated with the identified gene(s).

(ii) The DE genes can be used as inputs for further analysis using (a) supervised or (b) unsupervised learning approaches, depending on the availability of prior knowledge to identify a more refined set of biologically significant DE genes. Further analysis on the input initial set of DE genes may be (a) co-expression and regulatory network analysis, (b) statistical analysis or (c) analysis using supervised or unsupervised machine learning approaches to isolate a more acceptable set of DE genes. Among these, the network analysis approach is the most effective and hence commonly used. The other two approaches are mostly effective in finding co-expressed patterns, as discussed in *Chapter* 5.

Co-Expression Network Analysis

co-expression network analysis can be performed on the initial set of DE genes to extract biologically enriched co-expressed modules or subsets of DE genes to facilitate subsequent downstream analysis for identification of intertesting (i) disease biomarkers or (ii) gene-gene associations. There are five tasks in this process.

(i) *Preprocessing*, where tasks mostly such as (a) transformation into normal distribution (to enable some statistical analysis) or (b) removal of outlier or non-conforming or isolated gene(s) are performed on the initial set of DE genes to improve quality for the subsequent downstream processing.

(ii) *co-expression network construction*, where nodes of the network represent the DE genes and the connectivity between a pair of nodes represents an association when both are significantly co-expressed tissue samples. To instantiate an edge between a pair of nodes, one can use thresholding (either hard or soft) on pair-wise proximity values between nodes. Such networks can be of five types, signed, unsigned, weighted, unweighted, or hybrids.

(iii) *co-expressed module extraction*, where a module of a co-expression network is a subset of nodes or DE genes with tight connectivity or high cohesiveness. There is high functional similarity among the genes belonging to the module. Identification of such highly cohesive modules helps subsequent downstream analysis to identify interesting gene(s) or biomarker(s). One can extract the modules using (a) edge elimination or (b) clustering.

(iv) *Module-based analysis* which involves two types of analysis. (a) Hubgene-centric topological analysis, where the functional or structural behavior of a network module significantly depends on some gene(s). The shifting of such gene(s) from one module to another across states typically causes significant functional changes in addition to structural changes. Such genes are referred to as *hubgenes* and they play important roles in gene expression analysis. (b) Biological analysis, which extracts causality relationships among the DE genes which is not possible through hubgene-centric or gene-gene correlation-based association analysis. Causality information uncovers many interesting facts about Transcription Factors (TF) and Target Genes (TG) involved in a given disease in progression. A large number of tools extract such causality relationships. *ARACNE* [276] and *GENIE3* [415][17] are two tools that successfully construct regulatory networks from co-expression networks and support causality analysis. Although these tools offer a lot of good features, they have some drawbacks.

(v) *Validation* of the set of DE genes obtained from the above-mentioned analysis. It uses approaches such as the following: (a) *statistical*, where the significance of the DE genes is assessed using standard topological and statistical validity measures such as p-value, Q value, precision-recall, PPV, and accuracy. (b) *biological*, where significance of the DE genes is assessed using (i) regulatory network analysis which extracts causality relationships among the transcription factors and target genes and (ii) pathway analysis, to identify the association of a given DE gene with the related pathway of the given disease, (c) *literature mining*, where the validity of the identified genes is investigated against well-established facts (such as wet-lab results) related to the concerned biological question. A DE gene is generally considered biologically significant if all three validation approaches find it significant.

A conceptual framework for the network analysis and detailed discussion of each of these steps are presented in *Chapter* 5.

Regulatory Network Analysis

Co-expression network analysis identifies interesting groups of co-expressed genes that play an important role in disease progression and biological functions. However, gene-centric or gene-gene correlation-based approaches are not able to infer causality relationships among the DE genes. Causality information is useful in understanding interesting relationships among the Transcription Factors (TF) and Target Genes (TG) involved in a given disease. Based on statistically identified DE genes, one can apply algorithm or tool(s) to construct regulatory networks that identify causality relationships among the TFs, TGs, and miRNAs.

To construct a regulatory network, both transcriptional and post-transcriptional regulatory relationships are typically considered. Liu et al [257] consider both Transcription Factors and miRNAs as regulators in their

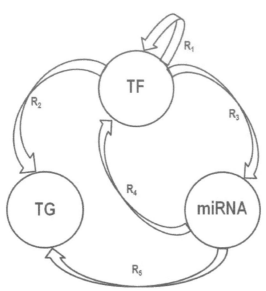

Figure 6.6: Five types of relationships among TF, TG and miRNA.

study. Unlike others, they emphasize the importance of miRNA as an important component in the causality behavior analysis among genes. They introduce different categories of relationships among TFs, TGs, and miRNAs (while creating GRNs) such as transcriptional, post-transcriptional and interplay between regulators. These relationships are

a. R_1: TF-TF relationship, which is a transcriptional regulatory relationship,

b. R_2: TF-gene relationship, which is also a transcriptional regulatory relationship,

c. R_3: TF-miRNA relationship, which indicates interplay between the regulators,

d. R_4: miRNA-TF relationship, which also indicates interplay between the regulators, and

e. R_5: miRNA-gene relationship, which is a post-transcriptional regulatory relationship.

Figure 6.6 shows a basic regulatory network with all these five relationships among these players.

Over the years, a large number of repositories have been created. These include regulatory information details for *Homo sapiens* and *Mus musculus*. From these sources, relevant information such as TFs, miRNAs, TFBS

(Transcription Factor Binding Sites) motifs, genes and their annotations can be collected and used during construction of the network by extracting transcriptional relationships and interplays. Some example repositories [257] are: (i) BioGrid, [1] (ii) Ensembl, [2] (iii) FANTOM, [3] (iv) GenBank, [4] (v) HPRD, [5] (vi) IntAct, [6] (vii) JASPAR, [7] (viii) KEGG, [8] (ix) Liftover, [9] and (x) MicroCosm. [10]

A number of available tools support causality relationship extraction activities. Two commonly used tools are: (i) *ARACNE* [276] and (ii) *GENIE3* [415][17]. Both tools construct regulatory networks from co-expression networks by extracting relevant causality information. *ARACNE* [276] filters out indirect associations among genes and presents only the strong and direct associations that are regulatory. On the other hand, *GENIE3* uses *TF* information where constructing the regulatory network by identifying expression patterns of *TF*s that best describe the expression of each of their *TG*s. A major constraint with this tool is that *TF* information used in the network perform better than random chance. A few other tools have also been proposed in [362], [172], [83]. The advantages and limitations of these tools are also reported in the literature.

6.3.5 Statistical Validation

Once a set of DE genes is identified, we need to validate them both statistically as well as biologically. Statistical validation uses a number of measures.

P-value is a measure which evaluates the quality of genes with reference to the genes already annotated in the Gene Ontology repository. One computes p-value using the hypergeometric test or Fisher's Exact Test [54]. For a given group of genes, this test assumes a null hypothesis that the genes included in the group represent a random sample from the population. A group of extracted genes with low p-value signifies that the genes belong to enriched functional categories found in the annotations of the Gene Ontology. Genes with significantly low p-values are biologically significant. A

---

[1]http://thebiogrid.org/
[2]http://www.ensembl.org
[3]http://fantom.gsc.riken.jp/
[4]http://www.ncbi.nlm.nih.gov/genbank/
[5]http://www.hprd.org
[6]http://www.ebi.ac.uk/intact/
[7]http://jaspar.genereg.net/
[8]http://www.genome.jp/kegg/
[9]http://genome.ucsc.edu/cgibin/hgLiftOver
[10]http://www.ebi.ac.uk/enright-srv/microcosm/

detailed discussion of this measure is available in Chapter 5.

Q-value is also an effective statistical validity measure to assess the significance of co-expressed gene groups extracted from gene expression data. The Q-value for a given gene is the proportion of *false positives* among all the genes that are as or more differentially expressed [344], [368]. In other words, it is nothing but the minimum *False Discovery Rate (FDR)* at which this gene appears biologically significant. One can determine the GO categories and Q-values from an *FDR* corrected hypergeometric test for enrichment using benchmark web-based tools such as GeneMANIA [288]. To estimate the Q-values for a given gene, the Benjamini Hochberg procedure [42] has been used. This procedure attempts to decrease the FDR. It is useful to avoid incorrect rejection of the null hypothesis by controlling the occurrence of smaller p-values by chance.

In addition to the above, there are several other statistical measures commonly used to validate the significance of DE genes. These are (i) Precision, Recall and F-measure, (ii) Positive Predictive Value (PPV), (iii) Sensitivity, and (iv) Accuracy. Precision is defined as the percentage of match between the predicted groups of genes and the known gene groups [368], whereas *recall* represents the fraction of known gene groups that has been detected within the predicted clusters. Another established validity index is F-measure, which is the harmonic mean of *Precision* and *Recall*, helps compare the effectiveness of a method with other methods. A method that extracts gene groups of high precision and recall values and consequently, a high F-measure, is considered effective. *Positive Predictive Value (PPV)* is another popular measure that assesses the amount of match of the predicted cluster set against the benchmark cluster set. The assessment of *PPV* for a gene group represents how closely the predicted co-expressed gene group resembles the best matching benchmark cluster. To estimate *PPV* over all gene groups with reference to an all-annotated complex, it is desirable to get an overall *PPV*. The overall *PPV* of a method is the average *PPVs* of all clusters. A high *PPV* value specifies a higher level of correspondence between the predicted cluster and the ground truth clusters, implying better quality results.

Sensitivity (S), another useful measure, indicates the fraction of the benchmark cluster set included in the predicted groups of genes. Sensitivity for a given gene group is defined as the maximum value of sensitivity obtained for the group over all real gene groups. *Accuracy* is a measure for unbiased evaluation of any co-expressed gene group finding method with reference to the ground truths. It is defined as the geometric mean of *PPV* and

S. Accuracy is considered a compromised yet effective evaluation measure for a gene group finding method compared to *PPV* and *S*. These measures have been discussed in detail in Chapter 5.

Biological Validation

To evaluate the biological significance of gene modules extracted from a network of DE genes, one can use analysis such as (i) preservation analysis, (ii) topological and differential connectivity analysis, and (iii) gene enrichment analysis. Preservation analysis evaluates whether an extracted module is preserved in the reference network or not [235]. A module may be preserved in test network due to (i) natural selection or (ii) changes in pathways or biological processes [87]. Two commonly used statistics for preservation analysis are (i) *Z-summary* and (ii) *medianRank*, which evaluate overall significance of the preservation of a test module based on density and connectivity [235].

Topological analysis investigates important network features of a module such as degree and connectivity, to identify interesting genes to characterize the network [235], [87]. This analysis also identifies closely connected genes in the network as hub genes. On the other hand, differential connectivity analysis finds genes which change significantly in terms of the degree or sum of connectivity in the test network [87].

Gene set enrichment analysis [358] identifies a set of genes that are overrepresented in a large collection of genes and may have an association with diseases [88], [208]. It uses the tool called Database for Annotation, Visualization Integrated Discovery (DAVID) to elucidate the biological insights of the non-preserved modules, based on the Gene ontology and pathway analysis [102]. It uses p-value to evaluate the significance of the enrichment terms. It considers those terms with $p-value < 0.05$. To better understand the associations with biological processes in which non-preserved module genes are involved. One can use the tool called *BiNGO* [268], a Cytoscape plugin [235] to find over-represented genes in a set of genes or a subgraph of biological network. It also maps the predominant function of a given gene set.

6.3.6 Discussion

The use of statistical measures on preprocessed gene expression data filters out unwanted genes and extracts relevant DE genes. However, a statistically significant DE gene may not be biologically significant. Network analysis is effective in obtaining a refined set of DE genes. Both co-expression network and regulatory network analysis help in extraction of significant DE genes and their biological assessment. A number of tools, systems and benchmark repositories have been developed for (i) extraction of topologically, statistically and biologically significant DE genes and (ii) significance analysis of the extracted DE genes based on (a) preservation analysis, (b) enrichment analysis, and (c) causality analysis.

To better illustrate the preprocessing, analysis and validation techniques discussed so far, we present a case study of neuropsychiatric disorder RNAseq count data analysis for potential biomarker identification using DE analysis approach.

6.4   Biomarker Identification Using DEA: A Case Study

This case study [358] deals with DE analysis of a neuropsychiatric disorder count dataset, generated using RNA-seq technology to identify differential regulation of neuroimmune system in schizophrenia (SCZ) and bipolar disorder (BPD). It eliminates biases and improves the quality of the raw count data using preprocessing techniques such as (i) removal of low read-count instances, (ii) normalization, and (iii) transformation. It extracts the preprocessed control and the corresponding SCZ and BPD instances for DE analysis. The DE genes generated are analyzed using co-expression analysis, topological and functional enrichment analysis for identification of unique and overlapping molecular signatures.

Differential expression analysis by Ramaker et al. [337] on molecular signatures across different disorders and brain regions observes that a majority of disease-associated expressions changes occur in the AnCg region. The AnCg region is associated with biological functions such as (i) cognition, (ii) error detection, (iii) conflict resolution, (iv) motivation, and (v) modulation of emotion [431], [321], [75]. It is also an observation that there is significant overlap in gene expression changes in BPD and SCZ of the AnCg region. This case study investigates further the overlapping and unique molecular signatures using DE analysis, followed by co-expression analysis, preservation analysis, topological analysis and gene set enrichment analysis on the DE genes identified from BPD and SCZ samples of the AnCg region.

An important observation of this study is that differential activation of immune system components and pathways can drive the common but unique pathogenesis of the BPD and SCZ.

6.4.1 Problem Definition

The problem is to analyze the preprocessed read-count data of (a) control and BPD samples and (b) control and SCZ samples of the AnCg region using DE analysis, followed by (i) co-expression analysis, (ii) preservation analysis, (iii) topological analysis, and (iv) gene set enrichment analysis on the generated DE genes for identification of unique and overlapping molecular signatures.

6.4.2 Dataset Used

The study uses GSE80655, a publicly available RNA seq dataset available at NCBI Gene Expression Omnibus (GEO) repository [337]. This dataset was generated from three different brain regions (i) AnCg, (ii) nucleus accumbens, and (iii) dorsolateral prefrontal cortex from 281 persons diagnosed with SCZ, BPD, major depressive disorder and normal (or control). This study analyses the count data of the AnCg region of two different disorders, BPD ($n = 24$), SCZ ($n = 24$) patients, and controls ($n = 24$). It applies preprocessings techniques such as low read-count instance removal, normalization, and transformation on raw count data to eliminate biases prior to downstream analysis.

6.4.3 Preprocessing

Genes with very low read counts in each sample are removed because such genes do not contribute to subsequent downstream analysis. It eliminates a gene from further consideration if it does not contain, on an average, 10 read-counts across all control and corresponding disorder samples [259]. Tools like DESeq2 [259] and edgeR [350] are used to remove low read-count genes.

To eliminate biases in the RNA-seq count data due to different sequencing depths and transcript sizes [88], it normalizes the data. Two normalization methods are used, DESeq and TMM, available in the DESeq2 and edgeR packages. To enable statistical modeling on discrete count data, transformation is necessary. The study uses *log2* transformation to convert the discrete count data into a normal like distribution prior to downstream analysis on the identified DE genes.

6.4.4   Framework of Analysis Used

Figure 6.7 presents the framework of the DE approach used to identify unique and overlapping molecular signatures from neuropsychiatric disorder count data generated using RNA-seq technology. The raw RNA-seq count data is initially preprocessed to (i) eliminate bias and (ii) transform into a ready-to-use form for statistical analysis. The preprocessed count data are input to a DE analysis module to detect an unbiased set of DE genes for subsequent network-based topological, functional and enrichment analyses. Detailed discussion of each module is reported next.

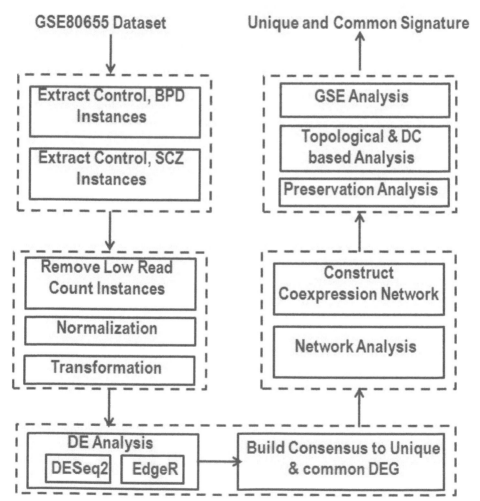

Figure 6.7: DE analysis of neuropsychiatric disorder count data: A framework

DEG Identification

There are several recently introduced tools for DEG identification. DEGs identified by these tools often have significant differences. To overcome this issue, two tools, namely *DESeq2* and *edgeR* are used to achieve an unbiased set of DE genes by eliminating individual bias. Consensus results given by these tools arc used for identification of a more accurate list of DEGs for subsequent downstream analysis. The study selects genes as differentially expressed if they are selected by both tools.

Co-Expressed Network Module Extraction and Analysis

Co-expression network analysis is conducted on the selected DE genes to identify highly coherent (tightly connected) modules of genes responsible for biological processes or pathways or phenotypic variations [87], [206], [208]. An established tool, called WGCNA, is used to extract interesting gene modules from the co-expression networks. To achieve unbiased inference(s), it is essential to perform *log2* transformation of the RNA-seq count data corresponding to the selected DE genes. In addition to that the authors cleaned all the non-conforming or outlier samples are removed, from control and disease datasets prior to further downstream analysis. The effectiveness of the extracted modules is assessed following topological, functional and enrichment analyses, as discussed below.

Preservation Analysis

Two commonly used statistics for preservation analysis are: (i) Z-summary and (ii) medianRank. These two statistics evaluate the overall significance of the preservation of a test module based on density and connectivity [235].

Topological and Differential Connectivity-Based Analysis

The degree differences and sum of connectivity difference are used as differential connectivity parameters to estimate the likelihood of a gene to be considered as central or border, in progressing from reference to test subnetwork [358].

Gene Set Enrichment Analysis

Further analysis considers terms with p-value ≤ 0.05 for subsequent analysis. To better understand the associations with biological processes in which non-preserved (NP) module genes are involved, the tool used is called *BiNGO* [268], a cytoscape plugin [235], which finds those genes that are over-represented in Gene Ontology in a set of genes or a subgraph of biological network. It also maps the predominant function of a given gene set.

6.4.5   Results

The results generated based on experimental study and their significance analysis are reported below.

Common and Unique Differentially Expressed Genes among BPD and SCZ

For a meaningful analysis of DE genes, genes with low read counts are removed [358]. After removal, the total number of genes for BPD and SCZ are 14,717 and 14,753 genes, respectively. Applying DESeq2 on these genes identifies 1,617 and 1,307 DE genes from BPD and SCZ, respectively. Application of edgeR, identifies 415 and 1,524 genes as significant DEGs for BPD and SCZ, respectively. It is necessary to identify the common DE genes for the two tools, *DESeq2* and *edgeR*. There are 787 common DEGs for BPD, and 325 common DEGs for SCZ based on intersection. Finally, (i) for *DE-Seq2*, 830 and 520 DEGs are found unique to BPD and SCZ, respectively, and (ii) for *edgeR*, 90 and 1,199 DEGs are found unique to BPD and SCZ, respectively. DEGs that are found common between BPD and SCZ given by DESeq2 and edgeR are 398 and 1,283, respectively.

Co-Expression Network Construction and Module-Finding

Suitable β power values of 16, 26 and 14 are used for control, BPD and SCZ datasets, respectively, to construct a weighted gene co-expression network for each type of disorder along with control. Two eigengenes are merged to form a single module, if the correlation value between the pair of eigengenes is ≥ 0.75. For the control network, 8 modules are considered before merging. After merging, 5 modules are obtained excluding the gray module of 177 genes for further analysis. Similarly, for BPD and SCZ, 14 modules are identified for each before merging, and after merging, 3 and 4 modules are found, respectively, excluding gray modules. In case of BPD and SCZ, the gray modules contain 55 genes and 13 genes, respectively.

Identification of NP Modules

Although a number of modules are identified, the focus is on modules that correspond to the control and BPD networks for this case study. A more detailed discussion on these results is found in [358]. Two modules in the control network are found to be non-preserved (NP) in BPD, as they are visible in Black and Red colors in Table 6.10. The Black and Red NP modules from the control network include 37 and 54 genes, respectively.

Topological Analysis of Non-Preserved Modules

Network topological analysis of all the identified NP modules identifies several interesting genes, and different network parameters for each of these genes are recorded. The top five prominent genes identified based on the sum of connectivity in control and disease from the NP modules are shown in Table 6.8. The table shows that the NP modules found in control and BPD networks have significantly lower sums of connectivities than the disease network. The authors of the study are of the opinion that this may be due to the activation or silencing of some genes in the disease condition. This analysis concludes that some genes of NP modules act abnormally due to external factors.

Table 6.8: Prominent Genes Based on Sum of Connectivity from each NP Module

Control vs BPD Black Module			Control vs BPD Red Module	
Gene Symbols	Control	BPD	Gene Symbols	control
TNFRSF1A	2.9708	6.196312	C8orf58	0.8782
C1R	2.7036	3.994709	TMEM150A	3.204
FZD4	2.6486	4.71635	IL17RC	2.267
DEPP1	2.5764	4.28416	HDAC7	3.295
ZNF395	2.5586	2.211981	SORBS3	3.173

GSEA of NP Modules

For gene set enrichment analysis, the Black and Red NP modules from Control are considered. The tools used are DAVID and BiNGO for analysis support. To assess the significance of the enriched GO terms, *p-value*-based

Table 6.9: Top 5 Enriched Biological Processes with Other Statistics

Term	P-value	Fold Enrichment	Gene symbols
Red Module			
cell surface receptor signaling pathway	0.022	6.451	PTH1R, ADGRF3, AGER, TSPAN4
osteoblast development	0.0368	51.987	PTH1R, LRP5
Black Module			
cell-cell adhesion	0.0091	8.851	PDLIM1, TES, F11R, SFN
cytokine-mediated signaling pathway	0.0186	13.733	TNFRSF1A, CISH, IL15RA
negative regulation of calcium ion transport	0.02	92.263	STC1, NOS3
inflammatory response	0.0223	6.329	TNFRSF1A, F11R, TNFRSF10D, CASP4
immune response	0.029	5.697	TNFRSF1A, C1R, TNFRSF10D, IFITM2

thresholding is applied. A gene with *p-value* < 0.05 is considered. Table 6.9 reports GSEA results that include the top 5 significantly enriched biological processes obtained from the DAVID tool for each NP module. Immunity-related biological processes are highly enriched in the NP modules.

Table 6.10: Modules with Preservation Statistics

	medianRank	medianRank Quality	Zsummary	Zsummary Quality
black	2	4	2.2	4.3
blue	3	2	9.2	16
brown	1	5	14	9.5
red	3	1	5	7.8
turquoise	4	3	15	20

6.4.6   Discussion

The following are the main findings of the experimental study discussed above [358].

- Co-expression analysis on DEGs, followed by non-preservation analysis, provides predictions of molecular signatures to produce molecular insights into neuropsychiatric disorders.

- Investigation of interactions among genes helps identify biologically interesting groups of genes.

- Topological and enrichment analysis of non-preserved modules helps infer molecular functions and signatures in BPD and SCZ patients in the AnCg brain region.

- Neuroinflammation is inflammation within the central nervous system with an important role in the etiology of these disorders.

- Eleven genes are enriched in BPD and SCZ networks. Some of these genes have relationships with autoimmunity and autophagy.

- An interesting observation is that DEPP1 acts as a critical modulator of FOXO3 induced autophagy via increased cellular ROS.

- Clear overrepresentation of pro-inflammatory cytokine receptors supports the conclusion that cytokine-mediated immunological responses play a vital role in the BPD.

- Investigation of the BPD can serve as a potential therapeutic strategy based on immune-neural system findings.

6.5 Summary and Recommendations

In this chapter, we have provided an extensive discussion on differential analysis, its background, importance, work flow, and applications in gene expression data. The following are some highlights of this chapter.

- Some genes exhibit significant change in the amount of mRNA during transition from one state or condition to another. Such genes are referred to as differentially expressed gene(s). DE genes often play crucial roles in the analysis of a disease in progression.

- Preprocessing technique(s) are necessary to achieve the desired goals of differential expression analysis. Although a number of preprocessing techniques have been developed, some are applicable to only specific type of gene expression data. For example, discretization may be relevant to microarray data, but not necessarily for RNA-seq data. Dimensionality reduction may be highly relevant for scRNA-seq data.

- Application of some preprocessing techniques may actually have negative impact on the overall DE analysis process. For example, discretization, normalization, and dimensionality reduction may sometimes lead to information loss.

- DE genes identified by various methods are often significantly different. To obtain an unbiased set of DE genes from a given gene expression dataset, one can use multiple DE analysis methods and build consensus to overcome the limitations or biases of the individual methods.

- Both parametric and non-parametric tests can be used in DE gene identification. A major weakness of parametric tests is that they assume the input gene expression data follows a normal distribution. Parametric tests can be used when the input data are expressed in (i) absolute numbers or (ii) values, rather than ranks.

- Non-parametric tests do not make such assumptions and hence are widely used more acceptability. Test statistics obtained from a single method are often not adequate for an unbiased assessment of significance. Knowledge of multiple testing techniques and their applicability is essential.

- Modeling based on Poisson distribution is widely used for read-count data, especially, when a small number of biological replicates are available.

- Co-expression analysis of DEGs followed by non-preservation analysis helps extract knowledge of molecular signatures to develop molecular insights to a critical disease.

- A differentially expressed gene of high statistical significance may not be biologically significant. Similarly, a topologically significant gene may not be biologically significant. Therefore, significance analysis of DE genes from statistical, topological and biological perspective is essential for better assessment of the relevance of DE genes.

- Topological and enrichment analysis of non-preserved modules help analyze, identify and establish the significance the DE genes.

Based on the discussions in this chapter we recommend the following to researchers and practitioners.

(a) Preprocessing prior to downstream analysis improves overall quality of gene expression data. Preprocessing steps include (a) normalization, (b) batch effect removal, (c) removal of outlier genes, (d) removal of genes with very low read-counts, (e) detection and filtration of low-quality cells in case of scRNA-seq, (f) quality control and quality improvement. Our observation is that reads of quality less than 30% do not contribute to effective analysis, hence can be discarded [91].

(b) Although a number of tools help differential expression analysis at various stages, most accept only a few limited data formats. A generic data format converter can help effective use of these tools. An example of such a data format converter is NGS-FC [451], which allows conversion among 14 different data formats.

(c) To empower differential expression of RNA-seq at low cost, one should consider lower sequential depth but a higher number of biological replicates because in DE analysis, the impact of the number of replicates is significantly higher than sequencing depth [234], [256].

(d) Most normalization tools are influenced by the presence of varying library sizes and varying sequencing depths. Tools like DESeq, TPM and RUV-seq have been found effective in the presence of such factors.

(e) For effective differential expression analysis of RNA-seq data towards unbiased inference generation, sample size plays an important role. With larger sample sizes, it is possible to achieve better measurements and precision, in spite of experimental variations. The analyst should achieve a better tradeoff between sequence depth and sample size for a specific dataset to generate unbiased inferences. Tools like RNAseqPS [154] and RNAseqPower [168] are useful in this regard.

(f) DEGs identified for a given disease using a single approach, method or tool based on a single source of data may not be authentic or unbiased. Hence, an integrative approach that uses multiple approaches, methods or tools supported by appropriate consensus function(s) over multiple sources of data is more promising.

(g) For a given disease, the causality (or strong associativity) of the DEGs identified by a differential expression analysis method can be analyzed using a multi-objective approach considering its topological characteristics, associated pathway information, and regulatory roles across progression states.

Chapter 7

Tools and Systems

7.1 Introduction

Metagenomics and behavioral analysis of microbiome population are currently high-priority thrust areas of research. Such research not only deals with the practice of sequencing DNA from the genomes of a wide spectrum of organisms, but also the study and analysis of the structural and functional behaviors of microbiome populations [263]. This interdisciplinary field of research has evolved with an increasing number of approaches, methods and techniques for storage, retrieval and analysis of voluminous, heterogenous, and constantly growing biological data. A major objective of such bioinformatics research is to develop light-weight, generic, and high-precision software tools to help extract interesting biological knowledge. Appropriately designed bioinformatics tools can help (i) sequence and annotate genomes and their observed mutations, (ii) mine textual biological literature to extract hidden knowledge, (iii) develop gene ontologies to support effective biological data querying, and (iv) understand and compare the structural and functional behaviors of gene and protein expressions and regulations towards extraction of interesting pattterns, characteristics, or biomarkers. [1]

7.1.1 Generic Characteristics of a Systems Biology Tool

A systems biology tool provides facilities to perform one or more pre-defined tasks on gene expression data at an initial or intermediate step for extraction of useful biological insights. Such a tool is expected to demosntrate the following software qualities to be widely accepted.

---

[1]http://www.informatics.sdsu.edu/bioinformatics

DOI: 10.1201/9780429322655-7

a) *Correctness*: The output generated by the tool must be correct. To ensure correctness, the developer needs to perform rigorous testing, covering as many scenarios as possible using synthetic as well as benchmark datasets, with reference to ground truths.

b) *Reliability*: The tool must perform reliably. In addition to the pre-specified cases or scenarios, the tool should perform well in some un-wanted or extended number of scenarios also.

c) *Recency*: The tool must perform well on recent gene expression data.

d) *Platform Independency*: The tool must not depend on a single platform; rather it should be able to function across multiple platforms with minimum customization.

e) *Generality*: The tool should be generic in nature. Its applicability should not be restricted to only microarray gene expression data or RNAseq data; rather it should apply to both types of data.

f) *Light-weight*: The tools should be developed using (i) efficient coding and (ii) appropriate language, so that it is light-weight and can be implemented on a wide range of devices including hand-held to high-end computing devices.

g) *Evolvability*: The tool should have the ability to easily evolve over time.

h) *Inter-operability*: The tool should have import-export abilities so that the output of the tool can be used by other tool(s) or system(s). Similarly, it should be able to import and use the output(s) of other tool(s).

7.1.2   Target Systems Biology Activities

A tool is developed to smoothly perform preprocessing and downstream activities on gene expression data for extraction of hidden knowledge in the form of interesting pattern(s), association(s), biomarker(s), or other forms of biological insights. This section presents six common activities that depend on efficient tools for effective analysis of gene expression data.

Pre-Processing

Prior to downstream gene expression data analysis, it is often necessary to apply pre-processing technique(s) on raw or intermediate gene expression data to transform the data into a useful, ready-to-use form. Based on technological as well as application-specific requirements, pre-processing techniques address various issues to support effective gene expression data analysis. These pre-processing techniques perform *eight* useful tasks, based on the application: (a) normalization, (b) missing value estimation, (c) discretization, (d) gene selection, (e) dimensionality reduction, (f) quality control, (g) quality improvement, and (h) batch effect removal. Each of these pre-processing tasks has a definite role to play in a cost-effective and high-quality analysis of the gene expression data. A well-developed pre-processing tool can ease the overall process substantially to achieve the desired outcome(s). The pre-processing tool developer must be careful to achieve wider acceptability of the tool(s).

Gene Expression Data Analysis

To analyze gene expression data, three approaches are commonly used: (i) co-expression analysis, (ii) differential co-expression analysis, and (iii) differential expression analysis. For each of the approaches, either in isolation or in collaboration, a large number of tools have been developed. Some tools can support both co-expression as well as differential co-expression analysis. In addition to statistical techniques, each of these approaches depends on three other types of techniques, (i) classification-based, (ii) clustering-based, and (iii) network-based techniques. These techniques have been successfully applied at the various levels of gene expression data analysis.

a) *Classification Technique*: A classification technique is a supervised machine learner, which identifies the subclass(es) (subcategory(ies) or subpopulation(s)), a new instance belongs to, based on prior knowledge. Such prior knowledge is acquired by training on a dataset where class membership of each instance is known [59]. Classification methods play a major role in supervised gene expression analysis. An appropriate classification tool helps gene expression analysis activities such as (i) gene selection, (ii) malignant instance identification, (iii) dimensionality reduction, (iv) feature selection, (v) tumor sub-type identification, and (vi) marker identification. A large number of classification tools have been developed. These tools usually consider only one type of gene expression data or aim to accomplish a specific set of objective(s).

b) *Clustering Techniques*: A cluster analysis tool partitions a set of objects or gene expression data instances into groups that are referred to as *clusters*, such that a pair of instances belonging to a cluster is more similar than pairs in different clusters [59]. A clustering algorithm does not require any supervision, and hence is a type of unsupervised machine learning algorithm. A clustering algorithm helps gene expression data analysis in (i) co-expressed pattern identification, (ii) biomarker or crucial gene identification, (iii) dimensionality reduction, (iv) feature selection, (v) gene selection, (vi) missing value estimation, and (vii) subtype identification. Like classification algorithms, a large number of clustering algorithms using various approaches have been developed to solve biological problems. Most tool designs are influenced by (i) technological advancements, (ii) constantly growing, heterogenous data processing requirements, and (iii) huge computational requirements. Like the classification tools, most clustering tools have been developed to achieve a specific set of objectives. Scalability and lack of ability to detect overlapped clusters are two major issues with most clustering tools.

c) *Network Analysis Techniques*: A gene-gene network can be of four possible types: (i) undirected and unweighted, (ii) weighted and undirected, (iii) unweighted and directed, and (iv) weighted and directed. A network analysis tool enables us to (i) construct networks, (ii) extract components, modules, complexes, or sub-networks with high functional coherence, (iii) identify crucial or hub genes from the modules or complexes, and (iv) perform hubgene-centric analysis for identification of biomarkers. Analysis of gene expression data uses various network representation approaches such as (i) co-expression networks, (ii) regulatory networks, and (iii) protein-protein interaction networks. Most network analysis tools also perform centrality analysis to help identify representatives or hubgenes for each module or complex for subsequent downstream analysis.

Visualization

A visualization tool helps view gene expression analysis results generated by (a) classification-based, (b) clustering-based, or (c) network-based approaches. Additionally, it helps interactive exploration of (i) gene-gene associativity, (ii) protein-protein interaction patterns, (iii) separability between a pair of gene clusters, (iv) grid approximations of data distributions over k-dimensional space, (v) spectral gene partitions, and (vi) hubgenes in a gene-gene network module. A visualization tool supports explorative interactions on a wide range devices, from smartphones to high-end computing workstations across various platforms.

Validation

Validation verifies the acceptability of a statistical or machine learning solution against a biological question. Validation helps the assessment of the performance of a gene expression analysis method at a given snapshot in time. The method needs to be revalidated after passage of time because of environmental or technological changes. Validation also helps develop strategies for refinement or re-engineering of the method to achieve better accuracy and performance. Gene expression analysis findings need validation both *statistically* as well as *biologically*. Statistical validation uses a set of measures or indices for the assessment of correctness, efficiency, and quality. A single statistical measure or index is often inadequate for an unbiased evaluation. Hence, the use of multiple measures or indices is recommended by eliminating the individiual biases. It is also essential to validate the significance of the findings biologically. Biological validation typically considers functional profiling, pathway sharing, regulatory behavior validation, and validation using heatmap biology.

7.2 Systems Biology Tools

Systems biology tools can be logically organized as shown in Figure 7.1. Initially, raw gene expression data are preprocessed with tools that support *Quality Control (QC)* or *Quality Improvement (QI)*. Next, based on requirements, one uses the tools in the third rectangular layer. The fourth layer shows various analysis approaches. Although the visualization tools arc shown in the fifth rectangular layer, the boundary between the fourth and fifth rectangular layers is thin. Many analysis tools are bundled with visualization tools. The validation tools are shown in the outer-most layer. A formal taxonomy of systems biology tools is presented next.

7.2.1 A Taxonomy

Figure 7.2 presents a taxonomy of systems biology tools. Tools are classified based on (i) requirements, (ii) salient features, and (iii) applications. The taxonomy categorizes the tools into four major categories and fifteen subcategories. The taxonomy also shows additional sub-categories based on approaches applied in the development of such tools. Next, we present some prominent tools, their capabilities, characteristics, and usage under four categories. We also present the conceptual frameworks for the four major categories of tools. Next, we discuss the capabilities of the four major categories of tools and present their ideal frameworks.

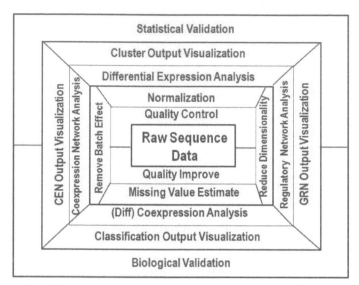

Figure 7.1: Organization of systems biology tools: A framework

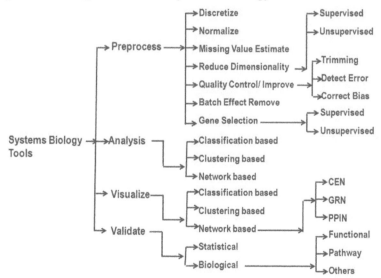

Figure 7.2: Systems biology tools: A taxonomy

7.2.2   Pre-Processing Tools

This section presents a set of popular tools that support preprocessing activities for microarray and / or RNAseq data prior to downstream analysis. A large number of preprocessing tools have been developed for qualitative gene expression analysis. Most tools have been introduced to achieve a focused set of objectives. For example, a normalization tool introduced for microarray data, may not be applicable to RNAseq data, unless it is adopted for it. The preprocessing requirements for a certain type of gene expression data may vary from others. For example, scRNAseq data requires dimensionality reduction, whereas it may be optional for microarray or bulk RNAseq data. There is still no integrated tool or system that supports all essential preprocessing activities for the three types of gene expression data. Figure 7.3 shows a generic framework of a preprocessing tool for all three types of gene expression data.

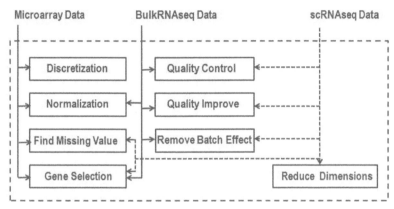

Figure 7.3: Gene expression data preprocessing tool: A framework

We present the tools under eight sub-categories, (i) *normalization*, (ii) *missing value estimation*, (iii) *discretization*, (iv) *gene selection*, (v) *dimensionality reduction*, (vi) *quality control*, (vii) *quality improvement*, and (viii) *batch effect removal*, and discuss their features and compare them based on characteristics, operational abilities, availabilities and other important features.

Normalization

Normalization is an essential task to be carried out on raw gene expression data to eliminate biases in data generation and to ensure unbiased inferences during downstream analysis. Undesired biases are often caused by systemic variations during the generation of the gene expression data, which subsequently can affect the expression levels [392]. We present some popular normalization tools used for microarray and RNAseq data. It is worth mentioning that some normalization tools support only a specific type of gene expression data. Some normalization tools are bundled with other preprocessing and analysis tools.

a) **DESeq:** This statistical tool [23], [264] tests, for a given gene, whether an observed difference in read counts is significant or not. It detects and corrects dispersion estimates that are too low. It models the number of reads in a sample j that are assigned to a particular gene i by a negative binomial (NB) distribution. $K_{ij} \longrightarrow NB(\mu_{ij}, \sigma_{ij}^2)$, where μ_{ij} is the mean and σ_{ij}^2 is the variance. The read count K_{ij} is a non-negative integer. It estimates the values of the mean and variance based on some assumptions.

b) **DESeq2:** It is an advanced version [264] of the DESeq tool. It facilitates quantitative and comparative analysis of RNA-seq data using shrinkage estimators for dispersion and fold change. It models the read counts K_{ij} using a negative binomial (NB) distribution with mean μ_{ij} and dispersion α_i. The mean μ_{ij} is taken as a quantity q_{ij}, proportional to the concentration of cDNA fragments of the gene in the sample scaled by a normalization factor s_{ij}.

c) **EdgeR:** This tool [350] normalizes the data by examining the differential expression of replicated count data using an over-dispersed Poisson distribution model. It uses empirical Bayes methods to moderate the degree of over-dispersion across transcripts to improve the inference. Testing for differential expression is supported for pairwise comparisons where the two groups to be compared are specified by the user.

d) **Limma:** This tool [349] helps overcome the problem of small sample sizes. It can perform traditional gene-wise expression analysis and can also analyze expression profiles in terms of higher-order expression signatures. Limma operates on a matrix of expression values where it fits a linear model to each row of data to handle experimental designs or to test hypotheses. It helps draw reliable statistical conclusions even with a small number of samples.

e) **Kallisto:** This tool [294] quantifies transcripts from both bulk and sc-RNAseq data using high- throughput sequencing reads. Kallisto depends on the novel idea of pseudo-alignment, and uses it to determine the compatibility of reads with targets, without the need for alignment of individual bases. It preserves key information for quantification, making the process faster. It is robust to errors in the reads.

f) **Sailfish:** Sailfish [320] quantifies abundance of previously annotated RNA isoforms from RNA-Seq data. This tool avoids mapping of reads and provides a faster quantification estimate. It facilitates frequent reanalysis of data and reduces the need to optimize parameters.

g) **Salmon:** It is a fast transcript quantification tool [319] that uses a set of target transcripts, references, or de novo assembly for quantification. It combines a dual-phase parallel inference algorithm and feature-rich bias models with an ultra-fast read mapping procedure. Salmon has two modes of operation: *mapping-based* and *alignment-based*. The mapping-based mode runs in two phases: *indexing* and *quantification*. Indexing is independent of reads, whereas quantification is specific to the set of RNA-seq reads. The alignment-based mode does not require indexing and can simply use a FASTA file of the transcripts and a SAM/BAM file containing the alignment that will be used for quantification.

h) **easyRNASeq:** This tool [100] simplifies the task of RNAseq data analysis by hiding the complexity of the required packages behind a single functionality. EasyRNASeq has four major steps: (i) read sequenced reads, (ii) retrieve annotations, (iii) summarize read counts by features of interest, and (iv) report the results in suitable formats for downstream analysis.

i) **ExpressionProfiler:** This web-based tool [209] is applicable to microarray gene expression data. ExpressionProfiler provides an extensible framework with a unified interface to facilitate users of different skill levels to interact with the tool effectively. In addition to normalization, this tool also performs annotation, transformation, and gene selection.

j) **CARMAweb:** It is a web-based tool that performs data processing such as background correction, quality control, and normalization. It also performs detection of differentially expressed genes, cluster analysis, dimension reduction, visualization, classification, and Gene-Ontology term analysis. It accepts raw data and uses Bioconductor packages along with the R function *Sweave* to generate analysis reports, which includes all R commands used to perform the analysis.

Table 7.1 lists some normalization tools for RNAseq data and Table 7.2 reports some normalization tools for microarray data.

Missing Value Estimation

Recent advances in sequencing technologies generate gene expression data matrices with very high or ultra-high numbers of instances as well as

Table 7.1: Normalization Tools for RNAseq Data

Name	Input	Output	Onl/ Off	WB/ CL	URL
DeSeq	FASTQ, SAM, BAM	Count data	Off	CL	bioconductor.org/ packages/DESeq/
DeSeq2	FASTQ, SAM, BAM	Count data	Off	CL	bioconductor.org/ packages/release/bioc/html/ DESeq2.html
EdgeR	SAM, BAM	FPKM, RPKM	Off	CL	bioconductor.org/ packages/release/bioc/html/ edgeR.html
Limma	SAM, BAM	Count data	Off	CL	bioconductor.org/ packages/release/bioc/html/ limma.html
Kallisto	FASTA, FASTQ	TPM	Off	WB	pachterlab.github.io/ kallisto/
Sailfish	FASTA	RPKM, KPKM, TPM	Off	WB	www.cs.cmu.edu/ ~ckingsf/software/sailfish
Salmon	FASTA, SAM, BAM	TPM	Off	WB	github.com/COMBINE-lab/Salmon
FeatureCounts	RPKM, FPKM, TPM	Count	Off	CL	subread.sourceforge.net/
HTSeq	FASTA, FASTQ, SAM, BAM	Count	Off	CL	htseq.readthedocs.io/
EasyRNASeq	BAM	RPKM	Off	CL	bioconductor.org/packages/release/bioc/
EMSAR	SAM, BAM	FPKM, TPM	Off	CL	github.com/parklab/ emsar
RSEM	FASTA, FASTQ	TPM	Off	CL	deweylab.biostat.wisc. edu/rsem
Cufflinks	FPKM		Off	CL	cole-trapnell-lab.github.io/cufflinks/
eXpress	SAM,BAM, FASTA	FPKM, count data	Off	CL	pachterlab.github.io/ eXpress/overview.html
NOISeq	RPKM, read counts	RPKM, TMM	Off	CL	bioconductor.org/ packages/release/bioc/html/ NOISeq.html
RUVSeq	read counts	read counts	Off	CL	bioconductor.org/ packages/release/bioc/html/ RUVSeq.html

dimensions. To handle such voluminous data matrices, the analysis algorithms need to be scalable as well as precise. Another challenge for such algorithms is that data matrices may include a significantly high number of missing values [410]. For qualitative analysis of such data to extract biologically interesting, hidden patterns, one needs to handle the missing values using imputing techniques. A number of imputing tools have been developed using statistical and machine learning techniques. We discuss some popular imputation solutions that are effective in handling missing values at various scales in a wide range of application domains.

Table 7.2: Normalization Tools for Microarray Gene Expression Data

Name	Input	Output	Onl/ Off	WB/ CL	URL
Expression Profiler	expression data	count data	Onl/ Off	WB/ CL	sourceforge.net/projects/ ep-sf/
CARMAweb	expression data	analysis report file	Onl	WB	carmaweb.genome.tugraz.at/ carma/
Limma	SAM, BAM	Count data	Off	CL	bioconductor.org/packages/release/ bioc/html/ limma.html
VSN			Off	CL	bioconductor.org/packages/release/ bioc/html/vsn.html

(a) **methyLImp:** This statistical imputation tool [312] uses linear regression. It assumes that there exists a high degree of inter-sample correlation among the methylation levels.

(b) **missForest:** This non-parametric imputation technique [386] handles missing values in mixed type data that may include both continuous and categorical values. It uses an iterative imputation approach based on the principle of the random forest classifier. The random forest constitutes a multiple imputation scheme that averages over many unpruned classification or regression trees.

(c) **KNNimpute** and **SVDimpute:** These two popular missing value estimation tools [410] are applicable to microarray gene expression data. They are designed based on the principles of weighted K-nearest neighbors and singular value decomposition, respectively. KNNimpute selects genes with expression profiles similar to the gene of interest to impute the missing values. The weighted average of K closest genes is used as an estimate for the missing value. SVDimpute uses SVD to obtain a set of mutually orthogonal expression patterns that is linearly combined to approximate the expression of all genes in the dataset.

Table 7.3 presents some missing value estimation tools for RNAseq data along with their features. Table 7.4 presents some missing value estimation tools for microarray data along with their features.

Discretization

Discretization techniques transform a complex, continuous-valued gene expression data matrix into a discrete-valued data matrix [283]. Gene expression values over a discrete domain are simple and easy to handle. Many statistical measures are more effective in discrete domains allowing analysis

Table 7.3: Missing Value Estimation Tools for RNAseq Data

Name	Input	Output	On/ Off	WB/ CL	URL
methyLImp	(CpGs x sam- ples) data matrix containing missing values	Returns im- puted values for the missing values	Offline	Command- line	github.com/pdilena/ methyLImp
missForest	n x p ma- trix	Imputed data matrix of same type as input matrix	Offline	Command- line	stat.ethz.ch/CRAN/

Table 7.4: Missing Value Estimation Tools for Microarray Data

Name	Input	Output	On/ Off	WB/ CL	URL
SVDimpute	$n \times p$ ma- trix, the rank of the SVD approx- imation, the con- vergence tolerence for the EM algo- rithm, the maximum number of EM steps to take	the com- puted version of the i/p matrix	off	CL	smi-web.stanford.edu/ projects/helix/pubs/impute/
KNNimpute	$n \times p$ matrix, nearest neighbour K				smi-web.stanford.edu/ projects/helix/pubs/impute/
Expression Profiler	expression data	count data	Onl/ Off	WB/ CL	sourceforge.net/ projects/ep- sf/

of gene expression data to extract interesting trends or patterns. Discretization also helps estimate missing values due to experimental errors in the original expression data. A detailed discussion on discretization approaches and methods can be found in [246], [262], [283], [267], [271], [73], [383]. This sub-section reports two commonly used discretization methods.

(a) **RefBool:** Most discretization methods [271] transform each gene expression measurement into two states, i.e., *active* or *inactive*. However, unlike these methods, RefBool [203] classifies the measurements with three states: *active, inactive,* and *intermediate.* This third state is interpreted as intermediate expression of genes. RefBool is a reference-based method that associates each measurement to a p and q-value indicating the significance of each classification [203].

(b) **BiTrinA:** This tool [292] integrates multiscale techniques for binarization and trinarization of one-dimensional data with methods for quality assessment and visualization of the results. BiTrinA identifies the measurements with significantly high variations over different time points or conditions, and determines the candidates that are related to specific experimental settings.

Table 7.5 presents some discretization tools for RNAseq data along with their features.

Table 7.5: Discretization Tools for RNAseq Data

Name	Input	Output	On/Off	WB/CL	URL
RefBool	expression matrix	discretized query expression	Off	CL	github.com/saschajung/ RefBool
BiTrinA	expression matrix		Off	CL	cran.r-project.org/web/ packages/BiTrinA/index.html
chiMerge			Off	CL	www.rdocumentation.org/ packages/dprep/versions/3.0.2/ topics/chiMerge
GABD	expression matrix	Discretized matrix	Off	CL	www.sampa.droppages. com/GABD.html

Gene Selection

Gene selection is a preprocessing task that identifies prominent or relevant genes for the subsequent downstream analysis. A common approach for selection of relevant genes is to use a gene ranking method. Such a method determines the relevance of a gene based on a statistical measure and selects them for downstream analysis with reference to a user-defined threshold [105]. Gene selection has significant impact on the cost-effective analysis of gene expression data for accurate prediction of source(s) of disease(s). Gene selection can be performed using both supervised and unsupervised approaches. Some commonly used supervised and unsupervised tools for gene selection are discussed next.

Table 7.6: Discretization Tools for Microarray Data

Name	Input	Output	Onl/ Off	WB/ CL	URL
RefBool	expression matrix	discretized query expression	Offline	Command-line	https://github.com/saschajung/RefBool
BiTrinA	expression matrix		Offline	Command-line	cran.r-project.org/web/ packages/BiTrinA/index.html
GABD	expression matrix	Discretized matrix	Off	CL	sampa.droppages.com/GABD.html
Bikmeans	(no. of genes)x(no. of time points) matrix	Gene Regulatory Network	off	CL	read.pudn.com/downloads749
Column K-means	(no of genes)x(no of time points) matrix	k intervals	off	CL	

We present four commonly used gene selection methods.

(i) *GeneSrF*: It is a web-based tool developed especially for exploratory work with microarray data. It is parallelized to achieve fast response without compromising accuracy. It selects a minimum subset of genes that ensures cost-effective classification with best possible accuracy.

(ii) *varSelRF*: It is an R package developed using parallelization for exploratory work with microarray data. It exploits multicore CPUs and a cluster of workstations to achieve fast response without compromising accuracy. It generates output with bootstrapped estimates of prediction error rates and evaluation of the stability of solutions.

(iii) *mAPKL*: This hybrid gene selection is an open-source R/Bioconductor package. It is able to select small number of gene exemplars and ensures better classification accuracy for microarray data. This package is accompanied with several other functionalities for preprocessing, classification, and validation.

(iv) *Expression Profiler*: This is a web-based tool accompanied with several other functionalities related to clustering, visualization, analysis, and interpretation with reference to benchmark microarray gene expression databases.

Table 7.7 presents some gene selection tools with their features. In addition to these, there are several other gene selection tools developed based on both supervised as well as unsupervised approaches.

Table 7.7: Gene Selection Tools

Name	Input	Output	On/Off	WB/CL	URL
GeneSrF	plain text files	PIF figures	On	WB	genesrf.iib.uam.es/
varSelRF	plain text files		Off	CL	cran.r-project.org/web/packages/varSelRF/index.html
mAPKL	expression data, class label		Off	CL	bioconductor.org/packages/3.1/bioc/html/mAPKL.html
Expression Profiler	expression data	count data	On/Off	WB/CL	sourceforge.net/ projects/ep-sf/

Dimensionality Reduction

Dimensionality reduction techniques play a significant role in handling high-dimensional gene expression data. They helps eliminate irrelevant dimensions and support cost-effective downstream processing over the relevant dimensional space. Arbitrary removal of uninformative genes may lead to inaccurate and biased estimates. However, with an appropriate use of dimensionality reduction, one can reduce the cost of handling numerous variables significantly [425, 226]. Such tools help achieve high-quality prediction at low cost [98]. Dimensionality reduction techniques can be categorized as *supervised* and *unsupervised*. There are several supervised dimensionality reduction techniques have been developed such as (i) CoRanking, (ii) DimRed, (iii) Expression Profiler, and (iv) CARMAweb. Table 7.8 presents these tools along with their characteristics. A popular unsupervised dimensionality reduction tools is PCAMethods. This offline tool performs in command-line mode and accepts microarray data as input to generate dimensionality reduced gene expression data. It is available as a Bioconductor package.

Table 7.8: Supervised Dimensionality Reduction Tools

Name	Input	Output	On/Off	WB/CL	URL
coRanking	microarray data	a matrix	Off	CL	cran.r-project.org/web/packages/coRanking/ index.html
dimRed	microarray data		Off	CL	cran.r-project.org/web/packages/dimRed/index.html
Expression Profiler	microarray data	count data	On/Off	WB/CL	sourceforge.net/ projects/ep-sf/
CARMAweb	expression data	analysis report file	On	WB	carmaweb.genome.tu-graz.at/carma/

Quality Control

Systematic variations in experimental setups often causes biases at various levels when generating gene expression data. It is essential to control such biases to achieve an unbiased downstream analysis. An appropriate quality control technique can help reduce such biases induced during (i) sequencing, (ii) alignment, (iii) de novo assembly, and (iv) expression quantification. To handle biases, it is necessary to use an appropriate quality checking mechanism to (i) identify the biases precisely and (ii) eliminate biases without information loss for improving the efficiency in the subsequent steps [88]. A number of quality control techniques have been developed, especially for RNAseq data. Some of them are: (i) fastQC, (ii) HTSeq, (iii) MultiQC, (iv) NGSQC, (v) PRINSEQ, (vi) RNA-SeQC, (vii) RSeQC, (viii) SAMStat, (ix) SolexaQA, (x) HTQC, (xi) QoRTs, (xii) Picard, (xiii) QuaCRS, (xiv) AlignerBoost, and (xv) NOISeq. In comparison to RNAseq data, the role of quality control tools for microarray data is significantly less. Two commonly used such tools for microarray data are: (i) CARMAweb and (ii) Limma. Table 7.9 and 7.10 list some quality control tools used for RNAseq data and microarray data, respectively.

Improving Quality

RNA sequencing technology-generated count data often include instances with a low read count. Such low read-count instances do not contribute to the subsequent downstream analysis. Therefore, a preprocessing technique is required to improve the quality of such count data by eliminating low-quality raw read instances. Three commonly used quality improving techniques are (i) trimming, (ii) error detection, and (iii) bias correction. We discuss each of these techniques and list the features of various trimming, error detection, and bias correction techniques.

(a) *Trimming:* This technique is useful in removing (i) adaptor sequences, (ii) ambiguous nucleotides, and (iii) instances with read quality scores less than 20 and length below 30 bp [88]. Table 7.11 presents some trimming tools that are commonly used in quality improvement of count data. Most of these tools perform other preprocessing activities also.

(b) *Error Detection:* Error detection is another important quality improving preprocessing technique that identifies and eliminates sequencing errors in low-quality count data. A large number of error detection techniques have been developed. Not all these tools are effective to the same extent. We present a few in Table 7.12.

(c) *Bias Correction:* This preprocessing technique improves the quality of read sequences by detecting and correcting spurious transcriptome. Table 7.14 presents some prominent tools for bias correction of RNAseq data.

Table 7.9: Some Quality Control Tools for RNAseq Data

Name	Input	Output	Onl/ Off	WB/ CL	URL
fastQC	FASTQ, SAM, BAM	HTML report	Off	CL	bioinformatics. babraham.ac.uk/ projects/fastqc/
HTSeq	FASTA, FASTQ, SAM, BAM	Count data	Off	CL	htseq.readthedocs.io/ en/release_0.11.1/
MultiQC	Log files	HTML report	Off	CL	github.com/ewels/ MultiQC
NGSQC	FASTA, FASTQ	HTML	Off	CL	nipgr.res.in/ngsqctoolkit.html
PRINSEQ	FASTA, FASTQ	FASTA, FASTQ, HTML	Both	Both	prinseq.sourceforge.net/
RNA-SeQC	BAM,GTF, RefGen	HTML	Off	CL	www.broadinstitute.org/rna-seqc/
RSeQC	SAM,BAM, FASTA, BED	HTML	Off	CL	rseqc.sourceforge.net/
SAMStat	SAM,BAM, FASTA, FASTQ	HTML	Off	CL	samstat.sourceforge.net/
SolexaQA	FASTQ	FASTQ, PNG	Off	CL	solexaqa.sourceforge.net/
HTQC	FASTQ	FASTQ	Off	CL	sourceforge.net/ projects/htqc/
QoRTs	SAM,BAM, FASTA,BED	PNG, PDF reports	Off	CL	hartleys.github.io/ QoRTs/doc/QoRTs-vignette.pdf
Picard	SAM,BAM, CRAM,VCF	SAM,BAM, Text file	Off	CL	broadinstitute.github.io/picard/
QuaCRS	FASTQ, BAM	CSV file	Offline	CL	bioserv.mps.ohio-state.edu/QuaCRS/
AlignerBoost	SAM,BAM, BED,VCF	SAM,BAM	Off	CL	github.com/Grice-Lab/AlignerBoost
NOISeq	RPKM, read counts		Off	CL	bioconductor.org/ packages/release/bioc/html/ NOISeq.html

Batch Effect Removal

Batch effect refers to the non-biological systemic biases or differences between batches of samples used as input for the generation of gene expression data. Such biases may often get included in the data due to factors such as (i) array type, (ii) experimental conditions of different batches, (iii) preservation protocols, and (iv) shipment conditions [192]. The inclusion of batch effects can significantly reduce the effectiveness statistical tests during downstream analysis [303]. We present some tools commonly used for batch effect removal of RNAseq and microarray data.

Table 7.10: Some Quality Control Tools for Microarray Data

Name	Input	Output	Onl/ Off	WB/ CL	URL
CARMAweb	expression data	analysis report file	Onl	WB	carmaweb.genome. tu-graz.at/carma/
Limma	SAM, BAM	Count data	Off	CL	bioconductor.org/ pack-ages/release/bioc/html/ limma.html

Table 7.11: Some Trimming Tools for RNAseq Data

Name	Input	Output	Onl/ Off	WB/ CL	URL
Cutadapt	FASTQ	FASTQ	Off	CL	cutadapt.readthedocs.io/en/ stable/
FASTX-Toolkit	FASTA, FASTQ	FASTA, FASTQ	Both	Both	hannonlab.cshl.edu/ fastx_toolkit/
PRINSEQ	FASTA, FASTQ	FASTA, FASTQ, HTML	Both	Both	prinseq.sourceforge.net/
Trimmomatic	FASTQ	FASTQ	Off	CL	www.usadellab.org/cms/ in-dex.php?page=trimmomatic
NGSQC	FASTA, FASTQ	HTML	Off	CL	nipgr.res.in/ngsqctoolkit.html
SolexaQA	FASTQ	FASTQ, PNG	Off	CL	solexaqa.sourceforge.net/
HTQC	FASTQ	FASTQ	Off	CL	sourceforge.net/ projects/htqc/

(a) *Batch Effect Removal Tools for RNAseq Data*: Batch effects occur in RNAseq data due to systemic variations. To handle such biases, a number of tools have been introduced such as (i) RUVSeq, (ii) DeSeq2, (iii) Limma, (iv) Swamp, (v) SVASeq, (vi) SVA, (vii) NOISeq, and (viii) PEER. Table 7.14 reports some batch effect removal tools for RNAseq data.

(b) *Batch Effect Removal Tools for Microarray Data*: A number of tools are also available for removal of batch effects in microarray data. A few are (i) COMBAT, (ii) ARSyN (microarray), (iii) gPCA, (iv) Swamp, and (v) Pamr. Table 7.15 presents some batch effect removal tools for microarray data.

7.3   Gene Expression Data Analysis Tools

A preprocessed gene expression dataset is analyzed for extraction of hidden knowledge using three approaches: (i) co-expression analysis, (ii) differential co-expression analysis, and (iii) differential expression analysis. Each approach handles the data with a specific set of goals.

Table 7.12: Some Error Detection Tools for RNAseq Data

Name	Input	Output	Onl/ Off	WB/ CL	URL
Rcorrector	FASTQ	FASTQ, FASTA	Off	CL	github.com/mourisl/ Rcorrector
SOAPdenovo	de novo transcriptome	Sequence file (contig, scaffold)	Off	CL	github.com/aquaskyline/ SOAPdenovo-Trans
UNOISE	FASTA	FASTA	off	CL	
UNOISE2	FASTA	FASTA	off	CL	
Bayes-Hammer	reads	reads	off	CL	
SEECER	FASTQ	FASTQ, FASTA	Offline	Command-line	https://github.com/haisonle/ SEECER
QoRTs	SAM,BAM, FASTA,BED	PNG, PDF reports	Offline	Command-line	https://hartleys.github.io/ QoRTs/doc/QoRTs-vignette.pdf
Oases	mapped reads, RPKM		Offline	Command-line	www.ebi.ac.uk/ zerbino/ oases/

Table 7.13: Bias Correction Tools for RNAseq Data

Name	Input	Output	Onl/ Off	WB/ CL	URL
GeneScissors	Standard pipeline - TopHat, MapSlice	GeneScissors pipeline	Off	CL	csbio.unc.edu/ genescissors/
EDASeq	genes-by-lanes counts, Gene GC-content	Genes-by-lanes Normalized counts	Off	CL	bioconductor.org/ packages/release/bioc/html/ EDASeq.html
SVASeq	FPKM, count data	FPKM	Off	CL	github.com/jtleek/ svaseq
RUVSeq	read counts	read counts	Off	CL	bioconductor.org/ packages/release/bioc/html/ RUVSeq.html
SVA	FPKM, count data	FPKM	Off	CL	bioconductor.org/ packages/release/bioc/html/ sva.html
PEER (2011)	CSV file	CSV file	Off	CL	github.com/PMBio/ peer/wiki/Tutorial

7.3.1 Co-Expression Analysis

A co-expression analysis approach handles the data using (a) a supervised or classification approach, (b) an unsupervised or clustering approach, or (c) a network-based approach.

Table 7.14: Some Batch Effect Removal Tools for RNAseq Data

Name	Input	Output	Onl/ Off	WB/ CL	URL
RUVSeq	read counts	read counts	Off	CL	bioconductor.org/packages/ release/bioc/html/ RU-VSeq.html
DeSeq2	FASTQ, SAM, BAM	Count	Off	CL	bioconductor.org/packages/ release/bioc/html/ DE-Seq2.html
Limma	SAM, BAM	Count	Off	CL	bioconductor.org/ packages/release/bioc/html/ limma.html
Swamp	RPKM	RPKM	Off	CL	cran.r-project.org/web/ packages/swamp/index.html
SVASeq	FPKM, count data	FPKM	Off	CL	github.com/jtleek/ svaseq
SVA	FPKM, count data	FPKM	Off	CL	bioconductor.org/packages/ release/bioc/html/ sva.html
NOISeq	RPKM, read counts		Off	CL	bioconductor.org/packages/ release/bioc/html/ NOISeq.html
PEER (2011)	CSV file	CSV file	Off	CL	github.com/PMBio/ peer/wiki/Tutorial

Table 7.15: Some Batch Effect Removal Tools for Microarray Data

Name	Input	Output	Onl/ Off	WB/ CL	URL
COMBAT	FPKM, count data	FPKM (expression matrix)	Off	CL	rdrr.io/bioc/sva/man/ ComBat.html
ARSyN (microarray)	TCM data, longitudinal TCM data	TCM	Off	CL	www.ua.es/personal/ mj.nueda/software.html
gPCA	expression data	expression data	Off	CL	cran.r-project.org/web/ packages/gPCA/index.html
Swamp	RPKM	RPKM	Off	CL	cran.r-project.org/web/ packages/swamp/index.html
Pamr			Off	CL	cran.r-project.org/web/ packages/pamr/

Co-Expression Analysis Using Classification

A classification approach can be supervised or semi-supervised depending upon the amount of prior knowledge used during downstream analysis. A classifier model is built based on training instances that include data samples and the corresponding correct class labels. The model predicts the class label of an unknown test sample or instance. In other words, a classification approach analyzes the training dataset and develops an accurate description or model for each class using the attributes present in the data for recognition of class membership of a given unknown test instance.

Many classification models have been developed including decision trees, support vector machines, neural networks, evolutionary models and other soft computing models and their hybrid versions. A major issue with most classification-based gene expression analysis tools is that the generated outputs lack consistency. For the same input dataset, different tools often generate significantly varied classification results. Figure 7.4 shows a framework of a generic classification-based co-expression analysis tool.

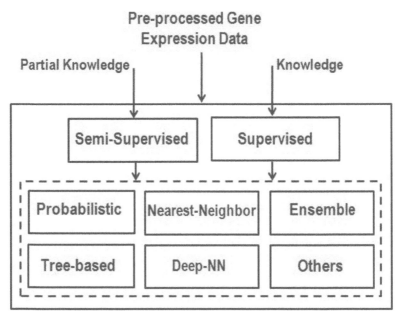

Figure 7.4: Supervised analysis: A framework

Co-Expression Analysis Using Clustering

A clustering approach partitions a dataset into disjoint groups or modules with high intra-group similarity or coherence. The formation of clusters or modules is largely dependent on the proximity measure used by the clustering algorithms. Based on the mode of operation and handling gene expression data, cluster analysis tools can be categorized as (i) 1-way clustering, (ii) 2-way clustering, and (iii) 3-way clustering. A major limitation of most clustering-based tools is lacking of generality. The application of the same clustering tool for all the three gene expression data types, microarray, bulkRNAseq or scRNAseq, may not be effective. Tools may require significant customization. Figure 7.5 shows a categorization of a generic clustering-based gene expression analysis tool. We present some cluster analysis tools developed for both microarray and RNAseq data.

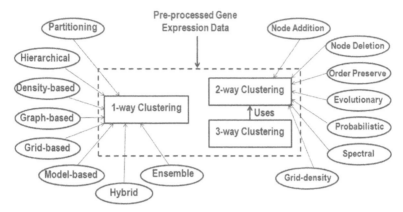

Figure 7.5: Clustering-based analysis: A framework

Co-Expression Analysis Using Network-Based Approach

A gene-gene network can be visualized as an undirected/directed and/or weighted/unweighted graph, where each node is a gene and a pair of nodes is connected with an edge if there exists significantly high functional or co-expression similarity. Network-based gene expression analysis is an established and promising approach for both microarray and RNseq data to extract biologically enriched, functionally similar co-expressed patterns. It can be carried out using various ways such as (i) co-expression network (CEN) analysis, (ii) regulatory network (GRN) analysis, and (iii) protein-protein interaction network (PPIN) analysis. Each of these approaches has its own merits and limitations. With the advent of advanced genomic data generation, all these three approaches have evolved with a large number of software tools. These supporting tools are used in the various stages of network-based gene expression data analysis. However, like the classification and clustering-based tools, most tools of this category also have limited scope. Figure 7.6 shows categorization of generic network-based gene expression analysis tools, including all three approaches on a single platform. Table 7.16 lists some tools for network-based analysis of RNAseq data. Table 7.17 presents some commonly used network analysis tools for microarray data.

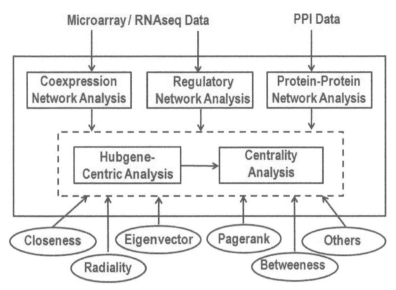

Figure 7.6: Network-based analysis

7.3.2 Differential Co-Expression Analysis

co-expression analysis discovers functionally similar or highly correlated groups of genes in a given gene expression dataset. If a set of co-expressed genes behave in a distinct way or responds in an unconventional way to biological changes (across the states i.e., normal to disease), we refer to it as a set of differentially co-expressed [403] genes. Differential co-expression network analysis helps the study of diseases and phenotypic variations in such co-expressed gene groups, which vary significantly across the conditions or states [87]. We list a set of tools for differential co-expression analysis of both microarray or RNAseq data. Tables 7.18 and 7.19 present some differential co-expression tools for RNAseq and microarray data, respectively.

7.3.3 Differential Expression Analysis

We define a set of genes as differentially expressed, if it shows significant differences across two states or conditions. Investigation of the behavior of such gene patterns helps extract hidden knowledge which may play a crucial role in the analysis of a disease in progression. DE analysis is a promising approach in Systems Biology research, and hence a large number of useful tools have been developed. Tables 7.20 and 7.21 present some useful tools for differential expression analysis of RNAseq data and microarray data, respectively.

Table 7.16: Some Network Analysis Tools for RNAseq Data

Name	Input	Output	Onl/ Off	WB/ CL	URL
WGCNA	RPKM, FPKM, normalized count data	RPKM, FPKM	Off	CL	cran.r-project.org/web/ packages/WGCNA/index.html
cMonkey2	genome data, gene annotations	expression data	Off	CL	github.com/baliga-lab/cmonkey2
ORTom	expression data	graphical visualization	Onl	WB	ortom.ivv.cnr.it/
FGMD	RPKM	RPKM	Off	CL	gcancer.org/FGMD/
CoNekT	FASTQ, TPM	heatmaps, matrices	Onl	WB	github.com/sepro/ CoNekT
WiPer			Off	CL	sourceforge.net/projects/ wipersoftware/
MRHCA	expression matrix	list of gene-hubs (text file)	Off	CL	github.com/zy26/mrct
EGAD	Gene label vectors	Performance metrics for each of the annotation sets tested	Off	CL	www.bioconductor.org/ packages/release/bioc/html/ EGAD.html
ARACNE	A text file, tab separated, with genes on rows and samples on columns	Text file containing every significant interaction	Off	CL	github.com/califano-lab/ARACNe-AP
wTO	RPKM, FPKM, TPM, rma, mas5, count data	graphical visualization	Off	CL	cran.r-project.org/web/ packages/wTO/index.html
CoRegNet	expression data	graphical visualization, heatmaps	Off	CL	bioconductor.org/ packages/CoRegNet
LSTrAP	FASTQ, FASTA, GFF3	graphical visualization	Off	CL	github.molgen.mpg.de/ proost/LSTrAP

7.4   Visualization

Visualization provides useful insights into the details of gene expressions, including co-expressed, differentially co-expressed, and differentially expressed patterns and other interaction patterns for a given gene expression dataset. Visualization tools also help load genomic datasets and visualize mapped reads in a graphical and interactive mode that allows zooming in and out of specific network regions to get a complete picture of the genomic data. Table 7.23 presents some useful tools for visualization.

Table 7.17: Some Network Analysis Tools for Microarray Data

Name	Input	Output	Onl/ Off	WB/ CL	URL
FGMD	RPKM	RPKM	Off	CL	gcancer.org/FGMD/
DHGA	microarray data		Off	CL	cran.r-project.org/web/ packages/dhga/index.html
GENIE3	microarray data	prediction scores	Off	CL	bioconductor.org/packages/ release/bioc/html/GENIE3.html
NICCE	Affymetrix Citrus Genome Array datasets	count data	Onl, Off	WB, CL	citrus.adelaide.edu.au/ nicce/home.aspx
ComPleX			Onl	WB	complex.plantgenie.org/
Expression Profiler	expression data	count data	Both	Both	sourceforge.net/ projects/ep-sf/

7.5 Validation

Validation is typically an assessment of the findings given by a gene expression analysis method for a given snapshot of time. With the passage of time, such a method needs to be re-validated or re-evaluated due to the changes in the (i) environmental conditions, (ii) technologies for data generation, and (iii) data characteristics such as volume, veracity, and velocity. Validation or evaluation not only gives an assessment of a method in terms of its accuracy and performance, but also helps refinement or re-engineering of the proposed method towards achievement of better acceptability by the society.

Any solution proposed against a biological question is necessary to be validated both (a) *statistically* and (b) *biologically*. A statistical validation is carried out considering three aspects: (i) correctness, (ii) efficiency, and (iii) data. The *correctness* of the gene expression analysis results can be validated using appropriate statistical measures. However, every statistical measure has its own advantages and limitations. Further, a measure alone may not be adequate to establish the correctness of a method. Hence, it is recommended to use multiple measures for unbiased evaluation (eliminating individual biasness) of a method in terms of accuracy. Similarly, to establish the *efficiency* of a method, one needs to assess the performance of the proposed method in terms of various quality aspects such as stability, completeness, scalability, etc. Also, one needs to consider three important facets of input gene expression data such as (i) quality, (ii) reliability, and (iii) completeness for appropriate validation of a method.

It is noteworthy that statistical measures alone are not adequate and also appropriate to establish a claim in solving a biological question, unless it is also validated biologically. Typical biological validations of gene expression analysis findings include (i) functional profiling, (ii) pathway sharing, (iii) regulatory behavior validation, and (iv) validation using heatmap biology.

Table 7.18: Some Differential Co-expression Analysis Tools for RNAseq Data

Name	Input	Output	Onl/ Off	WB/ CL	URL
DICER	expression data	gene similarity matrix	Off	CL	acgt.cs.tau.ac.il/dicer/
DifCoNet	expression data and metrics	graphical visualization	Off	CL	cran.rproject.org/web/ packages/difconet/index.html
DINGO	mRNA expression, DNA copy number, methylation, microRNA expression	differential scores for all pairs of genes	Off	CL	odin.mdacc.tmc.edu/ ~vbaladan
BFDCA	RNAseq, scRNAseq datasets	DCE analysis	Off	CL	data.mendeley.com/ datasets/jdz4vtvnm3/1
CEMiTool	expression data	HTML report/table/bar graph of significant pathways	Off	CL	Doi.org/10.18129/ B9.bioc.CEMiTool
CoDiNA	expression data	graphical visualization	Off	CL	rdrr.io/cran/CoDiNA/
DiffCoEx	Affymetrix gene expression profiles	heatmap	Off	CL	rdrr.io/github/ddeweerd/ MODi- fieRDev/man/diffcoex.html
MODA	gene expression profiles	heatmap	Off	CL	bioconductor.org/ packages/release/bioc/ html/MODA.html
DiffCorr	gene expression profiles	heatmap	Off	CL	cran.r- project.org/web/packages/ DiffCorr/index.html
THD-Module Extractor	gene expression profiles	GO pathways	Off	CL	pubmed.ncbi.nlm.nih.gov/ 27901073/

Next, we report some prominent tools commonly used in the validation of gene expression analysis results under two broad categories: *statistical* and *biological*.

7.5.1   Statistical Validation

The *correctness* of the gene expression analysis results can be established statistically using various measures such as (i) precision and recall, (ii) F measure, (iii) RoC curve, (iv) misclassification, (v) sensitivity and specificity, (vi) cluster validity indices, (vii) Rand index (R-index) and Adjusted Rand index (A-R index), (viii) p value, (ix) Q value, (x) overlap score, (xi) False Discovery Rate (FDR), and (xii) root mean square deviation (RMSD).

Table 7.19: Some Differential Co-expression Analysis Tools for Microarray
Data

Name	Input	Output	Onl/ Off	WB/ CL	URL
ROS-DET	expression data	gene pairs	Off	WB	www.bic.kyoto-u.ac.jp/pathway/kayano/ros-det.htm
WGCNA	RPKM, FPKM, normalized count data	RPKM, FPKM	Off	CL	cran.r-project.org/web/ packages/WGCNA/index.html
CoXpress	matrix with rows representing genes and columns representing samples	DCE genes	Off	CL	coxpress.sf.net
GSCA	gene expressions	FDR controlled genes	Off	CL	www.biostat.wisc.edu/ ~kendzior/GSCA/
Discordant	expression data	gene pairs	Off	CL	github.com/siskac/discordant
DiffCoMO	microarray data, scRNAseq data	graphical visualization			
MultiDCoX	expression data, attributes	graphical visualization	Off	CL	github.com/lianyh/ MultiDCoX
CEMiTool	expression data	statistical reports of singnificant pathways	Off	CL	doi.org/10.18129/ B9.bioc.CEMiTool
HO-GSVD	DNA copy number, DNA methylation and mRNA expression	graphical visualization	Off	CL	cran.r-project.org/web/packages/ hogsvdR/hogsvdR.pdf
EBcoexpress	expression matrix	pair-wise DCE analysis	Off	CL	bioconductor.org/ packages/release/bioc/html/ EBcoexpress.html
DiffCoEx	Affymetrix gene expression profiles	heatmap	Off	CL	rdrr.io/github/ddeweerd/ MODifieRDev/man/diffcoex.html

Similarly, one needs to consider the appropriate measures to establish *efficiency* of a method in terms of (i) stability, (ii) completeness, (iii) timeliness, (iv) scalability, and (v) ability to detect unknown interesting characteristics, biomarkers, genes, or patterns. Table 7.24 presents some useful tools for statistical validation of gene expression analysis results.

Table 7.20: Some Differential Expression Analysis Tools for RNAseq Data

Name	Input	Output	Onl/Off	WB/CL	URL
Limma-Voom	SAM, BAM	count	Off	CL	bioconductor.org/packages/release/bioc/html/limma.html
DeSeq	FASTQ, SAM, BAM	count	Off	CL	bioconductor.org/packages/DESeq/
DeSeq2	FASTQ, SAM, BAM	count	Off	CL	bioconductor.org/packages/release/bioc/html/DESeq2.html
EdgeR	SAM, BAM	FPKM, RPKM	Off	CL	bioconductor.org/packages/release/bioc/html/edgeR.html
CuffDiff2	raw counts	statistical reports	Off	CL	github.com/cole-trapnell-lab/cufflinks
DEApp	raw counts and metadata table	graphical visualization, pdf file	Onl	WB	yanli.shinyapps.io/ DEApp/
DSS	read counts	DEGs	Off	CL	bioconductor.org/packages/release/bioc/html/DSS.html
ImpulseDE	expression data	heatmaps, graphical visualization	Off	CL	bioconductor.org/packages/release/bioc/html/ImpulseDE.html
SARTools	count data file, target file describing experimental design	list of DE genes, HTML report	Off	CL	github.com/PF2-pasteur-fr/SARTools
baySeq	count data	statistical reports	Off	CL	bioconductor.org/packages/release/bioc/html/baySeq.html
NOISeq	RPKM, read counts	exploratory plots	Off	CL	bioconductor.org/packages/release/bioc/html/NOISeq.html
OASIS	FASTQ	interactive web reports	Onl	WB	oasis.dzne.de/

7.6 Biological Validation

It is essential to assess the results given by an analysis method in terms of biological enrichment. The findings need to be validated using appropriate functional analysis, pathway analysis, and crucial gene-centric analysis. A functional analysis validation determines the functional enrichments of the results (e.g., co-expressed gene groups) with reference to benchmark repositories or groundtruths. A pathway analysis attempts to validate the results

Table 7.21: Some Differential Expression Analysis Tools for Microarray Data

Name	Input	Output	Onl/ Off	WB/ CL	URL
CARMAweb	expression data	analysis report file	Onl	WB	carmaweb.genome.tugraz.at /carma/
Limma	SAM, BAM	Count data	Off	CL	bioconductor.org/packages/ release/bioc/html/ limma.html
vsn	microarray data	normalized microarray data	Off	CL	bioconductor.org/packages/ release/bioc/html/ vsn.html

(e.g. identified biomarker or crucial genes) from connectedness with the desired relevant pathway databases. Similarly, the findings related to crucial gene identification for a given disease can be validated using established (i) node centrality results and (ii) wet-lab results. Table 7.25 presents some popular functional analysis tools to validate gene expression analysis results.

7.7 Chapter Summary and Concluding Remarks

From the above discussion, we summarize as follows.

- Tools are useful for effective analysis of gene expression data. However, a single tool cannot help the entire process. So, we need an adequate knowledge on the existing tools, systems, and repositories for their appropriate use.

- Most raw gene expression datasets need appropriate preprocessing to transform it into more qualitative form for effective downstream processing.

- The preprocessing activities relevant or applicable for microarray data analysis, may not be relevant for RNAseq data analysis. An integrated tool or system that supports all kinds of preprocessing activities for both microarray and RNAseq data is useful.

- scRNAseq data needs special attention prior to downstream processing, especially to eliminate large occurrences of irrelevant cell information using appropriate dimensionality reduction technique. Similarly, the batch effect removal is mostly relevant for RNAseq data, but may not be for microarray data.

- 1-way cluster analysis tools are generally useful for microarray or bulk-RNAseq data analysis but not necessarily for scRNAseq data because of its high dimensionality.

- Biclustering or 2-way cluster analysis tools are effective in co-expressed pattern analysis or differentially co-expressed pattern analysis but cannot help in differentially expressed gene identification.

- Cluster visualization tools are generally not equipped for network analysis results visualization.

- Triclustering or 3-way cluster analysis tools are effective in identifying co-expressed patterns over Gene-Sample-Time space, but cannot help in DE analysis. Often such a tool consumes lot of computational resources.

- Graph-based algorithms are useful in unsupervised network analysis.

- A network analysis tool designed for co-expressed module(s) or association pattern(s) identification, may not be useful for gene regulatory behavior analysis, and vice versa.

- A statistically significant gene expression pattern or characteristics may not be biologically significant. Hence, such finding(s) need to be validated both statistically as well as biologically using appropriate and adequate number of indices, measures or benchmarks. Further, tools relevant for statistical validation of co-expression or differential co-expression analysis results may not be useful for differential expression analysis results.

- A set of genes identified as *elite* for a given disease is not necessarily static. It needs to be dynamically refreshed or updated with the recent sample data and validity indices or measures.

- Since a tool or system that enables analysis of only a single type of gene expression data (such as microarray or RNAseq data) may not be useful to generate the desired unbiased references. Therefore, we need tools or systems that allow handling of multiple gene expression data types in an integrated manner towards generation of more authentic inferences.

- An integrated tool/system that facilitates all types of preprocessing, analysis, visualization, and validation activities for all types of gene expression data is still missing.

Table 7.22: Some Prominent Tools for Visualization

Name	Input	Output	Onl/ Off	WB/ CL	URL
Artemis	BAM, SAM, VCF, BCF, FASTA	BAM, VCF, FASTA, graphs, heatmaps	Off	CL	sanger.ac.uk/ resources/software/artemis/
Integrated-Genome-Browser (IGB)	SAM, BAM, BED, PSL	Interactive genome visualization	Off	CL	igb.bioviz.org/ ; bitbucket.org/lorainelab/ integrated-genome-browser/src/ master/
UBiT2	FASTQ, expression data	PNG/ SVG images	Onl	WB	pklab.med.harvard.edu/ jean/ubit2/index.html
DEIVA	DGE statistical test results (text file)	SVG images	Onl	WB	github.com/Hypercubed/ DEIVA
iDEP	read counts, FPKM, RPKM, microarray (CSV/text)	graphical plots	Onl	WB	bioinformatics.sdstate. edu/idep/
Integrative-Genomics-Viewer (IGV)	SAM, BAM, VCF, GFF, BED, WIG, TDF	Interactive display	Both	Both	software.broadinstitute. org/software/igv/
GenomeView	SAM, BAM, FASTA, GFF, VCF, TDF	GFF, EMBL	Off	Stand-alone	genomeview.org/
ReadXplorer	Genbank, GFF, FASTA, SAM, BAM	mapped reads classification	Both	Both	readxplorer.org
RNASeq-Expression-Browser	gene expression matrix, annotation information	Interactive RNASeq visualization	Onl	WB	mips.helmholtz-muenchen.de/plant/ RNASeqExpression-Browser/
Tablet	ACE, AFG, MAQ, SOAP2, SAM,BAM, FASTA, FASTQ, GFF3	next generation sequence assemblies	Off	Stand-alone	ics.hutton.ac.uk/tablet/
JBrowse	FASTA, BED,GFF, BAM, WIG	Interactive genome display	Onl	WB	github.com/jbrowse/
Cascade	gene expression data	Interactive 3D forms	Onl	WB	bioinfo.iric.ca/ ~shifmana/Cascade/
Bambino	SAM,BAM	Assembly viewer display	Both	Both	cgwb.nci.nih.gov/ goldenPath/bamview/doc umentation/index.html
Savant	FASTA,BED, SAM,BAM, WIG,GFF	Interactive genome display	Off	Stand-alone	compbio.cs.toronto.edu/ savant
EagleView	ACE, READS, EGL,MAP, FASTA		Off	Stand-alone	bioinformatics.bc.edu/ marthlab/EagleView
HawkEye	FASTA, ACE	Graphical views of assembly	Off	Stand-alone	amos.sourceforge.net/wiki/ index.php?title= Hawkeye

Table 7.23: Some Tools for Visualization

Name	Input	Output	Onl/ Off	WB/ CL	URL
LookSeq	BAM, SAM	Interactive display	Off	Stand-alone	sanger.ac.uk/science/tools /lookseq
BamView	BAM, SAM		Off	Stand-alone	sanger.ac.uk/ science/tools/bamview
MagicViewer	BAM, SAM	Interactive display, VCF file	Off	Stand-alone	bioinformatics.zj.cn/ magicviewer/
Genome Data Viewer	FASTA	Interactive genome viewer	Onl	WB	ncbi.nlm.nih.gov/ genome/gdv/
SAMtools tview	SAM,BAM, CRAM, VCF, BCF	alignments in a per-position format	Off	Stand-alone	samtools.sourceforge.net/
ATGC transcriptomics	FASTA, VCF, GFF	HTML views	Onl	WB	sourceforge.net/projects/ atgcinta/
UCSC genome browser	genomic data	aligned annotation "tracks"	Onl	WB	genome.ucsc.edu/
GBrowse	SAM,BAM	PNG, SVG, PDF images	Both	Both	sourceforge.net/projects/ gmod/files/Generic 20Genome 20Browser/
Glimma	SAM, BAM	HTML views	Off	CL	bioconductor.org/ packages/release/bioc/html/ Glimma.html
DEGUST	count data (CSV file)	HTML views	Onl	WB	vicbioinformatics. com/software.degust.shtml
CARMAweb	expression data	analysis report file	Onl	WB	carmaweb.genome. tu-graz.at/carma/

Table 7.24: Some Tools for Statistical Validation

Name	Input	Output	Onl/ Off	WB/ CL	URL
VisRseq	Sequence data	graphical reports	Onl	WB	visrsoftware.github.io/
OASIS	FASTQ	interactive web reports	Onl	WB	oasis.dzne.de/
DEGUST	count data (CSV file)	HTML views	Onl	WB	vicbioinformatics. com/software.degust.shtml
NGSQC	FASTA, FASTQ	HTML	Off	CL	nipgr.ac.in/ ngsqc-toolkit.html

Table 7.25: Some Functional Analysis Tools

Name	Input	Output	Onl/ Off	WB/ CL	URL
ReactomePA	Entrez gene ID	graphical visualization	Off	CL	bioconductor.org/ packages/release/bioc/ html/ReactomePA.html
PAGE	expression data	graphical views, statistical indices	Off	Stand-alone	bit.ly/3tOnoCJ
GOMIner	expression data	DAG, statistical output files	Off	CL	discover.nci.nih.gov/ gominer/index.jsp
FatiGO	expression data	Graphical, statistical	Onl	WB	swmath.org/software/12318
GOTree-Machine	gene list	DAG	Onl	WB	genereg.ornl.gov/gotm/
GFINDer	expression data	graphical views, statistical indices	Onl	WB	medinfopoli. polimi.it/GFINDer/
FuncAssociate	Expression data	Statistical indices	Onl	WB	llama.mshri.on.ca/ funcassociate/
DAVID	expression data	graphical views, gene lists	Onl	WB	david.ncifcrf.gov/
GAGE	expression data	graphical views	Off	CL	bioconductor.org/ packages/release/bioc/html/ gage.html
GSEA	expression data	graphical views, statistical indices	Off	WB	gsea-msigdb.org/gsea/index.jsp
GO-Elite	expression data	statistical results	Off	Stand-alone	genmapp.org/ go_elite/
GSAASeqSP	raw count data	statistical results	Off	Stand-alone	gsaa.unc.edu/
GSVA	expression data matrix	score matrix	Off	CL	bioconductor.org/ packages/release/bioc/html/ GSVA.html
PathVisio	expression data	XML-based file	Off	Stand-alone	pathvisio.org
clusterProfiler	significant gene list and background gene list	CSV file (dotplot, enrichment plot, category netplot)	Off	CL	bioconductor.org/ packages/release/bioc/html/ clusterProfiler.html
CARMAweb	expression data	analysis report file	Onl	WB	carmaweb.genome. tu-graz.at/carma/

Chapter 8

Concluding Remarks and Research Challenges

8.1 Concluding Remarks

This book has focused on gene expression data analysis using three widely used approaches– co-expression analysis, differential co-expression analysis, and differential expression analysis using statistical and machine learning approaches. We have discussed recent high-throughput technologies used in the generation of large-scale gene expression data to provide a good understanding of how such data are generated as well as various issues involved in generating quality data. We have presented pre-processing requirements for both microarray and RNAseq data before such data can be used for analysis. To provide a strong background on machine learning techniques for use in gene expression analysis, this book has discussed a number of machine learning methods, systems and techniques under the categories of supervised learning, unsupervised learning, and combination learners in a simple, yet in-depth manner. To understand the effectiveness of a machine learning or statistical technique in analysing gene expression data, one needs to have an adequate knowledge of internal and external validity measures. This book, in a separate chapter, has introduced a number of internal and external validity measures, and also has enumerated the sources and repositories connected to the external validity measures. We believe it will help learn how to validate the outcomes of any existing method or their own method(s) using benchmark or real-life datasets. We have also presented pros and cons analysis of these methods so that learners can use them to solve bioinformatics problems to obtain high precision.

DOI: 10.1201/9780429322655-8

This book aims to impress upon readers that any gene expression data analysis problem can be solved with three approaches, co-expression analysis, differential co-expression analysis and differential expression analysis. In three dedicated chapters, we have introduced and analyzed a number of methods and techniques under these three approaches. We have also included case studies related to handling of critical diseases that identify and validate disease biomarkers. The study of real-life cases facilitates the development of a mental framework, to identify disease biomarkers with high precision.

This book has also introduced a broad array of practcal tools, systems and repositories for provide a hands-on experience with gene expression data pre-processing, analysis with clustering-based or network-based approaches, and to validate the outcomes of analysis. To locate tools easily for a bioinformatics task, the book has also introduced a taxonomy of tools. Finally, to motivate researchers and for the benefit of our society, we have enumerated a long list of issues and research challenges based on currently asked bioinformatics questions.

8.2   Some Issues and Research Challenges

A large number of techniques, methods and tools have been developed to assist microarray and RNA-seq data analysis. However, there are some important issues and research challenges [88] that still need attention.

(a) NVT [113] can select a suitable normalization technique from a set of 11 techniques for a specific dataset. These normalization techniques depend on properties of the dataset. It has been observed that the presence of a few highly expressed genes can repress the counts for all other genes. This may lead to false alarms, i.e., many genes may be falsely identified as differentially expressed. The use of an appropriate normalization technique can improve this issue in DE analysis. So, it will be neneficial to develop an effective statistical and/or computational method for RNA-seq data normalization, that works across datasets, to eliminate false alarms raised due to the misleading of DE analysis mentioned above.

(b) Often, RNA-seq experiments produce small samples which make further analysis difficult. More than 80% of RNA-seq count datasets available in public repositories, like recount2 and ARCHS4 [231], have very few samples. With low sample sizes, the ability of a method to detect DEGs decreases. To compensate, we need DE methods which not only work well for larger samples, but can also work with smaller sample sizes.,This is a challenging task.

(c) Although a large number of expression proximity and semantic proximity measures have been introduced for gene expression data analysis, there is no single proximity measure which captures both expression as well as semantic proximity at the same time. This is another important issue to investigate.

(d) No gene expression analysis method is consistent in performance over a wide range of disease datasets. Neither, is there any straightforward approach to identify an optimal method. So, developing an approach to evaluate a gene expression analysis method in terms of reproducibility, accuracy and robustness, remains a non-trivial task.

(e) Count data transformation is required for RNAseq data analysis using statistical and machine learning techniques. Such transformation is not starightforward. To avoid inconsistencies in the transformed data, one must consider statistical means appropriate for the data distribution. The development of a data transformation technique with wide applicability is another challenging task that need to be addressed.

(f) An important issue related to scRNA-seq data analysis is its low sensitivity. To distinguish biological variability with high precision in the presence of lowly expressed transcripts is challenging.

(g) The four major causes that often affect the analysis of scRNA-seq data are batch effects, zero inflation, presence of technical noise, and dropout events. It is necessary to develop accurate methods for imputation, normalization and batch effect removal to improve the performance of DE analysis of scRNA-seq data, especially in cell heterogeneity studies. However, there is no single, integrated framework to accurately deal with all such problems without compromising data quality.

(h) For scRNA-seq data analysis, cell-to-cell comparison is an essential for the tasks such as alignment and counting with high precision. With the increasing number of cells, such comparison becomes computationally intensive. Developing a robust and efficient parallel computing tool to support such activities without compromising quality is a need of the hour.

(i) Developing an integrated DE analysis framework to pre-process and analyze both bulk RNA-seq as well as scRNAseq data (for non-homogeneous tumor cells) for identification of interesting biomarkers is a challenging task that need to be investigated.

(j) For co-expression analysis of RNA-seq data, it is essential to choose optimal sample size and read depth. Non-optimal (too large or small) sample size or read depth leads to deviated co-expression analysis results. This remains a major research issue.

(k) Although a number of imputation techniques have been introduced to handle missing data, a robust and scalable imputation technique, applicable to all types of gene expression data is still missing.

(l) The performance of a normalization method generally depends on the particulars of a dataset. Different normalization methods have different impacts on co-expression analysis of RNA-seq data. There is need for an effective way to identify a suitable method from among alternatives for normalization of a particular scRNA-seq dataset with high precision.

(m) Although correlation measures have been commonly used in the co-expression analysis of microarray data, such measures are not suitable for RNA-seq because of the discrete nature of the data in RNA-seq data. Though logarithmic and variance stabilizing transformations have been used to address such problems, there is a chance of losing information. Thus, there is need for a robust proximity measure which can work well for RNA-seq data without transformation.

(n) To infer undirected networks, the selection of a proper co-expression measure and module-finding method is an issue. This is because the use of an appropriate unsupervised learning technique (e.g., clustering) to identify modules greatly influences downstream analysis.

(o) Co-expression modules are often not so modular because some members of modules overlap with members of other modules due to strong correlation. There is need for effective methods to address such perturbation effects from outside of a module.

(p) It is our experience that often a majority of the genes included in the network modules extracted from a CEN do not have annotation information. It creates difficulties in inferring association of such genes with traits. How to provide for such annotations or compensate for the lack of annotations is an important research issue.

(q) Developing a co-expression analysis method for splice variants is challenging if multiple splice variants are associated with the same exon. In such cases, we cannot predict which splice variant is expressed leading to biased results.

(r) Developing a user-friendly tool that can help identify unique and common features simultaneously for a number of species across two disease states based on topological comparison is a challenging task.

(s) The performance of differential co-expression analysis is largely dependent on the availability of quality data, species types, disease states, and the measures used. Developing a generic (diff) co-expression analysis framework that can help investigate a number of critical diseases to extract non-trivial biological knowledge, without depending on species type, will be beneficial.

(t) Statistical evaluation of pathway-based and network-based (co-expression or regulatory) methods in patient care needs further investigation. There is need for methods to bridge the gap between clinicians and patients to communicate genomic information.

(u) To compensate for biases (or limitations) of analysis results due to the use of a single source of data, a user-friendly framework to support integrative analysis that combines both transcriptomic and clinical data for a family of critical diseases is necessary. Such integrative analysis is useful to discover hidden biological knowledge.

(v) In differential co-expression analysis, often the modules obtained are not totally biologically enriched or pathway connected. As a consequence, such modules are usually discarded or not prioritized. However, in some such modules, some genes (a partial set of genes of a less prioritized module) may have high significance with reference to a given biological question. Appropriate strategies need to be developed to address this important research issue.

Bibliography

[1] CHOWDHURY, H. A., AHMED, H. A., BHATTACHARYYA, D. K. AND KALITA, J. K. NCBI: A Novel Correlation Based Imputing Technique using Biclustering. *Computational Intelligence in Pattern Recognition.* Springer, Singapore, (2020). 509–519.

[2] A. BARGHASH, T ARSLAN AND V HELMS. Detection of Outlier Samples and Genes in Expression Dataset. *Journal of Proteomics and Bioinformatics 9*, 2 (2016), 38–48.

[3] ABBOTT, D. W. Combining models to improve classifier accuracy and robustness. In *Proceedings of Second International Conference on Information Fusion, Fusion'99* (1999), vol. 1, Citeseer, pp. 289–295.

[4] AC'T HOEN, P., FRIEDLÄNDER, M. R., ALMLÖF, J., SAMMETH, M., PULYAKHINA, I., ANVAR, S. Y., LAROS, J. F., BUERMANS, H. P., KARLBERG, O., BRÄNNVALL, M., ET AL. Reproducibility of high-throughput mrna and small rna sequencing across laboratories. *Nature biotechnology 31*, 11 (2013), 1015–1022.

[5] ADAMS, M. D., KELLEY, J. M., GOCAYNE, J. D., DUBNICK, M., POLYMEROPOULOS, M. H., XIAO, H., MERRIL, C. R., WU, A., OLDE, B., MORENO, R. F., ET AL. Complementary DNA sequencing: expressed sequence tags and human genome project. *Science 252*, 5013 (1991), 1651–1656.

[6] AGARWAL, R., SRIKANT, R., ET AL. Fast algorithms for mining association rules. In *Proc. of the 20th VLDB Conference* (1994), pp. 487–499.

[7] AGARWAL, R. C., AGGARWAL, C. C., AND PRASAD, V. Depth first generation of long patterns. In *Proceedings of the sixth ACM*

SIGKDD international conference on Knowledge discovery and data mining (2000), pp. 108–118.

[8] AGGARWAL, C. C., WOLF, J. L., YU, P. S., PROCOPIUC, C., AND PARK, J. S. Fast algorithms for projected clustering. *ACM SIGMoD Record 28*, 2 (1999), 61–72.

[9] AGGARWAL, C. C., AND YU, P. S. Finding generalized projected clusters in high dimensional spaces. In *Proceedings of the 2000 ACM SIGMOD international conference on Management of data* (2000), pp. 70–81.

[10] AGRAWAL, R., GEHRKE, J. E., GUNOPULOS, D., AND RAGHAVAN, P. Automatic subspace clustering of high dimensional data for data mining applications, Dec. 14 1999. US Patent 6,003,029.

[11] AGRAWAL, R., IMIELINSKI, T., AND SWAMI, A. Mining association rules between sets of items in large databases. In ACM SIGMOD International Conference on Management of Data, pp. 206–216.

[12] AGRAWAL, R., IMIELINSKI, T., AND VU, A. Mining association rules with item constraints. In the Third International Conference on Knowledge Discovery in Databases and Data Mining, pp. 67–73.

[13] AGRAWAL, R., MANNILA, H., SRIKANT, R., TOIVONEN, H., VERKAMO, A. I., ET AL. Fast discovery of association rules. *Advances in knowledge discovery and data mining 12*, 1 (1996), 307–328.

[14] AHMADIAN, A., EHN, M., AND HOBER, S. Pyrosequencing: history, biochemistry and future. *Clinica chimica acta 363*, 1-2 (2006), 83–94.

[15] AHMED, H. A. Gene expression and protein interaction analysis using data mining techniques. *PhD Thesis*, TZ121399 (2012).

[16] AHMED, H. A., MAHANTA, P., BHATTACHARYYA, D. K., AND KALITA, J. K. Shifting-and-scaling correlation based biclustering algorithm. *IEEE/ACM transactions on computational biology and bioinformatics 11*, 6 (2014), 1239–1252.

[17] AIBAR, S., GONZÁLEZ-BLAS, C. B., MOERMAN, T., IMRICHOVA, H., HULSELMANS, G., RAMBOW, F., MARINE, J.-C., GEURTS, P., AERTS, J., VAN DEN OORD, J., ET AL. Scenic: single-cell regulatory network inference and clustering. *Nature methods 14*, 11 (2017), 1083–1086.

[18] ALBANESE, A., PAL, S. K., AND PETROSINO, A. Rough sets, kernel set, and spatiotemporal outlier detection. *IEEE Transactions on knowledge and data engineering 26*, 1 (2012), 194–207.

[19] ALBERT, R., JEONG, H., AND BARABÁSI, A.-L. Error and attack tolerance of complex networks. *nature 406*, 6794 (2000), 378–382.

[20] ALMAAS, E. Biological impacts and context of network theory. *Journal of Experimental Biology 210*, 9 (2007), 1548–1558.

[21] ALWINE, J. C., KEMP, D. J., AND STARK, G. R. Method for detection of specific RNAs in agarose gels by transfer to diazobenzyloxymethyl-paper and hybridization with DNA probes. *Proceedings of the National Academy of Sciences 74*, 12 (1977), 5350–5354.

[22] AMAR, D., YEKUTIELI, D., MARON-KATZ, A., HENDLER, T., AND SHAMIR, R. A hierarchical bayesian model for flexible module discovery in three-way time-series data. *Bioinformatics 31*, 12 (2015), i17–i26.

[23] ANDERS, S., AND HUBER, W. Differential expression analysis for sequence count data. *Nature Precedings* (2010), 1–1.

[24] ANKERST, M., BREUNIG, M. M., KRIEGEL, H.-P., AND SANDER, J. Optics: ordering points to identify the clustering structure. *ACM Sigmod record 28*, 2 (1999), 49–60.

[25] ARAVINDAN, M. An efficient approach for subspace clustering by using cat seeker. *Asian Journal of Computer Science and Technology (AJCST) 2*, 1 (2014), 49–52.

[26] ARDUI, S., AMEUR, A., VERMEESCH, J. R., AND HESTAND, M. S. Single molecule real-time (smrt) sequencing comes of age: applications and utilities for medical diagnostics. *Nucleic acids research 46*, 5 (2018), 2159–2168.

[27] ARIFOVIC, J. Genetic algorithm learning and the cobweb model. *Journal of Economic dynamics and Control 18*, 1 (1994), 3–28.

[28] ARTHUR, D., AND VASSILVITSKII, S. k-means++: The advantages of careful seeding. Tech. rep., Stanford, 2006.

[29] AUMANN, Y., FELDMAN, R., LIPSHTAT, O., AND MANILLA, H. Borders: An efficient algorithm for association generation in dynamic databases. *Journal of Intelligent Information Systems 12*, 1 (1999), 61–73.

[30] AWAD, M. M., AND DE JONG, K. Optimization of spectral signatures selection using multi-objective genetic algorithms. In *2011 IEEE Congress of Evolutionary Computation (CEC)* (2011), IEEE, pp. 1620–1627.

[31] AYAD, A., EL-MAKKY, N., AND TAHA, Y. Incremental mining of constrained association rules. In *Proceedings of the 2001 SIAM International Conference on Data Mining* (2001), SIAM, pp. 1–18.

[32] AYAD, H., AND KAMEL, M. Finding natural clusters using multi-clusterer combiner based on shared nearest neighbors. In *International Workshop on Multiple Classifier Systems* (2003), Springer, pp. 166–175.

[33] AYAN, N. F., TANSEL, A. U., AND ARKUN, E. An efficient algorithm to update large itemsets with early pruning. In *Proceedings of the fifth ACM SIGKDD international conference on Knowledge discovery and data mining* (1999), pp. 287–291.

[34] AZIZ, N. A. A., MOKHTAR, N. M., HARUN, R., MOLLAH, M. M. H., ROSE, I. M., SAGAP, I., TAMIL, A. M., NGAH, W. Z. W., AND JAMAL, R. A 19-gene expression signature as a predictor of survival in colorectal cancer. *BMC medical genomics 9*, 1 (2016), 58.

[35] AZIZI, E., PRABHAKARAN, S., CARR, A., AND PE'ER, D. Bayesian inference for single-cell clustering and imputing. *Genomics and Computational Biology 3*, 1 (2017).

[36] AZUAJE, F. J. Selecting biologically informative genes in co-expression networks with a centrality score. *Biology direct 9*, 1 (2014), 12.

[37] BADER, G. D., AND HOGUE, C. W. An automated method for finding molecular complexes in large protein interaction networks. *BMC Bioinformatics 4*, 1 (2003), 2.

[38] BALL, C. A., SPELLMAN, P. T., AND MILLER, M. MAGE-OM: An object model for the communication of microarray data. *Encyclopedia of Genetics, Genomics, Proteomics and Bioinformatics* (2004).

[39] BALLOUZ, S., VERLEYEN, W., AND GILLIS, J. Guidance for RNA-seq co-expression network construction and analysis: safety in numbers. *Bioinformatics 31*, 13 (2015), 2123–2130.

[40] BANDARU, S., NG, A. H., AND DEB, K. Data mining methods for knowledge discovery in multi-objective optimization: Part a-survey. *Expert Systems with Applications 70* (2017), 139–159.

[41] BANDYOPADHYAY, S., AND BHATTACHARYYA, M. A biologically inspired measure for coexpression analysis. *IEEE/ACM Transactions on Computational Biology and Bioinformatics 8*, 4 (2010), 929–942.

[42] BANDYOPADHYAY, S., MALLIK, S., AND MUKHOPADHYAY, A. A survey and comparative study of statistical tests for identifying differential expression from microarray data. *IEEE/ACM transactions on computational biology and bioinformatics 11*, 1 (2013), 95–115.

[43] BARABASI, A.-L., AND OLTVAI, Z. N. Network biology: understanding the cell's functional organization. *Nature reviews genetics 5*, 2 (2004), 101–113.

[44] BARCISZEWSKA, M. Z., PERRIGUE, P. M., AND BARCISZEWSKI, J. trna–the golden standard in molecular biology. *Molecular BioSystems 12*, 1 (2016), 12–17.

[45] BARRETT, T., BANYARD, A. C., AND DIALLO, A. Molecular biology of the morbilliviruses. In *Rinderpest and peste des petits ruminants*. Elsevier, 2006, pp. 31–IV.

[46] BARRETT, T., TROUP, D. B., WILHITE, S. E., LEDOUX, P., EVANGELISTA, C., KIM, I. F., TOMASHEVSKY, M., MARSHALL, K. A., PHILLIPPY, K. H., SHERMAN, P. M., ET AL. NCBI GEO: archive for functional genomics data sets—10 years on. *Nucleic Acids Research 39*, suppl_1 (2010), D1005–D1010.

[47] BASU, S., DAS, S., GHATAK, S., AND DAS, A. K. Strength pareto evolutionary algorithm based gene subset selection. In *2017 International Conference on Big Data Analytics and Computational Intelligence (ICBDAC)* (2017), IEEE, pp. 79–85.

[48] BAZLAMAÇCI, C. F., AND HINDI, K. S. Minimum-weight spanning tree algorithms a survey and empirical study. *Computers & Operations Research 28*, 8 (2001), 767–785.

[49] BECKER, E., ROBISSON, B., CHAPPLE, C. E., GUÉNOCHE, A., AND BRUN, C. Multifunctional proteins revealed by overlapping clustering in protein interaction network. *Bioinformatics 28*, 1 (2012), 84–90.

[50] BEN-DOR, A., CHOR, B., KARP, R., AND YAKHINI, Z. Discovering local structure in gene expression data: the order-preserving submatrix problem. In *Proceedings of the sixth annual international conference on Computational biology* (2002), pp. 49–57.

[51] BEN-DOR, A., SHAMIR, R., AND YAKHINI, Z. Clustering gene expression patterns. *Journal of computational biology 6*, 3-4 (1999), 281–297.

[52] BENJAMINI, Y., AND SPEED, T. P. Summarizing and correcting the gc content bias in high-throughput sequencing. *Nucleic acids research 40*, 10 (2012), e72–e72.

[53] BERGMANN, S., IHMELS, J., AND BARKAI, N. Iterative signature algorithm for the analysis of large-scale gene expression data. *Physical review E 67*, 3 (2003), 031902.

[54] BERRIZ, G. F., KING, O. D., BRYANT, B., SANDER, C., AND ROTH, F. P. Characterizing gene sets with FuncAssociate. *Bioinformatics 19*, 18 (2003), 2502–2504.

[55] BERTALANFFY, L. V. General system theory: Foundations, development, applications.

[56] BEZDEK, J. C., EHRLICH, R., AND FULL, W. Fcm: The fuzzy c-means clustering algorithm. *Computers & Geosciences 10*, 2-3 (1984), 191–203.

[57] BHAR, A., HAUBROCK, M., MUKHOPADHYAY, A., AND WINGENDER, E. Multiobjective triclustering of time-series transcriptome data reveals key genes of biological processes. *BMC bioinformatics 16*, 1 (2015), 200.

[58] BHATTACHARYYA, D. K., AND JUGAL K KALITA. *Network Anomaly Detection: A Machine Learning Perspective.* Chapman and Hall, CRC Press, 2013.

[59] BHATTACHARYYA, D. K., AND KALITA, J. K. *Network anomaly detection: A machine learning perspective.* Chapman and Hall/CRC, 2013.

[60] BLONDEL, V. D., GUILLAUME, J.-L., LAMBIOTTE, R., AND LEFEB-VRE, E. Fast unfolding of communities in large networks. *Journal of statistical mechanics: theory and experiment 2008*, 10 (2008), P10008.

[61] BORIAH, S., CHANDOLA, V., AND KUMAR, V. Similarity measure of categorical datas. In *Proceedings of SIAM Data Mining Conference, Atlanta, GA* (2008).

[62] BOTÍA, J. A., VANDROVCOVA, J., FORABOSCO, P., GUELFI, S., D'SA, K., HARDY, J., LEWIS, C. M., RYTEN, M., WEALE, M. E., CONSORTIUM, U. K. B. E., ET AL. An additional k-means clustering step improves the biological features of wgcna gene co-expression networks. *BMC systems biology 11*, 1 (2017), 47.

[63] BOUGUETTAYA, A. On-line clustering. *IEEE Transactions on knowledge and data engineering 8*, 2 (1996), 333–339.

[64] BOURGON, R., GENTLEMAN, R., AND HUBER, W. Independent filtering increases detection power for high-throughput experiments. *Proceedings of the National Academy of Sciences 107*, 21 (2010), 9546–9551.

[65] BRAZMA, A., KRESTYANINOVA, M., AND SARKANS, U. Standards for systems biology. *Nature Reviews Genetics 7*, 8 (2006), 593–605.

[66] BRAZMA, A., PARKINSON, H., SARKANS, U., SHOJATALAB, M., VILO, J., ABEYGUNAWARDENA, N., HOLLOWAY, E., KAPUSHESKY, M., KEMMEREN, P., LARA, G. G., ET AL. Arrayexpress—a public repository for microarray gene expression data at the ebi. *Nucleic acids research 31*, 1 (2003), 68–71.

[67] BREIMAN, L. Bagging predictors. *Machine learning 24*, 2 (1996), 123–140.

[68] BRENNER, S., JOHNSON, M., BRIDGHAM, J., GOLDA, G., LLOYD, D. H., JOHNSON, D., LUO, S., MCCURDY, S., FOY, M., EWAN, M., ET AL. Gene expression analysis by massively parallel signature sequencing (mpss) on microbead arrays. *Nature biotechnology 18*, 6 (2000), 630–634.

[69] BREUNIG, M. M., KRIEGEL, H.-P., NG, R. T., AND SANDER, J. Lof: identifying density-based local outliers. In *Proceedings of the 2000 ACM SIGMOD international conference on Management of data* (2000), pp. 93–104.

[70] BRIN, S., R.MOTWANI, J.D.ULLMAN, AND S.TSUR. Dynamic itemset counting and implication rules for market basket data. vol. 26, in Proc. of the 1997 ACM SIGMOD Int'n Conf. on Management of data, pp. 255–268.

[71] BUSTIN, S., BENES, V., NOLAN, T., AND PFAFFL, M. Quantitative real-time RT-PCR–a perspective. *Journal of Molecular Endocrinology 34*, 3 (2005), 597–601.

[72] BUSYGIN, S., JACOBSEN, G., KRÄMER, E., AND AG, C. Double conjugated clustering applied to leukemia microarray data. In *In 2nd SIAM ICDM, Workshop on clustering high dimensional data* (2002), Citeseer.

[73] C A GALLO, R L CECCHINI, J A CARBALLIDO, S MICHELETTO, I PONZONI. Discretization of gene expression data revised. *Briefings in Bioinformatics 17*, 5 (2015), 758–770.

[74] CAREY, M. F., PETERSON, C. L., AND SMALE, S. T. The RNase protection assay. *Cold Spring Harbor Protocols 2013*, 3 (2013), pdb–prot071910.

[75] CARTER, C. S., BRAVER, T. S., BARCH, D. M., BOTVINICK, M. M., NOLL, D., AND COHEN, J. D. Anterior cingulate cortex, error detection, and the online monitoring of performance. *Science 280*, 5364 (1998), 747–749.

[76] CHANG, J.-W., AND JIN, D.-S. A new cell-based clustering method for large, high-dimensional data in data mining applications. In *Proceedings of the 2002 ACM symposium on Applied computing* (2002), pp. 503–507.

[77] CHANG, T.-W. Binding of cells to matrixes of distinct antibodies coated on solid surface. *Journal of Immunological Methods 65*, 1-2 (1983), 217–223.

[78] CHEESEMAN, P., KELLY, J., SELF, M., STUTZ, J., TAYLOR, W., AND FREEMAN, D. Autoclass: A bayesian classification system. In *Machine Learning Proceedings 1988*. Elsevier, 1988, pp. 54–64.

[79] CHEN, Y., JIA, Z., MERCOLA, D., AND XIE, X. A gradient boosting algorithm for survival analysis via direct optimization of concordance index. *Computational and mathematical methods in medicine 2013* (2013).

[80] CHEN, Y.-J., KODELL, R., SISTARE, F., THOMPSON, K. L., MOR-RIS, S., AND CHEN, J. J. Normalization methods for analysis of microarray gene-expression data. *Journal of biopharmaceutical statistics 13*, 1 (2003), 57–74.

[81] CHENG, C.-H., FU, A. W., AND ZHANG, Y. Entropy-based subspace clustering for mining numerical data. In *Proceedings of the fifth ACM SIGKDD international conference on Knowledge discovery and data mining* (1999), pp. 84–93.

[82] CHENG, Y., AND CHURCH, G. M. Biclustering of expression data. In *Ismb* (2000), vol. 8, pp. 93–103.

[83] CHESLER, E. J., AND LANGSTON, M. A. Combinatorial genetic regulatory network analysis tools for high throughput transcriptomic data. In *Systems biology and regulatory genomics*. Springer, 2005, pp. 150–165.

[84] CHEUNG, D. W., LEE, S. D., AND KAO, B. A general incremental technique for maintaining discovered association rules. In *Database Systems For Advanced Applications' 97*. World Scientific, 1997, pp. 185–194.

[85] CHOE, S. E., BOUTROS, M., MICHELSON, A. M., CHURCH, G. M., AND HALFON, M. S. Preferred analysis methods for affymetrix genechips revealed by a wholly defined control dataset. *Genome biology 6*, 2 (2005), R16.

[86] CHOPRA-DEWASTHALY, R., KORB, M., BRUNTHALER, R., AND ERTL, R. Comprehensive RNA-Seq profiling to evaluate the sheep mammary gland transcriptome in response to experimental mycoplasma agalactiae infection. *PLoS One 12*, 1 (2017), e0170015.

[87] CHOWDHURY, H. A., BHATTACHARYYA, D. K., AND KALITA, J. K. (Differential) Co-Expression Analysis of Gene Expression: A Survey of Best Practices. *IEEE/ACM transactions on computational biology and bioinformatics* (2018).

[88] CHOWDHURY, H. A., BHATTACHARYYA, D. K., AND KALITA, J. K. Differential Expression Analysis of RNA-seq Reads: Overview, Taxonomy and Tools. *IEEE/ACM transactions on computational biology and bioinformatics* (2018).

[89] CHU, G., LI, J., NARASIMHAN, B., TIBSHIRANI, R., AND TUSHER, V. Significance analysis of microarrays users guide and technical document.

[90] CLOUGH, E., AND BARRETT, T. The gene expression omnibus database. In *Statistical Genomics*. Springer, 2016, pp. 93–110.

[91] CONESA, A., MADRIGAL, P., TARAZONA, S., GOMEZ-CABRERO, D., CERVERA, A., MCPHERSON, A., SZCZEŚNIAK, M. W., GAFFNEY, D. J., ELO, L. L., ZHANG, X., ET AL. A survey of best practices for rna-seq data analysis. *Genome biology 17*, 1 (2016), 13.

[92] COSTA-SILVA, J., DOMINGUES, D., AND LOPES, F. M. Rna-seq differential expression analysis: An extended review and a software tool. *PloS one 12*, 12 (2017).

[93] CRICK, F. Central dogma of molecular biology. *Nature 227*, 5258 (1970), 561.

[94] D. JIANG, J. PEI, M. RAMANATHANY, C. TANG AND A. ZHANG. Mining coherent gene clusters from gene-sample-time microarray data. In *Proc of the 10 th ACM SIGKDD Conference(KDD04)* (2004), ACM, pp. 430–439.

[95] DAS, A., AND BHATTACHARYYA, D. K. Rule mining for dynamic databases. In *International Workshop on Distributed Computing* (2004), Springer, pp. 46–51.

[96] DAS, R., MITRA, S., BANKA, H., AND MUKHOPADHYAY, S. Evolutionary biclustering with correlation for gene interaction networks. In *International Conference on Pattern Recognition and Machine Intelligence* (2007), Springer, pp. 416–424.

[97] DAS, S., MEHER, P. K., RAI, A., BHAR, L. M., AND MANDAL, B. N. Statistical approaches for gene selection, hub gene identification and module interaction in gene co-expression network analysis: an application to aluminum stress in soybean (glycine max l.). *PloS one 12*, 1 (2017).

[98] DE SOUZA, J. T., DE FRANCISCO, A. C., AND DE MACEDO, D. C. Dimensionality reduction in gene expression data sets. *IEEE Access 7* (2019), 61136–61144.

[99] DEFAYS, D. An efficient algorithm for a complete link method. *The Computer Journal 20*, 4 (1977), 364–366.

[100] DELHOMME, N., PADIOLEAU, I., AND ET AL. Easyrnaseq: a bioconductor package for processing rna-seq data. *Bioinformatics 28*, 19 (2012), 2532–2533.

[101] DEMBÉLÉ, D., AND KASTNER, P. Fuzzy c-means method for clustering microarray data. *bioinformatics 19*, 8 (2003), 973–980.

[102] DENNIS, G., SHERMAN, B. T., HOSACK, D. A., YANG, J., GAO, W., LANE, H. C., AND LEMPICKI, R. A. David: database for annotation, visualization, and integrated discovery. *Genome biology 4*, 9 (2003), R60.

[103] DI, Y., SCHAFER, D. W., CUMBIE, J. S., AND CHANG, J. H. The nbp negative binomial model for assessing differential gene expression from rna-seq. *Statistical applications in genetics and molecular biology 10*, 1 (2011).

[104] DIAZ, A., LIU, S. J., SANDOVAL, C., POLLEN, A., NOWAKOWSKI, T. J., LIM, D. A., AND KRIEGSTEIN, A. Scell: integrated analysis of single-cell rna-seq data. *Bioinformatics 32*, 14 (2016), 2219–2220.

[105] DÍAZ-URIARTE, R., AND DE ANDRES, S. A. Gene selection and classification of microarray data using random forest. *BMC bioinformatics 7*, 1 (2006), 3.

[106] DIETTERICH, T. G. Machine-learning research. *AI magazine 18*, 4 (1997), 97–97.

[107] DILLIES, M.-A., RAU, A., AUBERT, J., ET AL. A comprehensive evaluation of normalization methods for Illumina high-throughput RNA sequencing data analysis. *Briefings in Bioinformatics 14*, 6 (2013), 671–683.

[108] DO, C. B., AND BATZOGLOU, S. What is the expectation maximization algorithm? *Nature biotechnology 26*, 8 (2008), 897.

[109] DOVICHI, N. J. Dna sequencing by capillary electrophoresis. *Electrophoresis 18*, 12-13 (1997), 2393–2399.

[110] DUBOIA, D., AND PRADE, H. Rough-fuzzy sets and fuzzy-rough sets. *International Journal of General Systems 17* (1990), 191–209.

[111] DUFFY, D. E., ET AL. A permutation-based algorithm for block clustering. *Journal of Classification 8*, 1 (1991), 65–91.

[112] DÜNDAR, F., SKRABANEK, L., AND ZUMBO, P. Introduction to differential gene expression analysis using RNA-seq. *Appl. Bioinformatics* (2015), 1–67.

[113] EDER, T., GREBIEN, F., AND RATTEI, T. Nvt: a fast and simple tool for the assessment of rna-seq normalization strategies. *Bioinformatics 32*, 23 (2016), 3682–3684.

[114] EDWARDS, D. Non-linear normalization and background correction in one-channel cdna microarray studies. *Bioinformatics 19*, 7 (2003), 825–833.

[115] ELDER, J., AND PREGIBON, D. A statistical perspective on knowledge discovery in databases. advances in knowledge discovery and data mining. um fayyad, g. piatetsky-shapiro, p. smyth, and r. uthurusamy, editors. aaai, 1995.

[116] ELYASIGOMARI, V., LEE, D., SCREEN, H. R., AND SHAHEED, M. H. Developmenmt of a two-stage gene selection method that incorporates a novel hybrid approach using the cuckoo optimization algorithm and harmony search for cancer classification. *Journal of Biomedical Informatics 67* (2017), 11–20.

[117] ERDAL, S., OZTURK, O., ARMBRUSTER, D., FERHATOSMANOGLU, H., AND RAY, W. C. A time series analysis of microarray data. In *Proceedings. Fourth IEEE Symposium on Bioinformatics and Bioengineering* (2004), IEEE, pp. 366–375.

[118] ESTER, M., KRIEGEL, H.-P., SANDER, J., XU, X., ET AL. A density-based algorithm for discovering clusters in large spatial databases with noise. In *Kdd* (1996), vol. 96, pp. 226–231.

[119] ESTIVILL-CASTRO, V., AND LEE, I. Amoeba: Hierarchical clustering based on spatial proximity using delaunay diagram. In *Proceedings of the 9th International Symposium on Spatial Data Handling. Beijing, China* (2000), pp. 1–16.

[120] ESTIVILL-CASTRO, V., AND LEE, I. Autoclust: Automatic clustering via boundary extraction for mining massive point-data sets. In *In Proceedings of the 5th International Conference on Geocomputation* (2000), Citeseer.

[121] EZEIFE, C. I., AND SU, Y. Mining incremental association rules with generalized fp-tree. In *Conference of the Canadian Society for Computational Studies of Intelligence* (2002), Springer, pp. 147–160.

[122] FANG, Q., NG, W., FENG, J., AND LI, Y. Mining order-preserving submatrices from probabilistic matrices. *ACM Transactions on Database Systems (TODS) 39*, 1 (2014), 1–43.

[123] FELIX BRECHTMANN, C MERTES, A MATUSEVICIUTE, ET AL. OUTRIDER: A Statistical Method for Detecting Aberrantly Expressed Genes in RNA Sequencing Data. *AJHG 103*, 6 (2018), 907–917.

[124] FELL, D. A. Metabolic control analysis: a survey of its theoretical and experimental development. *Biochemical Journal 286*, Pt 2 (1992), 313.

[125] FILTEAU, M., PAVEY, S. A., ST-CYR, J., ET AL. Gene coexpression networks reveal key drivers of phenotypic divergence in lake whitefish. *Molecular Biology and Evolution 30*, 6 (2013), 1384–1396.

[126] FLOOD, R. L., AND CARSON, E. R. *Dealing with complexity: an introduction to the theory and application of systems science.* Springer Science & Business Media, 2013.

[127] FOSS, A., AND ZAÏANE, O. R. A parameterless method for efficiently discovering clusters of arbitrary shape in large datasets. In *2002 IEEE International Conference on Data Mining, 2002. Proceedings.* (2002), IEEE, pp. 179–186.

[128] FREUND, Y., SCHAPIRE, R. E., ET AL. Experiments with a new boosting algorithm. In *icml* (1996), vol. 96, Citeseer, pp. 148–156.

[129] FRIEDL, M. A., AND BRODLEY, C. E. Decision tree classification of land cover from remotely sensed data. *Remote sensing of environment 61*, 3 (1997), 399–409.

[130] FRIEDMAN, J. H., AND MEULMAN, J. J. Clustering objects on subsets of attributes (with discussion). *Journal of the Royal Statistical Society: Series B (Statistical Methodology) 66*, 4 (2004), 815–849.

[131] FU, L., AND MEDICO, E. Flame, a novel fuzzy clustering method for the analysis of dna microarray data. *BMC bioinformatics 8*, 1 (2007), 3.

[132] GAMAZON, E. R., WHEELER, H. E., SHAH, K. P., MOZAFFARI, S. V., AQUINO-MICHAELS, K., CARROLL, R. J., EYLER, A. E., DENNY, J. C., NICOLAE, D. L., COX, N. J., ET AL. A gene-based association method for mapping traits using reference transcriptome data. *Nature genetics 47*, 9 (2015), 1091.

[133] GANTI, V., GEHRKE, J., AND RAMAKRISHNAN, R. Cactus—clustering categorical data using summaries. In *Proceedings of the fifth ACM SIGKDD international conference on Knowledge discovery and data mining* (1999), pp. 73–83.

[134] GARRAGHAN, P., TOWNEND, P., AND XU, J. An analysis of the server characteristics and resource utilization in google cloud. In *2013 IEEE International Conference on Cloud Engineering (IC2E)* (2013), IEEE, pp. 124–131.

[135] GE, S. X., SON, E. W., AND YAO, R. idep: an integrated web application for differential expression and pathway analysis of rna-seq data. *BMC bioinformatics 19*, 1 (2018), 534.

[136] GERSTNER, W. Hebbian learning and plasticity. *From neuron to cognition via computational neuroscience* (2011), 0–25.

[137] GETZ, G., LEVINE, E., AND DOMANY, E. Coupled two-way clustering analysis of gene microarray data. *Proceedings of the National Academy of Sciences 97*, 22 (2000), 12079–12084.

[138] GEURTS, P., ERNST, D., AND WEHENKEL, L. Extremely randomized trees. *Machine learning 63*, 1 (2006), 3–42.

[139] GHAFFARI, N., ABANTE, J., SINGH, R., BLOOD, P. D., AND JOHNSON, C. D. Computational considerations in transcriptome assemblies and their evaluation, using high quality human rna-seq data. In *Proceedings of the XSEDE16 Conference on Diversity, Big Data, and Science at Scale* (2016), pp. 1–4.

[140] GHASEMI, A., AND ZAHEDIASL, S. Normality tests for statistical analysis: a guide for non-statisticians. *International journal of endocrinology and metabolism 10*, 2 (2012), 486.

[141] GHOSH, A., AND NATH, B. Multi-objective rule mining using genetic algorithms. *Information Sciences 163*, 1-3 (2004), 123–133.

[142] GIBBONS, F. D., AND ROTH, F. P. Judging the quality of gene expression-based clustering methods using gene annotation. *Genome research 12*, 10 (2002), 1574–1581.

[143] GOGOI, P., BHATTACHARYYA, D., BORAH, B., AND KALITA, J. K. A survey of outlier detection methods in network anomaly identification. *The Computer Journal 54*, 4 (2011), 570–588.

[144] GOH, W. W. B., WANG, W., AND WONG, L. Why batch effects matter in omics data, and how to avoid them. *Trends in biotechnology 35*, 6 (2017), 498–507.

[145] GOIL, S., NAGESH, H., AND CHOUDHARY, A. Mafia: Efficient and scalable subspace clustering for very large data sets. In *Proceedings of the 5th ACM SIGKDD International Conference on Knowledge Discovery and Data Mining* (1999), vol. 443, ACM, p. 452.

[146] GOLUB, T. R., SLONIM, D. K., TAMAYO, P., HUARD, C., GAASENBEEK, M., MESIROV, J. P., COLLER, H., LOH, M. L., DOWNING, J. R., CALIGIURI, M. A., ET AL. Molecular classification of cancer: class discovery and class prediction by gene expression monitoring. *science 286*, 5439 (1999), 531–537.

[147] GRABHERR, M. G., HAAS, B. J., YASSOUR, M., LEVIN, J. Z., THOMPSON, D. A., AMIT, I., ADICONIS, X., FAN, L., RAYCHOWDHURY, R., ZENG, Q., ET AL. Full-length transcriptome assembly from rna-seq data without a reference genome. *Nature biotechnology 29*, 7 (2011), 644–652.

[148] GRAHNE, G., AND ZHU, J. Efficiently using prefix-trees in mining frequent itemsets. In *FIMI* (2003), vol. 90, p. 65.

[149] GRÜN, D., LYUBIMOVA, A., KESTER, L., WIEBRANDS, K., BASAK, O., SASAKI, N., CLEVERS, H., AND VAN OUDENAARDEN, A. Single-cell messenger rna sequencing reveals rare intestinal cell types. *Nature 525*, 7568 (2015), 251–255.

[150] GUHA, S., RASTOGI, R., AND SHIM, K. Cure: an efficient clustering algorithm for large databases. *ACM Sigmod record 27*, 2 (1998), 73–84.

[151] GUHA, S., RASTOGI, R., AND SHIM, K. Rock: A robust clustering algorithm for categorical attributes. *Information systems 25*, 5 (2000), 345–366.

[152] GUO, M., WANG, H., POTTER, S. S., WHITSETT, J. A., AND XU, Y. Sincera: a pipeline for single-cell rna-seq profiling analysis. *PLoS computational biology 11*, 11 (2015).

[153] GUO, Y., SHENG, Q., LI, J., YE, F., SAMUELS, D. C., AND SHYR, Y. Large scale comparison of gene expression levels by microarrays and rnaseq using tcga data. *PloS one 8*, 8 (2013), e71462.

[154] GUO, Y., ZHAO, S., LI, C.-I., SHENG, Q., AND SHYR, Y. Rnaseqps: a web tool for estimating sample size and power for rnaseq experiment. *Cancer informatics 13* (2014), CIN–S17688.

[155] GUSENLEITNER, D., HOWE, E. A., BENTINK, S., QUACKENBUSH, J., AND CULHANE, A. C. ibbig: iterative binary bi-clustering of gene sets. *Bioinformatics 28*, 19 (2012), 2484–2492.

[156] GUTIÉRREZ-AVILÉS, D., GIRÁLDEZ, R., GIL-CUMBRERAS, F. J., AND RUBIO-ESCUDERO, C. Triq: a new method to evaluate triclusters. *BioData mining 11*, 1 (2018), 15.

[157] GUTIÉRREZ-AVILÉS, D., RUBIO-ESCUDERO, C., MARTÍNEZ-ÁLVAREZ, F., AND RIQUELME, J. C. Trigen: A genetic algorithm to mine triclusters in temporal gene expression data. *Neurocomputing 132* (2014), 42–53.

[158] H. JIANG, S. ZHOU, J. GUAN AND Y. ZHENG. gtricluster: A more general and effective 3d clustering algorithm for gene-sample-time microarray data. In *Proc. BioDM'06* (2006), ACM, pp. 48–59.

[159] HA, M. J., BALADANDAYUTHAPANI, V., AND DO, K.-A. DINGO: differential network analysis in genomics. *Bioinformatics 31*, 21 (2015), 3413–3420.

[160] HAGHVERDI, L., LUN, A. T., MORGAN, M. D., AND MARIONI, J. C. Batch effects in single-cell rna-sequencing data are corrected by matching mutual nearest neighbors. *Nature biotechnology 36*, 5 (2018), 421–427.

[161] HAMERLY, G., AND DRAKE, J. Accelerating lloyd's algorithm for k-means clustering. In *Partitional clustering algorithms*. Springer, 2015, pp. 41–78.

[162] HAN, J., PEI, J., AND KAMBER, M. *Data mining: concepts and techniques*. Elsevier, 2011.

[163] HAN, J., PEI, J., AND YIN, Y. Mining frequent patterns without candidate generation. In *ACM sigmod record* (2000), vol. 29, ACM, pp. 1–12.

[164] HANDHAYANI, T., AND WASITO, I. Fully unsupervised clustering in nonlinearly separable data using intelligent kernel k-means. In *2014 International Conference on Advanced Computer Science and Information System* (2014), IEEE, pp. 450–453.

[165] HANSEN, K. D., IRIZARRY, R. A., AND WU, Z. Removing technical variability in rna-seq data using conditional quantile normalization. *Biostatistics 13*, 2 (2012), 204–216.

[166] HANTAO, L. B. S. Research of association rules incremental updating algorithm [j]. *Computer Engineering and Applications 23* (2002).

[167] HARATI, S., PHAN, J. H., AND WANG, M. D. Investigation of factors affecting rna-seq gene expression calls. In *2014 36th Annual International Conference of the IEEE Engineering in Medicine and Biology Society* (2014), IEEE, pp. 5232–5235.

[168] HART, S. N., THERNEAU, T. M., ZHANG, Y., POLAND, G. A., AND KOCHER, J.-P. Calculating sample size estimates for rna sequencing data. *Journal of computational biology 20*, 12 (2013), 970–978.

[169] HARTIGAN, J. A. Direct clustering of a data matrix. *Journal of the american statistical association 67*, 337 (1972), 123–129.

[170] HAWKINS, D. M. *Identification of outliers*, vol. 11. Springer, 1980.

[171] HE, S. L., AND GREEN, R. Northern Blotting. In *Methods in Enzymology*, vol. 530. Elsevier, 2013, pp. 75–87.

[172] HECKER, M., LAMBECK, S., TOEPFER, S., VAN SOMEREN, E., AND GUTHKE, R. Gene regulatory network inference: data integration in dynamic models—a review. *Biosystems 96*, 1 (2009), 86–103.

[173] HEIMBERG, G., BHATNAGAR, R., EL-SAMAD, H., AND THOMSON, M. Low dimensionality in gene expression data enables the accurate extraction of transcriptional programs from shallow sequencing. *Cell systems 2*, 4 (2016), 239–250.

[174] HENRIQUES, R., AND MADEIRA, S. C. Biclustering with flexible plaid models to unravel interactions between biological processes. *IEEE/ACM Transactions on computational biology and bioinformatics 12*, 4 (2015), 738–752.

[175] HENRIQUES, R., AND MADEIRA, S. C. Triclustering algorithms for three-dimensional data analysis: a comprehensive survey. *ACM Computing Surveys (CSUR) 51*, 5 (2018), 1–43.

[176] HICKS, S. C., TENG, M., AND IRIZARRY, R. A. On the widespread and critical impact of systematic bias and batch effects in single-cell rna-seq data. *BioRxiv* (2015), 025528.

[177] HICKS, S. C., TOWNES, F. W., TENG, M., AND IRIZARRY, R. A. Missing data and technical variability in single-cell rna-sequencing experiments. *Biostatistics 19*, 4 (2018), 562–578.

[178] HINNEBURG, A., AND GABRIEL, H.-H. Denclue 2.0: Fast clustering based on kernel density estimation. In *International symposium on intelligent data analysis* (2007), Springer, pp. 70–80.

[179] HOCHREITER, S., BODENHOFER, U., HEUSEL, M., MAYR, A., MITTERECKER, A., KASIM, A., KHAMIAKOVA, T., VAN SANDEN, S., LIN, D., TALLOEN, W., ET AL. Fabia: factor analysis for bicluster acquisition. *Bioinformatics 26*, 12 (2010), 1520–1527.

[180] HOQUE, N., NATH, B., AND BHATTACHARYYA, D. A new approach on rare association rule mining. *International Journal of Computer Applications 53*, 3 (2012), 1–6.

[181] HOQUE, N., NATH, B. AND BHATTACHARYYA, D.K. An efficient approach on rare association rule mining. *In Proceedings of Seventh International Conference on Bio-Inspired Computing: Theories and Applications* (BIC-TA 2012), Springer, India, pp. 193–203.

[182] HOSSEINI, B., AND KIANI, K. Fwcmr: A scalable and robust fuzzy weighted clustering based on mapreduce with application to microarray gene expression. *Expert Systems with Applications 91* (2018), 198–210.

[183] HOWELLS, T., SMIELEWSKI, P., DONNELLY, J., CZOSNYKA, M., HUTCHINSON, P. J., MENON, D. K., ENBLAD, P., AND ARIES, M. J. Optimal cerebral perfusion pressure in centers with different treatment protocols. *Critical care medicine 46*, 3 (2018), e235–e241.

[184] HSU, C.-C., AND HUANG, Y.-P. Incremental clustering of mixed data based on distance hierarchy. *Expert systems with applications 35*, 3 (2008), 1177–1185.

[185] HUANG, M., WANG, J., TORRE, E., ET AL. Gene expression recovery for single cell RNA sequencing. *BioRxiv* (2017), 138677.

[186] HUSSAIN, H. M., BENKRID, K., SEKER, H., AND ERDOGAN, A. T. Fpga implementation of k-means algorithm for bioinformatics application: An accelerated approach to clustering microarray data. In *2011 NASA/ESA Conference on Adaptive Hardware and Systems (AHS)* (2011), IEEE, pp. 248–255.

[187] HWANG, B., LEE, J. H., AND BANG, D. Single-cell RNA sequencing technologies and bioinformatics pipelines. *Experimental & Molecular Medicine 50*, 8 (2018), 1–14.

[188] HWANG, S., RHEE, S. Y., MARCOTTE, E. M., AND LEE, I. Systematic prediction of gene function in arabidopsis thaliana using a probabilistic functional gene network. *Nature protocols 6*, 9 (2011), 1429.

[189] ILICIC, T., KIM, J. K., KOLODZIEJCZYK, A. A., BAGGER, F. O., MCCARTHY, D. J., MARIONI, J. C., AND TEICHMANN, S. A. Classification of low quality cells from single-cell rna-seq data. *Genome biology 17*, 1 (2016), 29.

[190] IVLIEV, A. E., AC'T HOEN, P., AND SERGEEVA, M. G. Coexpression network analysis identifies transcriptional modules related to proastrocytic differentiation and sprouty signaling in glioma. *Cancer Research 70*, 24 (2010), 10060–10070.

[191] J HAN, J PEI AND M KAMBER. *Data Mining: Concepts and Techniques.* Elsevier, 2011.

[192] J LUO, M SCHUMACHER, A SCHERER, ET AL. A comparison of batch effect removal methods for enhancement of prediction performance using MAQC-II microarray gene expression data. *The Pharmacogenomics Journal 10* (2010), 278–291.

[193] J Y H YANG, S DUDOIT, P LUU, T P SPEED. Normalization for cdna microarray data. In *Proc of the SPIE Digital Library 4266, 4 June* (2001), SPIE, USA, pp. 1–12.

[194] JAISWAL, B., UTKARSH, K., AND BHATTACHARYYA, D. Pnme–a gene-gene parallel network module extraction method. *Journal of Genetic Engineering and Biotechnology 16*, 2 (2018), 447–457.

[195] JI, L., AND TAN, K.-L. Mining gene expression data for positive and negative co-regulated gene clusters. *Bioinformatics 20*, 16 (2004), 2711–2718.

[196] JIANG, D., PEI, J., AND ZHANG, A. Dhc: a density-based hierarchical clustering method for time series gene expression data. In *Bioinformatics and Bioengineering, 2003. Proceedings. Third IEEE Symposium on* (2003), IEEE, pp. 393–400.

[197] JIANG, P., THOMSON, J. A., AND STEWART, R. Quality control of single-cell rna-seq by sinqc. *Bioinformatics 32*, 16 (2016), 2514–2516.

[198] JIN, D., AND LEE, H. FGMD: A novel approach for functional gene module detection in cancer. *PloS One 12*, 12 (2017), e0188900.

[199] JOHNSON, W. E., LI, C., AND RABINOVIC, A. Adjusting batch effects in microarray expression data using empirical Bayes methods. *Biostatistics 8*, 1 (2007), 118–127.

[200] JOTHI, R., MOHANTY, S. K., AND OJHA, A. Dk-means: a deterministic k-means clustering algorithm for gene expression analysis. *Pattern Analysis and Applications 22*, 2 (2019), 649–667.

[201] JUAN, X., ANJUN, M., ANNE, F., QIN, M., AND JING, Z. It is time to apply biclustering: a comprehensive review of biclustering applications in biological and biomedical data. *Briefings in bioinformatics 20*, 4 (2019), 1450–1465.

[202] JUNG, I., JO, K., KANG, H., AHN, H., YU, Y., AND KIM, S. Timesvector: a vectorized clustering approach to the analysis of time series transcriptome data from multiple phenotypes. *Bioinformatics 33*, 23 (2017), 3827–3835.

[203] JUNG, S., HARTMANN, A., AND SOL, A. D. RefBool: A reference based algorithm for discretizing gene expression data. *Bioinformatics 33*, 13 (2017), 1953–1962.

[204] KAISER, S., AND LEISCH, F. A toolbox for bicluster analysis in r.

[205] KAKATI, T., AHMED, H. A., BHATTACHARYYA, D. K., AND KALITA, J. K. Thd-tricluster: A robust triclustering technique and its application in condition specific change analysis in hiv-1 progression data. *Computational biology and chemistry 75* (2018), 154–167.

[206] KAKATI, T., BHATTACHARYYA, D. K., BARAH, P., AND KALITA, J. K. Comparison of methods for differential co-expression analysis for disease biomarker prediction. *Computers in biology and medicine 113* (2019), 103380.

[207] KAKATI, T., BHATTACHARYYA, D. K., AND KALITA, J. K. X-module: A novel fusion measure to associate co-expressed gene modules from condition-specific expression profiles. *Journal of Biosciences 45*, 1 (2020), 33.

[208] KAKATI, T., KASHYAP, H., AND BHATTACHARYYA, D. K. THD-module extractor: an application for CEN module extraction and interesting gene identification for Alzheimer's disease. *Scientific Reports 6* (2016), 38046.

[209] KAPUSHESKY, M., KEMMEREN, P., CULHANE, A. C., AND ET AL. Expression profiler: Next generation – an online platform for analysis of microarray data. *Nucleic Acids Research 32* (2004), 465–470.

[210] KARMAKAR, B., DAS, S., BHATTACHARYA, S., SARKAR, R., AND MUKHOPADHYAY, I. Tight clustering for large datasets with an application to gene expression data. *Scientific reports 9*, 1 (2019), 3053.

[211] KARYPIS, G., HAN, E.-H., AND KUMAR, V. Chameleon: Hierarchical clustering using dynamic modeling. *Computer 32*, 8 (1999), 68–75.

[212] KASIM, A., SHKEDY, Z., KAISER, S., HOCHREITER, S., AND TALLOEN, W. The xmotif algorithm. In *Applied Biclustering Methods for Big and High-Dimensional Data Using R*. Chapman and Hall/CRC, 2016, pp. 75–86.

[213] KAUFMAN, L., AND ROUSSEEUW, P. J. *Finding groups in data: an introduction to cluster analysis*, vol. 344. John Wiley & Sons, 2009.

[214] KCHOUK, M., GIBRAT, J.-F., AND ELLOUMI, M. Generations of sequencing technologies: from first to next generation. *Biology and Medicine 9*, 3 (2017).

[215] KEPLER, T. B., CROSBY, L., AND MORGAN, K. T. Normalization and analysis of dna microarray data by self-consistency and local regression. *Genome Biology 3*, 7 (2002), research0037–1.

[216] KILGER, C., KRINGS, M., POINAR, H., AND PÄÄBO, S. "colony sequencing": direct sequencing of plasmid dna from bacterial colonies. *BioTechniques 22*, 3 (1997), 412–418.

[217] KIRAN, R. U., AND REDDY, P. K. Mining rare association rules in the datasets with widely varying items' frequencies. The 15th International Conference on Database Systems for Advanced Applications Tsukuba, Japan, April 1-4,.

[218] KISELEV, V. Y., ANDREWS, T. S., AND HEMBERG, M. Challenges in unsupervised clustering of single-cell rna-seq data. *Nature Reviews Genetics 20*, 5 (2019), 273–282.

[219] KISELEV, V. Y., KIRSCHNER, K., SCHAUB, M. T., ANDREWS, T., YIU, A., CHANDRA, T., NATARAJAN, K. N., REIK, W., BARAHONA, M., GREEN, A. R., ET AL. Sc3: consensus clustering of single-cell rna-seq data. *Nature methods 14*, 5 (2017), 483.

[220] KLUGER, Y., BASRI, R., CHANG, J. T., AND GERSTEIN, M. Spectral biclustering of microarray data: coclustering genes and conditions. *Genome research 13*, 4 (2003), 703–716.

[221] KNORR, E. M., AND NG, R. T. Algorithms for mining distance-based outliers in large datasets. *Proc of the 24th Int. Conf. on Very Large Data Bases, New York USA* (1998), 392–403.

[222] KNOX, D. P., AND SKUCE, P. J. SAGE and the quantitative analysis of gene expression in parasites. *TRENDS in Parasitology 21*, 7 (2005), 322–326.

[223] KODAMA, Y., SHUMWAY, M., AND LEINONEN, R. The Sequence Read Archive: explosive growth of sequencing data. *Nucleic Acids Research 40*, D1 (2012), D54–D56.

[224] KOLODZIEJCZYK, A. A., KIM, J. K., SVENSSON, V., MARIONI, J. C., AND TEICHMANN, S. A. The technology and biology of single-cell RNA sequencing. *Molecular Cell 58*, 4 (2015), 610–620.

[225] KOZIK, E. U., AND WEHNER, T. C. A single dominant gene ch for chilling resistance in cucumber seedlings. *Journal of the American Society for Horticultural Science 133*, 2 (2008), 225–227.

[226] KRAEMER, G., REICHSTEIN, M., AND MAHECHA, M. D. dimred and coranking—unifying dimensionality reduction in r. *R Journal 10*, 1 (2018), 342–358.

[227] KRAUS, J. M., MÜSSEL, C., PALM, G., AND KESTLER, H. A. Multi-objective selection for collecting cluster alternatives. *Computational Statistics 26*, 2 (2011), 341–353.

[228] KUKURBA, K. R., AND MONTGOMERY, S. B. RNA sequencing and analysis. *Cold Spring Harbor Protocols 2015*, 11 (2015), pdb–top084970.

[229] KUMARI, S., NIE, J., CHEN, H.-S., ET AL. Evaluation of gene association methods for coexpression network construction and biological knowledge discovery. *PloS One 7*, 11 (2012), e50411.

[230] L. ZHAO AND M. J. ZAKI. Tricluster: An effective algorithm for mining coherent clustersin 3d microarray data. In *Proc of SIGMOD'05* (2005), ACM, pp. 694–705.

[231] LACHMANN, A., TORRE, D., KEENAN, A. B., JAGODNIK, K. M., LEE, H. J., WANG, L., SILVERSTEIN, M. C., AND MA'AYAN, A. Massive mining of publicly available rna-seq data from human and mouse. *Nature communications 9*, 1 (2018), 1–10.

[232] LAGARIAS, J. Euler's constant: Euler's work and modern developments. *Bulletin of the American Mathematical Society 50*, 4 (2013), 527–628.

[233] LAM, F., AND LONGNECKER, M. A modified wilcoxon rank sum test for paired data. *Biometrika 70*, 2 (1983), 510–513.

[234] LAMARRE, S., FRASSE, P., ZOUINE, M., LABOURDETTE, D., SAINDERICHIN, E., HU, G., BERRE-ANTON, L., BOUZAYEN, M., MAZA, E., ET AL. Optimization of an rna-seq differential gene expression analysis depending on biological replicate number and library size. *Frontiers in plant science 9* (2018), 108.

[235] LANGFELDER, P., AND HORVATH, S.   Wgcna: an r package for weighted correlation network analysis.   BMC bioinformatics 9, 1 (2008), 559.

[236] LANGFELDER, P., LUO, R., OLDHAM, M. C., AND HORVATH, S. Is my network module preserved and reproducible? PLoS computational biology 7, 1 (2011).

[237] LANGFELDER, P., MISCHEL, P. S., AND HORVATH, S. When is hub gene selection better than standard meta-analysis? PloS one 8, 4 (2013).

[238] LAW, C. W., CHEN, Y., SHI, W., AND SMYTH, G. K. voom: Precision weights unlock linear model analysis tools for rna-seq read counts. Genome biology 15, 2 (2014), R29.

[239] LAZAREVIC, A., AND KUMAR, V.   Feature bagging for outlier detection. In Proceedings of the eleventh ACM SIGKDD international conference on Knowledge discovery in data mining (2005), pp. 157–166.

[240] LEEK, J. T.   SVAseq: removing batch effects and other unwanted noise from sequencing data. Nucleic Acids Research 42, 21 (2014), e161–e161.

[241] LEEK, J. T., SCHARPF, R. B., BRAVO, H. C., SIMCHA, D., LANGMEAD, B., JOHNSON, W. E., GEMAN, D., BAGGERLY, K., AND IRIZARRY, R. A.   Tackling the widespread and critical impact of batch effects in high-throughput data. Nature Reviews Genetics 11, 10 (2010), 733–739.

[242] LEINONEN, R., AKHTAR, R., BIRNEY, E., BOWER, L., CERDENO-TÁRRAGA, A., CHENG, Y., CLELAND, I., FARUQUE, N., GOODGAME, N., GIBSON, R., ET AL.   The european nucleotide archive. Nucleic Acids Research 39, suppl_1 (2010), D28–D31.

[243] LEVINE, J. H., SIMONDS, E. F., BENDALL, S. C., DAVIS, K. L., EL-AD, D. A., TADMOR, M. D., LITVIN, O., FIENBERG, H. G., JAGER, A., ZUNDER, E. R., ET AL. Data-driven phenotypic dissection of aml reveals progenitor-like cells that correlate with prognosis. Cell 162, 1 (2015), 184–197.

[244] LI, W. V., AND LI, J. J. scImpute: accurate and robust imputation for single cell RNA-seq data. *BioRxiv* (2017), 141598.

[245] LI, W. V., AND LI, J. J. An accurate and robust imputation method scimpute for single-cell rna-seq data. *Nature communications 9*, 1 (2018), 1–9.

[246] LI, Y., LIU, L., BAI, X., CAI, H., JI, W., GUO, D., AND ZHU, Y. Comparative study of discretization methods of microarray data for inferring transcriptional regulatory networks. *BMC bioinformatics 11*, 1 (2010), 520.

[247] LIAW, A., WIENER, M., ET AL. Classification and regression by randomforest. *R news 2*, 3 (2002), 18–22.

[248] LIN, D.-I., AND KEDEM, Z. M. Pincer-search: A new algorithm for discovering the maximum frequent set. In *International conference on Extending database technology* (1998), Springer, pp. 103–119.

[249] LIN, P., TROUP, M., AND HO, J. W. Cidr: Ultrafast and accurate clustering through imputation for single-cell rna-seq data. *Genome biology 18*, 1 (2017), 59.

[250] LIU, B., HSU, W., AND MA, Y. Mining association rules with multiple minimum supports. ACM Special Interest Group on Knowledge Discovery and Data Mining Explorations, pp. 337–341.

[251] LIU, B., XIA, Y., AND YU, P. S. Clustering through decision tree construction. In *Proceedings of the ninth international conference on Information and knowledge management* (2000), pp. 20–29.

[252] LIU, G., LU, H., YU, J. X., WANG, W., AND XIAO, X. Afopt: An efficient implementation of pattern growth approach. In *FIMI* (2003), Citeseer.

[253] LIU, Q., AND MARKATOU, M. Evaluation of methods in removing batch effects on RNA-seq data. *Infectious Diseases and Translational Medicine 2*, 1 (2016), 3–9.

[254] LIU, X., AND WANG, L. Computing the maximum similarity biclusters of gene expression data. *Bioinformatics 23*, 1 (2007), 50–56.

[255] LIU, Y., YANG, T., AND FU, L. A partitioning based algorithm to fuzzy tricluster. *Mathematical Problems in Engineering 2015* (2015).

[256] LIU, Y., ZHOU, J., AND WHITE, K. P. Rna-seq differential expression studies: more sequence or more replication? *Bioinformatics 30*, 3 (2014), 301–304.

[257] LIU, Z.-P., WU, C., MIAO, H., AND WU, H. Regnetwork: an integrated database of transcriptional and post-transcriptional regulatory networks in human and mouse. *Database 2015* (2015).

[258] LOBO, I. Biological complexity and integrative levels of organization. *Nature Education 1*, 1 (2008), 141.

[259] LOVE, M. I., HUBER, W., AND ANDERS, S. Moderated estimation of fold change and dispersion for rna-seq data with deseq2. *Genome biology 15*, 12 (2014), 550.

[260] LOWE, R., SHIRLEY, N., BLEACKLEY, M., DOLAN, S., AND SHAFEE, T. Transcriptomics technologies. *PLoS Computational Biology 13*, 5 (2017), e1005457.

[261] LUECKEN, M. D., AND THEIS, F. J. Current best practices in single-cell rna-seq analysis: a tutorial. *Molecular systems biology 15*, 6 (2019).

[262] LUSTGARTEN, J. L., VISWESWARAN, S., GOPALAKRISHNAN, V., AND COOPER, G. F. Application of an efficient bayesian discretization method to biomedical data. *BMC bioinformatics 12*, 1 (2011), 309.

[263] M, O. I., G, G. J., J, M. S., A, S. E., J, S. A., NAGESH, A., L, A. A., AND YU, J.-P. J. Gut microbiome populations are associated with structure-specific changes in white matter architecture. *Translational psychiatry 8*, 1 (2018), 1–11.

[264] M. I. LOVE, W. HUBER, AND S. ANDERS. Moderated estimation of fold change and dispersion for RNA-seq data with DESeq2. *Genome Biology 15*, 12 (2014), 550.

[265] MACQUEEN, J., ET AL. Some methods for classification and analysis of multivariate observations. In *Proceedings of the fifth Berkeley symposium on mathematical statistics and probability* (1967), vol. 1, Oakland, CA, USA, pp. 281–297.

[266] MADEIRA, S. C., AND OLIVEIRA, A. L. Biclustering algorithms for biological data analysis: a survey. *IEEE/ACM transactions on computational biology and bioinformatics 1*, 1 (2004), 24–45.

[267] MADEIRA, S. C., AND OLIVEIRA, A. L. An evaluation of discretization methods for non-supervised analysis of time-series gene expression data. *Instituto de Engenharia de Sistemas e Computadores Investigacao e Desenvolvimento, Technical Report 42* (2005).

[268] MAERE, S., HEYMANS, K., AND KUIPER, M. BiNGO: a Cytoscape plugin to assess overrepresentation of gene ontology categories in biological networks. *Bioinformatics 21*, 16 (2005), 3448–3449.

[269] MAHANTA, P., AHMED, H., BHATTACHARYYA, D., AND KALITA, J. K. Triclustering in gene expression data analysis: a selected survey. In *Emerging Trends and Applications in Computer Science (NCETACS), 2011 2nd National Conference on* (2011), IEEE, pp. 1–6.

[270] MAHANTA, P., AHMED, H. A., BHATTACHARYYA, D. K., AND KALITA, J. K. An effective method for network module extraction from microarray data. *BMC bioinformatics 13*, S13 (2012), S4.

[271] MAHANTA, P., AHMED, H. A., KALITA, J. K., AND BHATTACHARYYA, D. K. Discretization in gene expression data analysis: a selected survey. In *Proceedings of the Second International Conference on Computational Science, Engineering and Information Technology* (2012), pp. 69–75.

[272] MAHANTA, P., AHMED, H.A., BHATTACHARYYA, D.K. ET AL. An effective method for network module extraction from microarray data. *BMC Bioinformatics 13*, S4 (2012).

[273] MAHMOUDI, S., AND MENHAJ, M. B. A new hybrid evolutionary biclustring algorithm based on transposed virtual error. In *2015 2nd International Conference on Knowledge-Based Engineering and Innovation (KBEI)* (2015), IEEE, pp. 533–538.

[274] MALLIK, S., AND ZHAO, Z. ConGEMs: Condensed gene co-expression module discovery through rule-based clustering and its application to carcinogenesis. *Genes 9*, 1 (2017), 7.

[275] MANDAL, K., SARMAH, R., AND BHATTACHARYYA, D. K. Biomarker identification for cancer disease using biclustering approach: an empirical study. *IEEE/ACM transactions on computational biology and bioinformatics 16*, 2 (2018), 490–509.

[276] MARGOLIN, A. A., NEMENMAN, I., BASSO, K., ET AL. ARACNE: an algorithm for the reconstruction of gene regulatory networks in a mammalian cellular context. *BMC Bioinformatics 7*, 1 (2006), S7.

[277] MARIONI, J. C., MASON, C. E., MANE, S. M., STEPHENS, M., AND GILAD, Y. Rna-seq: an assessment of technical reproducibility and comparison with gene expression arrays. *Genome research 18*, 9 (2008), 1509–1517.

[278] MARLER, R. T., AND ARORA, J. S. Survey of multi-objective optimization methods for engineering. *Structural and multidisciplinary optimization 26*, 6 (2004), 369–395.

[279] MATSUMURA, H., URASAKI, N., YOSHIDA, K., KRÜGER, D. H., KAHL, G., AND TERAUCHI, R. SuperSAGE: powerful serial analysis of gene expression. In *RNA Abundance Analysis*. Springer, 2012, pp. 1–17.

[280] MAULIK, U., AND BANDYOPADHYAY, S. Performance evaluation of some clustering algorithms and validity indices. *IEEE Transactions on pattern analysis and machine intelligence 24*, 12 (2002), 1650–1654.

[281] MAX, B., DANIEL, E., ANDREAS, S., STEFAN, J., THOMAS, M., AND JOHAN, T. Orthogonal projections to latent structures as a strategy for microarray data normalization. *BMC bioinformatics 8*, 1 (2007), 207.

[282] MI, H., POUDEL, S., MURUGANUJAN, A., CASAGRANDE, J. T., AND THOMAS, P. D. PANTHER version 10: expanded protein families and functions, and analysis tools. *Nucleic Acids Research 44*, D1 (2015), D336–D342.

[283] MISRA, S., AND RAY, S. S. Finding optimum width of discretization for gene expressions using functional annotations. *Computers in biology and medicine 90* (2017), 59–67.

[284] MITRA, A., HARDING, T., MUKHERJEE, U., JANG, J., LI, Y., HONGZHENG, R., JEN, J., SONNEVELD, P., KUMAR, S., KUEHL, W., ET AL. A gene expression signature distinguishes innate response and resistance to proteasome inhibitors in multiple myeloma. *Blood cancer journal 7*, 6 (2017), e581–e581.

[285] MITRA, S., AND BANKA, H. Multi-objective evolutionary biclustering of gene expression data. *Pattern Recognition 39*, 12 (2006), 2464–2477.

[286] MOH'D BELAL, A.-Z., AL-DAHOUD, A., AND YAHYA, A. A. New outlier detection method based on fuzzy clustering. *WSEAS transactions on information science and applications 7* (2010).

[287] MONICA, B., JOEL, P., QUAN, D., JUNYUAN, W., DONG, X., M, P. C., AND STEPHEN, M. J. Adjustment of systematic microarray data biases. *Bioinformatics 20*, 1 (2004), 105–114.

[288] MOSTAFAVI, S., RAY, D., WARDE-FARLEY, D., GROUIOS, C., AND MORRIS, Q. Genemania: a real-time multiple association network integration algorithm for predicting gene function. *Genome biology 9*, S1 (2008), S4.

[289] MOUSSA, M., AND MĂNDOIU, I. I. Single cell rna-seq data clustering using tf-idf based methods. *BMC genomics 19*, 6 (2018), 569.

[290] MUNRO, S. A., LUND, S. P., PINE, P. S., BINDER, H., CLEVERT, D.-A., CONESA, A., DOPAZO, J., FASOLD, M., HOCHREITER, S., HONG, H., ET AL. Assessing technical performance in differential gene expression experiments with external spike-in rna control ratio mixtures. *Nature communications 5* (2014), 5125.

[291] MURALI, T., AND KASIF, S. Extracting conserved gene expression motifs from gene expression data. In *Biocomputing 2003*. World Scientific, 2002, pp. 77–88.

[292] MUSSEL, C., SCHMID, F., TJ BLATTE, M. H., AND ET AL. BiTrinA: A multiscale binarization and trinarization with quality analysis. *Bioinformatics 32*, 3 (2016), 465–468.

[293] MUTZ, K.-O., HEILKENBRINKER, A., LÖNNE, M., WALTER, J.-G., AND STAHL, F. Transcriptome analysis using next-generation sequencing. *Current opinion in biotechnology 24*, 1 (2013), 22–30.

[294] N. L. BRAY, H. PIMENTEL, E. A. Near-optimal probabilistic rna-seq quantification. *Nature Biotechnology 34* (2016), 525–527.

[295] NAGHIEH, E., AND PENG, Y. Microarray gene expression data mining: clustering analysis review. *Aug 20* (2009), 2009.

[296] NAGI, S., AND BHATTACHARYYA, D. K. Classification of microarray cancer data using ensemble approach. *Network Modeling Analysis in Health Informatics and Bioinformatics 2*, 3 (2013), 159–173.

[297] NAGI, S., BHATTACHARYYA, D. K., AND KALITA, J. K. A preview on subspace clustering of high dimensional data. *INTERNATIONAL JOURNAL OF COMPUTERS & TECHNOLOGY 6*, 3 (2013), 441–448.

[298] NASSIRI, I., LOMBARDO, R., LAURIA, M., MORINE, M. J., MOYSEOS, P., VARMA, V., NOLEN, G. T., KNOX, B., SLOPER, D., KAPUT, J., ET AL. Systems view of adipogenesis via novel omics-driven and tissue-specific activity scoring of network functional modules. *Scientific reports 6* (2016), 28851.

[299] NATH, B., BHATTACHARYYA, D., AND GHOSH, A. Incremental association rule mining: a survey. *Wiley Interdisciplinary Reviews: Data Mining and Knowledge Discovery 3*, 3 (2013), 157–169.

[300] NATH, B., BHATTACHARYYA, D. K., AND GOSH, A. Faster generation of association rules. vol. 2, IJITKM, pp. 267–279.

[301] NG, R. T., AND HAN, J. Clarans: A method for clustering objects for spatial data mining. *IEEE transactions on knowledge and data engineering 14*, 5 (2002), 1003–1016.

[302] NOVIKOFF, A. B. The concept of integrative levels and biology. *Science 101*, 2618 (1945), 209–215.

[303] NUEDA, M. J., FERRER, A., AND CONESA, A. ARSyN: a method for the identification and removal of systematic noise in multifactorial time course microarray experiments. *Biostatistics 13*, 3 (2012), 553–566.

[304] OCONE, A., HAGHVERDI, L., MUELLER, N. S., AND THEIS, F. J. Reconstructing gene regulatory dynamics from high-dimensional single-cell snapshot data. *Bioinformatics 31*, 12 (2015), i89–i96.

[305] ORTIZ, R. A., GOLUB, A., LUGOVOY, O., MARKANDYA, A., AND WANG, J. Dicer: A tool for analyzing climate policies. *Energy Economics 33* (2011), S41–S49.

[306] OSHLACK, A., ROBINSON, M. D., AND YOUNG, M. D. From rna-seq reads to differential expression results. *Genome biology 11*, 12 (2010), 220.

[307] OTOO, E. J., SHOSHANI, A., AND HWANG, S.-W. Clustering high dimensional massive scientific datasets. *Journal of Intelligent Information Systems 17*, 2-3 (2001), 147–168.

[308] OYELADE, J., ISEWON, I., OLADIPUPO, F., AROMOLARAN, O., UWOGHIREN, E., AMEH, F., ACHAS, M., AND ADEBIYI, E. Clustering algorithms: Their application to gene expression data. *Bioinformatics and Biology insights 10* (2016), BBI–S38316.

[309] OZEL, S. A., AND GUVENIR, H. An algorithm for mining association rules using perfect hashing and database pruning. In *10th Turkish Symposium on Artificial Intelligence and Neural Networks* (2001), Citeseer, pp. 257–264.

[310] OZSOLAK, F., PLATT, A. R., JONES, D. R., REIFENBERGER, J. G., SASS, L. E., MCINERNEY, P., THOMPSON, J. F., BOWERS, J., JAROSZ, M., AND MILOS, P. M. Direct RNA sequencing. *Nature 461*, 7265 (2009), 814–818.

[311] ÖZTUNA, D., ELHAN, A. H., AND TÜCCAR, E. Investigation of four different normality tests in terms of type 1 error rate and power under different distributions. *Turkish Journal of Medical Sciences 36*, 3 (2006), 171–176.

[312] P, D. L., C, S., A, P., AND C, N. Missing value estimation methods for dna methylation data. *Bioinformatics 35*, 19 (2019), 3786–3793.

[313] P., M., K., B. D., AND A., G. Mipce: An mi-based protein complex extraction technique. *Journal of biosciences 40*, 4 (2015), 701–708.

[314] PAN, Q., SHAI, O., LEE, L. J., FREY, B. J., AND BLENCOWE, B. J. Deep surveying of alternative splicing complexity in the human transcriptome by high-throughput sequencing. *Nature Genetics 40*, 12 (2008), 1413–1415.

[315] PANDIT, A. A., SHAH, R. A., AND HUSAINI, A. M. Transcriptomics: A time-efficient tool with wide applications in crop and animal biotechnology. *Journal of Pharmacognosy and Phytochemistry 7*, 2 (2018), 1701–1704.

[316] PAPIEZ, A., MARCZYK, M., POLANSKA, J., AND POLANSKI, A. Batchi: Batch effect identification in high-throughput screening data using a dynamic programming algorithm. *Bioinformatics 35*, 11 (2019), 1885–1892.

[317] PARSONS, L., HAQUE, E., AND LIU, H. Subspace clustering for high dimensional data: a review. *Acm Sigkdd Explorations Newsletter 6*, 1 (2004), 90–105.

[318] PATNAIK, A. K., BHUYAN, P. K., AND RAO, K. K. Divisive analysis (diana) of hierarchical clustering and gps data for level of service criteria of urban streets. *Alexandria Engineering Journal 55*, 1 (2016), 407–418.

[319] PATRO, R., DUGGAL, G., LOVE, M. I., ET AL. Salmon provides fast and bias-aware quantification of transcript expression. *Nature Methods* (2017).

[320] PATRO, R., MOUNT, S. M., AND KINGSFORD, C. Sailfish enables alignment-free isoform quantification from rna-seq reads using lightweight algorithms. *Nature Biotechnology 32* (2014), 462–464.

[321] PAUS, T. Primate anterior cingulate cortex: where motor control, drive and cognition interface. *Nature reviews neuroscience 2*, 6 (2001), 417–424.

[322] PEI, J., HAN, J., LU, H., NISHIO, S., TANG, S., AND YANG, D. H-mine: Hyper-structure mining of frequent patterns in large databases. In *proceedings 2001 IEEE international conference on data mining* (2001), IEEE, pp. 441–448.

[323] PENG, L. Estimating the mean of a heavy tailed distribution. *Statistics & Probability Letters 52*, 3 (2001), 255–264.

[324] PÉREZ-ALONSO, M.-M., CARRASCO-LOBA, V., MEDINA, J., VICENTE-CARBAJOSA, J., AND POLLMANN, S. When transcriptomics and metabolomics work hand in hand: a case study characterizing plant cdf transcription factors. *High-throughput 7*, 1 (2018), 7.

[325] PIERSON, E., KOLLER, D., BATTLE, A., ET AL. Sharing and specificity of co-expression networks across 35 human tissues. *PLoS Computational Biology 11*, 5 (2015), e1004220.

[326] PONTES, B., DIVINA, F., GIRÁLDEZ, R., AND AGUILAR-RUIZ, J. S. Virtual error: a new measure for evolutionary biclustering. In *European Conference on Evolutionary Computation, Machine Learning and Data Mining in Bioinformatics* (2007), Springer, pp. 217–226.

[327] PONTES, B., GIRÁLDEZ, R., AND AGUILAR-RUIZ, J. S. Measuring the quality of shifting and scaling patterns in biclusters. In *IAPR International Conference on Pattern Recognition in Bioinformatics* (2010), Springer, pp. 242–252.

[328] PONTES, B., GIRÁLDEZ, R., AND AGUILAR-RUIZ, J. S. Biclustering on expression data: A review. *Journal of Biomedical Informatics 57* (2015), 163–180.

[329] PRELIĆ, A., BLEULER, S., ZIMMERMANN, P., WILLE, A., BÜHLMANN, P., GRUISSEM, W., HENNIG, L., THIELE, L., AND ZITZLER, E. A systematic comparison and evaluation of biclustering methods for gene expression data. *Bioinformatics 22*, 9 (2006), 1122–1129.

[330] PRIM, R. C. Shortest connection networks and some generalizations. *The Bell System Technical Journal 36*, 6 (1957), 1389–1401.

[331] PRIYANTINI, D. T., WARDHANA, Y., ALHAMIDI, M. R., PURNOMO, D. M., MURSANTO, P., JATMIKO, W., ET AL. An efficient implementation of generalized extreme studentized deviate (gesd) on field programmable gate array (fpga). In *2015 International Conference on Computers, Communications, and Systems (ICCCS)* (2015), IEEE, pp. 14–18.

[332] PROCOPIUC, C. M., JONES, M., AGARWAL, P. K., AND MURALI, T. A monte carlo algorithm for fast projective clustering. In *Proceedings of the 2002 ACM SIGMOD international conference on Management of data* (2002), pp. 418–427.

[333] PRUGEL-BENNETT, A., AND SHAPIRO, J. L. Statistical mechanics of unsupervised hebbian learning. *Journal of Physics A: Mathematical and General 26*, 10 (1993), 2343.

[334] PUSHKAREV, D., NEFF, N. F., AND QUAKE, S. R. Single-molecule sequencing of an individual human genome. *Nature biotechnology 27*, 9 (2009), 847–850.

[335] QI, R., MA, A., MA, Q., AND ZOU, Q. Clustering and classification methods for single-cell rna-sequencing data. *Briefings in bioinformatics* (2019).

[336] QIN, Z. S. Clustering microarray gene expression data using weighted chinese restaurant process. *Bioinformatics 22*, 16 (2006), 1988–1997.

[337] RAMAKER, R. C., BOWLING, K. M., LASSEIGNE, B. N., HAGENAUER, M. H., HARDIGAN, A. A., DAVIS, N. S., GERTZ, J., CARTAGENA, P. M., WALSH, D. M., VAWTER, M. P., ET AL. Post-mortem molecular profiling of three psychiatric disorders. *Genome medicine 9*, 1 (2017), 72.

[338] RAPAPORT, F., KHANIN, R., LIANG, Y., PIRUN, M., KREK, A., ZUMBO, P., MASON, C. E., SOCCI, N. D., AND BETEL, D. Comprehensive evaluation of differential gene expression analysis methods for rna-seq data. *Genome biology 14*, 9 (2013), 1–13.

[339] RAU, A., AND MAUGIS-RABUSSEAU, C. Transformation and model choice for rna-seq co-expression analysis. *Briefings in bioinformatics 19*, 3 (2017), 425–436.

[340] RAU, A, MAUGIS-RABUSSEAU, C AND MARTIN-MAGNIETTE, M AND CELEUX, G. Coexpression Analysis of High-throughput Transcriptome Sequencing Data with Possion Mixture Model. *Bioinformatics 31*, 9 (2015), 1420–1427.

[341] RAVASZ, E., SOMERA, A. L., MONGRU, D. A., OLTVAI, Z. N., AND BARABÁSI, A.-L. Hierarchical organization of modularity in metabolic networks. *Science 297*, 5586 (2002), 1551–1555.

[342] REBHAN, M., CHALIFA-CASPI, V., PRILUSKY, J., AND LANCET, D. GeneCards: integrating information about genes, proteins and diseases. *Trends in Genetics 13*, 4 (1997), 163.

[343] REESE, S. E., ARCHER, K. J., THERNEAU, T. M., ET AL. A new statistic for identifying batch effects in high-throughput genomic data that uses guided principal component analysis. *Bioinformatics 29*, 22 (2013), 2877–2883.

[344] REINER, A., YEKUTIELI, D., AND BENJAMINI, Y. Identifying differentially expressed genes using false discovery rate controlling procedures. *Bioinformatics 19*, 3 (2003), 368–375.

[345] REISS, D. J., PLAISIER, C. L., WU, W.-J., ET AL. cMonkey2: Automated, systematic, integrated detection of co-regulated gene modules for any organism. *Nucleic Acids Research 43*, 13 (2015), e87–e87.

[346] RISSO, D. Edaseq: Exploratory data analysis and normalization for rna-seq. *R package version 1*, 0 (2011).

[347] RISSO, D., SCHWARTZ, K., SHERLOCK, G., AND DUDOIT, S. Gc-content normalization for rna-seq data. *BMC bioinformatics 12*, 1 (2011), 480.

[348] RISTEVSKI, B. A survey of models for inference of gene regulatory networks. *Nonlinear Anal Model Control 18*, 4 (2013), 444–465.

[349] RITCHIE, M. E., PHIPSON, B., AND D. WU, E. A. Limma powers differential expression analyses for rna-sequencing and microarray studies. *Nucleic Acid Research 43* (2015), 7.

[350] ROBINSON, M. D., MCCARTHY, D. J., AND SMYTH, G. K. edger: a bioconductor package for differential expression analysis of digital gene expression data. *Bioinformatics 26*, 1 (2010), 139–140.

[351] ROBINSON, M. D., AND OSHLACK, A. A scaling normalization method for differential expression analysis of rna-seq data. *Genome biology 11*, 3 (2010), R25.

[352] ROBINSON, M. D., AND SMYTH, G. K. Moderated statistical tests for assessing differences in tag abundance. *Bioinformatics 23*, 21 (2007), 2881–2887.

[353] ROBINSON, M. D., AND SMYTH, G. K. Small-sample estimation of negative binomial dispersion, with applications to sage data. *Biostatistics 9*, 2 (2008), 321–332.

[354] ROSTOM, R., SVENSSON, V., TEICHMANN, S. A., AND KAR, G. Computational approaches for interpreting scrna-seq data. *FEBS letters 591*, 15 (2017), 2213–2225.

[355] RUSSELL, S. J., AND NORVIG, P. *Prentice Hall Series in Artificial Intelligence.* Prentice Hall Englewood Cliffs, NJ:, 1995.

[356] RUSSO, P. S., FERREIRA, G. R., CARDOZO, L. E., BÜRGER, M. C., ARIAS-CARRASCO, R., MARUYAMA, S. R., HIRATA, T. D., LIMA,

D. S., PASSOS, F. M., FUKUTANI, K. F., ET AL. Cemitool: a bioconductor package for performing comprehensive modular co-expression analyses. *Bmc Bioinformatics 19*, 1 (2018), 56.

[357] S., B. J., AND G., P. A. *Random data: analysis and measurement procedures*, vol. 729. John Wiley & Sons., 2011.

[358] SAHU, A., CHOWDHURY, H. A., GAIKWAD, M., CHONGTHAM, C., TALUKDAR, U., PHUKAN, J. K., BHATTACHARYYA, D. K., AND BARAH, P. Integrative network analysis identifies differential regulation of neuroimmune system in schizophrenia and bipolar disorder. *Brain, Behavior, & Immunity-Health 2* (2020), 100023.

[359] SAN SEGUNDO-VAL, I., AND SANZ-LOZANO, C. S. Introduction to the gene expression analysis. In *Molecular Genetics of Asthma*. Springer, 2016, pp. 29–43.

[360] SANDER, J., ESTER, M., KRIEGEL, H.-P., AND XU, X. Density-based clustering in spatial databases: The algorithm gdbscan and its applications. *Data mining and knowledge discovery 2*, 2 (1998), 169–194.

[361] SANGER, F., NICKLEN, S., AND COULSON, A. R. Dna sequencing with chain-terminating inhibitors. *Proceedings of the National Academy of Sciences 74*, 12 (1977), 5463–5467.

[362] SANGUINETTI, G., ET AL. Gene regulatory network inference: an introductory survey. In *Gene Regulatory Networks*. Springer, 2019, pp. 1–23.

[363] SATIJA, R., FARRELL, J. A., GENNERT, D., SCHIER, A. F., AND REGEV, A. Spatial reconstruction of single-cell gene expression data. *Nature biotechnology 33*, 5 (2015), 495–502.

[364] SEVERIN, A. J., WOODY, J. L., BOLON, Y.-T., JOSEPH, B., DIERS, B. W., FARMER, A. D., MUEHLBAUER, G. J., NELSON, R. T., GRANT, D., SPECHT, J. E., ET AL. Rna-seq atlas of glycine max: a guide to the soybean transcriptome. *BMC plant biology 10*, 1 (2010), 160.

[365] SHA, Y., PHAN, J. H., AND WANG, M. D. Effect of low-expression gene filtering on detection of differentially expressed genes in rna-seq data. In *2015 37th Annual International Conference of the IEEE*

Engineering in Medicine and Biology Society (EMBC) (2015), IEEE, pp. 6461–6464.

[366] SHAO, C., AND HÖFER, T. Robust classification of single-cell transcriptome data by nonnegative matrix factorization. *Bioinformatics 33*, 2 (2017), 235–242.

[367] SHARAN, R., AND SHAMIR, R. Click: a clustering algorithm with applications to gene expression analysis. In *Proc Int Conf Intell Syst Mol Biol* (2000), vol. 8, p. 16.

[368] SHARMA, P. Development of effective algorithms for analyzing protein protein interaction and gene expression data to support disease biomarker identification. *PhD Thesis*, TZ167043 (2016).

[369] SHARMA, P., BHATTACHARYYA, D. K., AND KALITA, J. K. Centrality analysis in ppi networks. In *2016 International Conference on Accessibility to Digital World (ICADW)* (2016), IEEE, pp. 135–140.

[370] SHEIKHOLESLAMI, G., CHATTERJEE, S., AND ZHANG, A. Wavecluster: A multi-resolution clustering approach for very large spatial databases. In *VLDB* (1998), vol. 98, pp. 428–439.

[371] SHI, F., AND HUANG, H. Identifying cell subpopulations and their genetic drivers from single-cell rna-seq data using a biclustering approach. *Journal of Computational Biology 24*, 7 (2017), 663–674.

[372] SI, Y., LIU, P., LI, P., AND BRUTNELL, T. P. Model-based clustering for rna-seq data. *Bioinformatics 30*, 2 (2013), 197–205.

[373] SIMA, C., HUA, J., AND JUNG, S. Inference of gene regulatory networks using time-series data: a survey. *Current genomics 10*, 6 (2009), 416–429.

[374] SINHA, D., KUMAR, A., KUMAR, H., BANDYOPADHYAY, S., AND SENGUPTA, D. dropclust: efficient clustering of ultra-large scrna-seq data. *Nucleic acids research 46*, 6 (2018), e36–e36.

[375] SISKA, C., BOWLER, R., AND KECHRIS, K. The discordant method: a novel approach for differential correlation. *Bioinformatics 32*, 5 (2016), 690–696.

[376] SMITH, R. Aristotle's logic.

[377] SMITH-UNNA, R., BOURSNELL, C., PATRO, R., HIBBERD, J. M., AND KELLY, S. Transrate: reference-free quality assessment of de novo transcriptome assemblies. *Genome research 26*, 8 (2016), 1134–1144.

[378] SOMLO, G. L., AND HOWE, A. E. Incremental clustering for profile maintenance in information gathering web agents. In *Proceedings of the fifth international conference on Autonomous agents* (2001), pp. 262–269.

[379] SONESON, C., AND DELORENZI, M. A comparison of methods for differential expression analysis of rna-seq data. *BMC bioinformatics 14*, 1 (2013), 91.

[380] SPIELMAN, R. S., BASTONE, L. A., BURDICK, J. T., MORLEY, M., EWENS, W. J., AND CHEUNG, V. G. Common genetic variants account for differences in gene expression among ethnic groups. *Nature genetics 39*, 2 (2007), 226–231.

[381] SPIES, D., RENZ, P. F., BEYER, T. A., AND CIAUDO, C. Comparative analysis of differential gene expression tools for rna sequencing time course data. *Briefings in bioinformatics 20*, 1 (2019), 288–298.

[382] SRIKANT, R., VU, Q., AND AGRAWAL, R. Mining association rules with item constraints. In *Kdd* (1997), vol. 97, pp. 67–73.

[383] SRIWANNA, K., BOONGOEN, T., AND IAM-ON, N. Graph clustering-based discretization approach to microarray data. *Knowledge and Information Systems 60*, 2 (2019), 879–906.

[384] STANFORD, N. J., WOLSTENCROFT, K., GOLEBIEWSKI, M., KANIA, R., JUTY, N., TOMLINSON, C., OWEN, S., BUTCHER, S., HERMJAKOB, H., LE NOVÈRE, N., ET AL. The evolution of standards and data management practices in systems biology. *Molecular systems biology 11*, 12 (2015), 851.

[385] STEINBERG, D., AND COLLA, P. Cart users manual. *Salford Systems, San Diego, CA* (1997).

[386] STEKHOVEN, D. J., AND BUHLMANN, P. Missforest- non-parametric missing value imputation for mixed type data. *Bioinformatics 28*, 1 (2012), 112–118.

[387] STUART, J. M., SEGAL, E., KOLLER, D., ET AL. A gene-coexpression network for global discovery of conserved genetic modules. *Science 302*, 5643 (2003), 249–255.

[388] SUN, Z., AND ZHU, Y. Systematic comparison of rna-seq normalization methods using measurement error models. *Bioinformatics 28*, 20 (2012), 2584–2591.

[389] SZATHMARY, L., NAPOLI, A., AND VALTCHEV, P. Towards rare itemset mining. In *19th IEEE International Conference on Tools with Artificial Intelligence (ICTAI 2007)* (2007), vol. 1, IEEE, pp. 305–312.

[390] SZATHMARY, L., VALTCHEV, P., AND NAPOLI, A. Generating rare association rules using the minimal rare itemsets family. vol. 4, International Journal on Software Informatics, pp. 219–238.

[391] DONDERS, A. R. T., VAN DER HEIJDEN, G. J. M. G., STIJNEN, T. AND MOONS, K. G. M. Review: a gentle introduction to imputation of missing values. *Journal of clinical epidemiology* 59, no. 10 (2006): 1087–1091.

[392] T PARK, S YI, S KANG, ET AL. Evaluation of normalization methods for microarray data. *BMC Bioinformatics 4*, 33 (2003).

[393] TAFT, R. J., PANG, K. C., MERCER, T. R., DINGER, M., AND MATTICK, J. S. Non-coding RNAs: regulators of disease. *The Journal of Pathology: A Journal of the Pathological Society of Great Britain and Ireland 220*, 2 (2010), 126–139.

[394] TAN, K. M., KILLOURHY, K. S., AND MAXION, R. A. Undermining an anomaly-based intrusion detection system using common exploits. In *International Workshop on Recent Advances in Intrusion Detection* (2002), Springer, pp. 54–73.

[395] TAN, P.-N., STEINBACH, M., AND KUMAR, V. Data mining cluster analysis: basic concepts and algorithms. *Introduction to data mining* (2013).

[396] TAN, P.-N., STEINBACH, M., AND KUMAR, V. *Introduction to Data Mining*. Pearson, 2016.

[397] TANAY, A., SHARAN, R., AND SHAMIR, R. Discovering statistically significant biclusters in gene expression data. *Bioinformatics 18*, suppl_1 (2002), S136–S144.

[398] TANAY, A., SHARAN, R., AND SHAMIR, R. Biclustering algorithms: A survey. *Handbook of computational molecular biology 9*, 1-20 (2005), 122–124.

[399] TANG, C., AND ZHANG, A. Interrelated two-way clustering and its application on gene expression data. *International Journal on Artificial Intelligence Tools 14*, 04 (2005), 577–597.

[400] TCHAGANG, A. B., PHAN, S., FAMILI, F., SHEARER, H., FOBERT, P., HUANG, Y., ZOU, J., HUANG, D., CUTLER, A., LIU, Z., ET AL. Mining biological information from 3d short time-series gene expression data: the optricluster algorithm. *BMC bioinformatics 13*, 1 (2012), 54.

[401] TEMIN, H. M. The dna provirus hypothesis. *Science 192*, 4244 (1976), 1075–1080.

[402] TENG, L., AND CHAN, L. Discovering biclusters by iteratively sorting with weighted correlation coefficient in gene expression data. *Journal of Signal Processing Systems 50*, 3 (2008), 267–280.

[403] TESSON, B. M., BREITLING, R., AND JANSEN, R. C. DiffCoEx: a simple and sensitive method to find differentially coexpressed gene modules. *BMC Bioinformatics 11*, 1 (2010), 497.

[404] THORNLOW, B. P., HOUGH, J., ROGER, J. M., GONG, H., LOWE, T. M., AND CORBETT-DETIG, R. B. Transfer RNA genes experience exceptionally elevated mutation rates. *Proceedings of the National Academy of Sciences 115*, 36 (2018), 8996–9001.

[405] TIAN, L., SU, S., AMANN-ZALCENSTEIN, D., BIBEN, C., NAIK, S. H., AND RITCHIE, M. E. scpipe: a flexible data preprocessing pipeline for single-cell rna-sequencing data. *bioRxiv* (2017), 175927.

[406] TIBSHIRANI, R., HASTIE, T., EISEN, M., ROSS, D., BOTSTEIN, D., BROWN, P., ET AL. Clustering methods for the analysis of dna microarray data. *Dept. Statist., Stanford Univ., Stanford, CA, Tech. Rep* (1999).

[407] TONG, W., CAO, X., HARRIS, S., SUN, H., FANG, H., FUSCOE, J., HARRIS, A., HONG, H., XIE, Q., PERKINS, R., ET AL. Arraytrack–supporting toxicogenomic research at the us food and drug administration national center for toxicological research. *Environmental Health Perspectives 111*, 15 (2003), 1819.

[408] TORIBIO, A. L., ALAKO, B., AMID, C., CERDEÑO-TARRÁGA, A., CLARKE, L., CLELAND, I., FAIRLEY, S., GIBSON, R., GOODGAME, N., TEN HOOPEN, P., ET AL. European nucleotide archive in 2016. *Nucleic Acids Research 45*, D1 (2017), D32–D36.

[409] TRAPNELL, C., CACCHIARELLI, D., GRIMSBY, J., POKHAREL, P., LI, S., MORSE, M., LENNON, N. J., LIVAK, K. J., MIKKELSEN, T. S., AND RINN, J. L. The dynamics and regulators of cell fate decisions are revealed by pseudotemporal ordering of single cells. *Nature biotechnology 32*, 4 (2014), 381.

[410] TROYANSKAYA, O., CANTOR, M., SHERLOCK, G., BROWN, P., HASTIE, T., TIBSHIRANI, R., BOTSTEIN, D., AND ALTMAN, R. B. Missing value estimation methods for dna microarrays. *Bioinformatics 17*, 6 (2001), 520–525.

[411] TSENG, G. C., OH, M.-K., ROHLIN, L., LIAO, J. C., AND WONG, W. H. Issues in cdna microarray analysis: quality filtering, channel normalization, models of variations and assessment of gene effects. *Nucleic acids research 29*, 12 (2001), 2549–2557.

[412] TUSHER, V. G., TIBSHIRANI, R., AND CHU, G. Significance analysis of microarrays applied to the ionizing radiation response. *Proceedings of the National Academy of Sciences 98*, 9 (2001), 5116–5121.

[413] UPTON, G. J. Fisher's exact test. *Journal of the Royal Statistical Society. Series A (Statistics in Society)* (1992), 395–402.

[414] VALLEJOS, C. A., RISSO, D., SCIALDONE, A., DUDOIT, S., AND MARIONI, J. C. Normalizing single-cell RNA sequencing data: challenges and opportunities. *Nature Methods 14*, 6 (2017), 565.

[415] VÂN ANH HUYNH-THU, A. I., WEHENKEL, L., AND GEURTS, P. Inferring regulatory networks from expression data using tree-based methods. *PloS one 5*, 9 (2010).

[416] VAN DAM, S., CRAIG, T., AND DE MAGALHÃES, J. P. GeneFriends: a human RNA-seq-based gene and transcript co-expression database. *Nucleic Acids Research 43*, D1 (2014), D1124–D1132.

[417] VAN DIJK, D., NAINYS, J., SHARMA, R., ET AL. MAGIC: A diffusion-based imputation method reveals gene-gene interactions in single-cell RNA-sequencing data. *BioRxiv* (2017), 111591.

[418] VAN DIJK, E. L., JASZCZYSZYN, Y., NAQUIN, D., AND THERMES, C. The third revolution in sequencing technology. *Trends in Genetics 34*, 9 (2018), 666–681.

[419] VEGARD NYGAARD, EINAR ANDREAS RODLAND. Methods that remove batch effects while retaining group differences may lead to exaggerated confidence in downstream analyses. *Biostatistics 17*, 1 (2016), 29–39.

[420] VELCULESCU, V. E., VOGELSTEIN, B., AND KINZLER, K. W. Analysing uncharted transcriptomes with SAGE. *Trends in Genetics 16*, 10 (2000), 423–425.

[421] VIDMAN, L., KÄLLBERG, D., AND RYDÉN, P. Cluster analysis on high dimensional rna-seq data with applications to cancer research-an evaluation study. *BioRxiv* (2019), 675041.

[422] GONG, W., KWAK, I.-Y., POTA, P., KOYANO-NAKAGAWA, N. AND GARRY, D. J. DrImpute: imputing dropout events in single cell RNA sequencing data. *BMC bioinformatics* 19, no. 1 (2018): 1–10.

[423] WANG, B., ZHU, J., PIERSON, E., RAMAZZOTTI, D., AND BATZOGLOU, S. Visualization and analysis of single-cell rna-seq data by kernel-based similarity learning. *Nature methods 14*, 4 (2017), 414.

[424] WANG, C., TONG, T., CAO, L., AND MIAO, B. Non-parametric shrinkage mean estimation for quadratic loss functions with unknown covariance matrices. *Journal of Multivariate Analysis 125* (2014), 222–232.

[425] WANG, H., AND VAN DER LAAN, M. J. Dimension reduction with gene expression data using targeted variable importance measurement. *BMC bioinformatics 12*, 1 (2011), 312.

[426] WANG, L., WANG, S., AND LI, W. Rseqc: quality control of rna-seq experiments. *Bioinformatics 28*, 16 (2012), 2184–2185.

[427] WANG, W., YANG, J., AND MUNTZ, R. Sting+: An approach to active spatial data mining. In *Proceedings 15th International Conference on Data Engineering (Cat. No. 99CB36337)* (1999), IEEE, pp. 116–125.

[428] WANG, W., YANG, J., MUNTZ, R., ET AL. Sting: A statistical information grid approach to spatial data mining. In *VLDB* (1997), vol. 97, pp. 186–195.

[429] WANG, Y., LU, J., LEE, R., GU, Z., AND CLARKE, R. Iterative normalization of cdna microarray data. *IEEE Transactions on Information Technology in Biomedicine 6*, 1 (2002), 29–37.

[430] WATSON, M. Coxpress: differential co-expression in gene expression data. *BMC bioinformatics 7*, 1 (2006), 509.

[431] WEISSMAN, D. H., GOPALAKRISHNAN, A., HAZLETT, C., AND WOLDORFF, M. Dorsal anterior cingulate cortex resolves conflict from distracting stimuli by boosting attention toward relevant events. *Cerebral cortex 15*, 2 (2005), 229–237.

[432] WHETZEL, P. L., PARKINSON, H., CAUSTON, H. C., FAN, L., FOSTEL, J., FRAGOSO, G., GAME, L., HEISKANEN, M., MORRISON, N., ROCCA-SERRA, P., ET AL. The MGED Ontology: a resource for semantics-based description of microarray experiments. *Bioinformatics 22*, 7 (2006), 866–873.

[433] WITTEN, D. M., ET AL. Classification and clustering of sequencing data using a poisson model. *The Annals of Applied Statistics 5*, 4 (2011), 2493–2518.

[434] WOLF, F. A., ANGERER, P., AND THEIS, F. J. Scanpy: large-scale single-cell gene expression data analysis. *Genome biology 19*, 1 (2018), 15.

[435] WOLPERT, D. H. Stacked generalization. *Neural networks 5*, 2 (1992), 241–259.

[436] WOO, K.-G., LEE, J.-H., KIM, M.-H., AND LEE, Y.-J. Findit: a fast and intelligent subspace clustering algorithm using dimension voting. *Information and Software Technology 46*, 4 (2004), 255–271.

[437] WORKMAN, C., JENSEN, L. J., JARMER, H., BERKA, R., GAUTIER, L., NIELSER, H. B., SAXILD, H.-H., NIELSEN, C., BRUNAK, S., AND KNUDSEN, S. A new non-linear normalization method for reducing variability in dna microarray experiments. *Genome biology 3*, 9 (2002), research0048–1.

[438] WU, X., AND KUMAR, V. *The top ten algorithms in data mining.* CRC press, 2009.

[439] XIE, J., MA, A., ZHANG, Y., LIU, B., CAO, S., WANG, C., XU, J., ZHANG, C., AND MA, Q. Qubic2: a novel and robust biclustering algorithm for analyses and interpretation of large-scale rna-seq data. *Bioinformatics 36*, 4 (2020), 1143–1149.

[440] XU, B., YI, T., WU, F., AND CHEN, Z. An incremental updating algorithm for mining association rules. *Journal of Electronics (China) 19*, 4 (2002), 403–407.

[441] XU, C., AND SU, Z. Identification of cell types from single-cell transcriptomes using a novel clustering method. *Bioinformatics 31*, 12 (2015), 1974–1980.

[442] XU, C., AND SU, Z. Identification of cell types from single-cell transcriptomes using a novel clustering method. *Bioinformatics 31*, 12 (2015), 1974–1980.

[443] XU, X., ESTER, M., KRIEGEL, H.-P., AND SANDER, J. A distribution-based clustering algorithm for mining in large spatial databases. In *Proceedings 14th International Conference on Data Engineering* (1998), IEEE, pp. 324–331.

[444] XU, Y., LIN, Z., TANG, C., TANG, Y., CAI, Y., ZHONG, H., WANG, X., ZHANG, W., XU, C., WANG, J., ET AL. A new massively parallel nanoball sequencing platform for whole exome research. *BMC bioinformatics 20*, 1 (2019), 1–9.

[445] XU, Y., OLMAN, V., AND XU, D. Clustering gene expression data using a graph-theoretic approach: an application of minimum spanning trees. *Bioinformatics 18*, 4 (2002), 536–545.

[446] Y. S. KOH AND N. ROUNTREE. Finding Sporadic Rules Using Apriori-Inverse. Springer-Verlag Berlin Heidelberg, pp. 97–106.

[447] Y WANG AND R REKAYA. LSOSS: Detection of Cancer Outlier Differential Gene Expression. *Biomarker Insights 5* (2010), BMI–55175.

[448] YANG, J., WANG, H., WANG, W., AND YU, P. Enhanced biclustering on expression data. In *Third IEEE Symposium on Bioinformatics and Bioengineering, 2003. Proceedings.* (2003), IEEE, pp. 321–327.

[449] YANG, J., WANG, W., WANG, H., AND YU, P. /spl delta/-clusters: capturing subspace correlation in a large data set. In *Proceedings 18th international conference on data engineering* (2002), IEEE, pp. 517–528.

[450] YIP, A. M., AND HORVATH, S. Gene network interconnectedness and the generalized topological overlap measure. *BMC bioinformatics 8*, 1 (2007), 22.

[451] YU, C., WU, W., WANG, J., LIN, Y., YANG, Y., CHEN, J., ZHU, F., AND SHEN, B. Ngs-fc: A next-generation sequencing data format converter. *IEEE/ACM transactions on computational biology and bioinformatics 15*, 5 (2017), 1683–1691.

[452] YU, F. Y., YANG, Z. H., TANG, N., LIN, H. F., WANG, J., AND YANG, Z. W. Predicting protein complex in protein interaction network-a supervised learning based method. *BMC Systems Biology 8*, 3 (2014), S4.

[453] YU, G., LI, F., QIN, Y., BO, X., WU, Y., AND WANG, S. GOSemSim: an R package for measuring semantic similarity among GO terms and gene products. *Bioinformatics 26*, 7 (2010), 976–978.

[454] YUN, H., HA, D., HWANG, B., AND RYU, K. H. Mining association rules on significant rare data using relative support. vol. 67, The Journal of Systems and Software, pp. 181–191.

[455] ZAKI, M. J. Parallel and distributed association mining: A survey. *IEEE concurrency 7*, 4 (1999), 14–25.

[456] ZEISEL, A., ANA B. MUNOZ MANCHADO, AND SIMONE CODELUPPI, E. A. Cell types in the mouse cortext and hippocampus revealed by single-cell rna-seq. *SCIENCE 347*, 6226 (2015), 1138–1142.

[457] ZHANG, B., AND HORVATH, S. A general framework for weighted gene co-expression network analysis. *Statistical applications in genetics and molecular biology 4*, 1 (2005).

[458] ZHANG, L., GU, S., LIU, Y., WANG, B., AND AZUAJE, F. Gene set analysis in the cloud. *Bioinformatics 28*, 2 (2012), 294–295.

[459] ZHANG, L., AND ZHANG, S. Comparison of computational methods for imputing single-cell rna-sequencing data. *IEEE/ACM transactions on computational biology and bioinformatics* (2018).

[460] ZHANG, T., RAMAKRISHNAN, R., AND LIVNY, M. Birch: A new data clustering algorithm and its applications. *Data Mining and Knowledge Discovery 1*, 2 (1997), 141–182.

[461] ZHANG, Z., ZHANG, Y., EVANS, P., CHINWALLA, A., AND TAYLOR, D. Rna-seq 2g: online analysis of differential gene expression with comprehesive options of statistical methods. *BioRxiv* (2017), 122747.

[462] ZHOU, H., FENG, B., LV, L., HUI, Y., CHENG, C., WAICHEE, A., ZHANG, F., KAILING, K., KRIEGEL, H., KROGER, P., ET AL. delta-clusters: Capturing subspace correlation in a large dataset. *Information Technology Journal 6*, 2 (1999), pp–84.

[463] ZIEGENHAIN, C., VIETH, B., PAREKH, S., REINIUS, B., GUILLAUMET-ADKINS, A., SMETS, M., LEONHARDT, H., HEYN, H., HELLMANN, I., AND ENARD, W. Comparative analysis of single-cell RNA sequencing methods. *Molecular Cell 65*, 4 (2017), 631–643.

[464] ZYPRYCH-WALCZAK, J., SZABELSKA, A., HANDSCHUH, L., GÓRCZAK, K., KLAMECKA, K., FIGLEROWICZ, M., AND SIATKOWSKI, I. The impact of normalization methods on rna-seq data analysis. *BioMed research international 2015* (2015).

Glossary

ArrayExpress One of the major public repositories for functional genomic datasets.. 8

ArrayTrack A tool for managing, analyzing and interpreting microarray gene expression data. It was created by Dr. Weida Tong, director of the Division of Biostatistics and Bioinformatics, FDA (USA).. 8

Autophagy Natural self-destruction of the body's own tissues.. 255

base pairings Chemical cross-links between two strands in DNA by complementary pairs of bases; thymine with adenine (T–A) and guanine with cytosine (G–C).. 39

bioinformatics The science of collecting and analysing complex biological data.. 11

biological domains The highest taxonomic rank in the hierarchical biological classification system, above the kingdom level.. 31

biosphere The regions of the surface and atmosphere of the earth or another planet occupied by living organisms.. 31

Bipolar disorder A disorder with episodes of depression and extreme highs.. 248

blood The red liquid that circulates in the arteries and veins of animals carrying oxygen to and carbon dioxide from the tissues of the body.. 182

cardiovascular system Organ system comprising the blood, heart, and blood vessels.. 31

cell Basic structural and functional unit of an organism.. 30

cell lineages Development of cells and tissues from a fertilized embryo.. 173

cell proliferation Increase in the total mass of a cell by its growth and division.. 183

cell wall The outermost membrane of a plant cell which differentiates it from an animal cell.. 30

Central Dogma The scientific concept of a gene being transcribed to an mRNA which is in turn translated to protein.. 4

chloroplast A cell organelle where photosynthesis takes place.. 30

coexpression analysis The study to derive genes whose expression is correlated with each other during a condition.. 261

cytokine Glycoproteins secreted by the cells of the immune system.. 255

Danio rerio embryos Embryo of zebrafish.. 175

de novo assembly Construction of a genome from a large number of DNA fragments without knowledge of correct sequence or order of those fragments.. 267

DNA A coiled double helix molecule composed of two polynucleotide chains that carry genetic instructions.. 30, 44

DNA helicase An essential enzyme of DNA replication that separates double-stranded DNA into single strands.. 33

DNA nanoball 160,000 to 200,000-bp-long single-stranded replicated DNA fragments made of the original library DNA molecules.. 7

DNA polymerase A type of enzyme that synthesizes new copies of DNA.. 34

DNA provirus RNA-directed DNA synthesis for the formation of the provirus.. 36

dNTPs A molecule containing a nitrogenous base bound to a 5-carbon deoxyribose sugar, with three phosphate groups bound to the sugar.. 44

ecosystem An environment made up of all the biotic and abiotic components.. 31

endoplasmic reticulum The cell organelle responsible for the synthesis, modification and transport of proteins.. 30

enrichment analysis A computational method to determine a statistically significant set of genes among samples.. 182

enzyme Proteins that act as biological catalysts in living organisms.. 33

epigenome Chemical changes in the genome of an organism.. 167

evolution A process by which different kinds of living organisms have developed from earlier forms during the history of the earth.. 54

exogenous RNA spike-ins Incidental complementary binding of external RNA to DNA sequences in a reaction concentration other than the target sequence.. 54

fertilized egg A cell that results from the union of a female gamete with a male gamete.. 41

fetal A developing organism that is not yet born.. 177

fibroblasts Connective tissue cell of collagen and other fibres.. 183

flow-activated cell sorting (FACS) Specialized type of flow cytometry that sorts heterogeneous mixture of cells into two or more containers.. 50

fluxes Discharge of fluid from a body.. 30

gene Basic physical and functional unit of heredity.. 31

gene annotation Study of finding coding regions along the genome of an organism and identifying them.. 45

gene expression Expression of a gene into its functional products.. 41

Gene Ontology A major bioinformatics initiative to unify the representation of gene and gene product attributes across all species .. 211

Genecard A database of human genes developed and maintained by the Crown Human Genome Centre of Weizmann Institute of Science.. 13

genetic code The sequence of nucleotides in deoxyribonucleic acid (DNA) and ribonucleic acid (RNA) that determines the amino acid sequence of proteins.. 32

genetic material Any material of an organism that carries genetic information and passes from one generation to the next.. 37

genome All the genetic material of an organism.. 213

genotype Genetic makeup of an organism describing an organism's complete set of genes.. 21

germ cells Reproductive cells of the body.. 32

Golgi apparatus A membrane-bound cell organelle that is involved in transportation of biomolecules.. 30

. 7

homeostasis The capacity of the body to maintain the stability of diverse internal variables, such as temperature, acidity, and water level, in the face of constant environmental disturbance.. 30

hormones The chemical messengers secreted directly into the blood.. 30

hybrid Combination of more than one entity or process.. 37

hybridization Combining two complementary single-stranded DNA or RNA molecules to form a single double-stranded molecule.. 42

in-vitro Refers to any work that occurs outside the living organism.. 43

inflammation The body's reaction to injury or infection.. 254

interactomic Pertaining to a process/phenomenon/entity occurring between two atoms.. 31

intra-species variability Variability occurring among the different populations of a species.. 32

ionomic It is the complete set of ions in a living organism and is measured quantitatively.. 31

long ncRNAs RNA/transcripts with lengths exceeding 200 nucleotides that do not overlap into protein coding genes.. 48

macroscopic level Anything seen with a measurable scale large enough to be visible by the naked eye.. 30

metabolism Process that takes place in a cell.. 30

metabolites The intermediate products of metabolic reactions occurring within the cells.. 30

metabolomic Pertaining to a collection of all the metabolites in an organism.. 31

metagenomics Study of the genetic material recovered from environmental samples directly.. 259

microbial colonies Visible cluster of microorganisms.. 45

microbiome Collection of all the genetic material of the microorganism.. 259

micromanipulation Manipulation of cell organelles at a macroscopic level.. 49

microtiter plates A plate with multiple small wells that can be used for different experiments.. 45

mitochondria The cell organelle which provides energy to the cell.. 30

mRNA Messenger RNA is a single-stranded molecule of RNA that consists of the sequence of a gene .. 45

multicellular Organisms which are made up of more than one cell.. 30

mutations Alteration in the nucleotide sequence of the genome of an organism, virus, or extrachromosomal DNA.. 259

myoblasts Progenitors of muscle cells.. 176

neuroimmune system Neurological disorders that share the inflammatory involvement of the central nervous system.. 248

neuroinflammation Inflammation of the neurons.. 254

neuron The structural and functional unit of a neuron.. 105

neuropsychiatric disorder Disorders of affect, cognition, and behavior that arise from abnormalities in cerebral function.. 248

nitrocellulose papers A paper made of nitrocellulose that is used for protein immobilization because of its protein binding affinity.. 45

non-coding RNAs RNAs which are not translated into proteins.. 46

nuclease digestion Cleavage of phosphodiester bonds between nucleotides of nucleic acids by the enzyme nuclease.. 44

nucleotides Organic molecule of a nucleoside and phosphate that act as monomers of nucleic acid polymers.. 275

nucleus The cell organelle where DNA replication and transcription takes place.. 30

oligo primer Small 2'-deoxyribonucleotides molecules used to amplify a small amount of DNA.. 44

omics Total collection of all biomolecules such as proteins, nucleic acids, vitamins, etc.. 31

organelle Organelles are subunits of cells that are specialized in cellular function.. 30

organs Group of tissues which are similar in their functions and act as a unit.. 30

phenomic Set of all phenotypes expressed by a cell or organism.. 31

phenotype Observable features of an organism.. 120

phenotypic Expression of genes in an observable and measurable way.. 251

polyadenylation Addition of adenine nucleotides at the end of mRNA.. 48

probes A single-stranded sequence of DNA or RNA used to search for its complementary sequence in a sample genome.. 43

prognostic Prediction of likely development of a disease.. 70

proteins Large biomolecules, or macromolecules, that are an essential part of living organisms, especially as structural components of body tissues.. 48

proteomic Collection of all the proteins produced in a sample.. 31

reference genomes The genomes whose sequences are considered the standard genome of an organism.. 47

reverse transcriptase enzyme An enzyme used to generate complementary DNA (cDNA) from an RNA template.. 36

ribosome biogenesis The process of formation of ribosomes.. 183

ribosomes Cell organelles where proteins are synthesized.. 30

RNA High-molecular-weight compound involved in cellular protein synthesis that replaces DNA as a carrier of genetic codes in some viruses.. 30, 183

RNA interference The process where a RNA is hybridized to mRNA and its translation is prevented.. 48

RNA polymerase Enzyme that actively copies a DNA sequence into an RNA sequence, during transcription.. 4

RNA synthetase An enzyme that catalyzes the synthesis of RNA in cells infected with RNA viruses.. 38

RNA templates Template for protein synthesis that carries information from DNA in the nucleus to ribosome sites of protein synthesis in the cell.. 44

RNA viruses A virus that has RNA as its genetic material.. 36

RNAseq Examining the quantity and sequences of RNA in a sample using next-generation sequencing (NGS).. 152

Schizophrenia A mental disorder in which the patient interprets reality differently.. 248

sequenced concatemers A molecule that contains multiple copies of the same sequence in series.. 44

single-cell qPCR A technique used to measure the quantitative amount of gene in a single cell using PCR.. 49

single-cellular Refers to those organisms which are made up of a single cell.. 30

single-molecule FISH A process to detect individual RNA molecules within cells via fluorescence microscopy.. 49

somatosensory cortex Primary receptor of the general bodily sensations in the brain.. 185

splice junctions The site of a former intron in a mature mRNA in splicing.. 54

tissue Collection of cells that are similar in function and work as a unit.. 46

trait An observable characteristic which is passed on as hereditary.. 84

transcript mRNA produced from a gene.. 46

transcript isoform Different types of mRNA produced from a single locus.. 47

transcription factor Proteins that bind to the DNA to regulate the expression of a gene.. 10

transcriptome Collection of cells that are similar in function and work as a unit.. 46

transfer RNAs (tRNA) The subtype of RNA that adds new amino to a growing peptide chain to form protein.. 46

veins The blood vessels that carry deoxygenated blood from the organ to the heart.. 31

viral RNA The RNA which encodes the genetic information in a virus.. 36

Watson-Crick base pairing DNA consists of adenine, guanine, cytosine and thymine nucleotide bases. Adenine from one strand always binds with thymine of other complementary strand and cytosine from one strand always binds with guanine from another strand. These pairs are known as base pairs.. 35

. 42

wrapper A process to find the optimal features of a model by employing a specific machine learning algorithm.. 18

Index

Milton Keynes UK
Ingram Content Group UK Ltd.
UKHW031533071024
449327UK00005B/80